Metabolomics, Metabonomics and Metabolite Profiling

RSC Biomolecular Sciences

Editorial Board:

This Series is devoted to coverage of the interface between the chemical and biological sciences, especially structural biology, chemical biology, bio- and chemo-informatics, drug discovery and development, chemical enzymology and biophysical chemistry. Ideal as reference and state-of-the-art guides at the graduate and post-graduate level.

Titles in the Series:

Visit our website on www.rsc.org/biomolecularsciences

For further information please contact:
Sales and Customer Care, Royal Society of Chemistry, Thomas Graham House, Science Park, Milton Road, Cambridge, CB4 0WF, UK
Telephone: +44 (0)1223 432360, Fax: +44 (0)1223 426017, Email: sales@rsc.org

Metabolomics, Metabonomics and Metabolite Profiling

Edited by

William J. Griffiths
University of London, The School of Pharmacy, University of London, London, UK

RSCPublishing

ISBN: 978-0-85404-299-9

A catalogue record for this book is available from the British Library

Published by The Royal Society of Chemistry,
Thomas Graham House, Science Park, Milton Road,
Cambridge CB4 0WF, UK

Registered Charity Number 207890

For further information see our web site at www.rsc.org

Preface

With the completion of gene sequencing projects, scientific interest is shifting to the investigation of the proteome and metabolome. Broadly speaking, the metabolome can be considered as the dynamic complement of metabolites formed by, or found within, a cell type, tissue, body fluid, or organism; and the study of the metabolome can be viewed as metabonomics or metabolomics, both of which are applications of metabolite profiling. Metabolite profiling is not a new concept, and has been successfully used for the diagnosis of specific diseases for many years. However, with the advent of high-throughput technology, metabolite profiling is now being used extensively in the quest for the discovery of markers for a wide range of diseases.

In this volume, we draw together experts in the fields of metabolite profiling and identification. The main techniques used in metabolite profiling are mass spectrometry and NMR, and an introduction to these techniques is covered. There follow chapters on the current application of metabolite profiling for the diagnosis of disease. Specific classes of metabolites are the subject of further chapters, and these are followed by chapters on plant metabolomics and metabolite data mining. Finally, we conclude with a chapter on global systems biology.

Contents

Chapter 3   Steroids, Sterols and the Nervous System
Yuqin Wang and William J. Griffiths

Chapter 4   Phospholipid Profiling
Anthony D. Postle

**Chapter 5   New Developments in Multi-dimensional Mass Spectrometry
                     Based Shotgun Lipidomics**
Xianlin Han and Richard W. Gross

Chapter 6   Neutral Lipidomics and Mass Spectrometry
Robert C. Murphy, Mark Fitzgerald and Robert M. Barkley

Chapter 7 Bioinformatics of Lipids
Eoin Fahy

Chapter 8 Mass Spectrometry in Glycobiology
João Rodrigues, Carla Antonio, Sarah Robinson and Jane Thomas-Oates

**Chapter 9   Matrix Assisted Laser Desorption Ionisation Mass
               Spectrometric Imaging – Principles and Applications**
*Caroline J. Earnshaw, Sally J. Atkinson, Michael Burrell and
Malcolm R. Clench*

Chapter 10   Plant Metabolomics
Thomas Moritz and Annika I. Johansson

Chapter 11 Data Mining for Metabolomics
Anders Nordström

Chapter 12 Metabonomics and Global Systems Biology
Ian D Wilson and Jeremy K. Nicholson

CHAPTER 1

Mass Spectrometry for Metabolite Identification

YUQIN WANG AND WILLIAM J. GRIFFITHS

The School of Pharmacy, University of London, 29–39 Brunswick Square, London WC1N 1AX, UK

1.1 Introduction

Mass spectrometry (MS) and nuclear magnetic resonance (NMR) constitute the two major pillars upon which the disciplines of metabolomics and metabolite profiling are built. Both these techniques have their advantages and disadvantages, but the fundamental difference in the nature of their spectroscopy means that they provide complementary information to the analytical scientist. NMR spectroscopy is discussed in detail in Chapter 2, while this chapter will concentrate on mass spectrometry and associated methodologies appropriate in metabolomics research. The principles of mass spectrometry will be described and examples of metabolite analysis given. As the range of metabolite structures present in biology (almost) exceeds the imagination, we will take many examples from the class of biomolecules that we have been most intimately involved with, *i.e.* sterols and steroids. However, many of the references quoted are equally applicable to other area of metabolite profiling and metabolomics.

1.2 Mass Spectrometry

1.2.1 Principles

Simplistically, a mass spectrometer consists of an ion source, a mass analyser, a detector and a data system (Figure 1.1). Sample molecules are admitted to the ion source, where they become ionised. The ions, which are now in the gas phase, are separated according to their mass-to-charge ratio (m/z) in the mass analyser and are then detected. The resulting signals are transmitted to the data

1

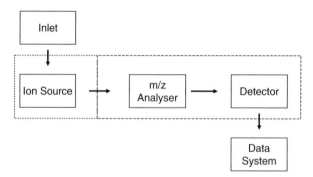

Figure 1.1   Simplistic view of a mass spectrometer. The ion source may be under vacuum or at atmospheric pressure. The analyser and detector are under vacuum. The acceptable pressure in the analyser region depends on the analyser, ranging from 10^{-3} mbar to 10^{-11} mbar for quadrupole and ion cyclotron resonance (ICR) analysers respectively.

system and a plot of ion abundance against *m/z* corresponds to a mass spectrum. In many cases, a separating inlet device precedes the ion source, so that complex mixtures can be separated prior to admission to the mass spectrometer. Today, the separating inlet device is usually either a capillary gas chromatography (GC) column or a high-performance liquid chromatography (HPLC) column, although capillary electrophoresis and thin-layer chromatography can be interfaced with mass spectrometry.

For metabolite analysis a number of different types of ionisation methods are used to generate gas-phase ions and these include: electron ionisation (EI), chemical ionisation (CI), electrospray (ES), atmospheric pressure chemical ionisation (APCI), atmospheric pressure photoionisation (APPI), and the recently introduced, desorption electrospray ionisation (DESI) technique. Other ionisation techniques used, but to a lesser extent, are liquid secondary ion mass spectrometry (LSIMS) and fast atom bombardment (FAB), or for more specific applications, matrix-assisted laser desorption/ionisation (MALDI) (see Chapter 9) and desorption ionisation on silicon (DIOS). The most widely used ionisation modes are discussed below, as are the chromatographic devices to which they are interfaced.

1.2.2   Ionisation

1.2.2.1   *Electron Ionisation (EI) and Chemical Ionisation (CI)*

Historically, the most important method of ionisation of small biomolecules ($< \sim 500$ Da) is EI. The effluent from a GC column is readily transferred to an EI source, thereby allowing the combination of the high separating power of a GC column with mass analysis. EI involves the bombardment of gas-phase sample molecules (M) with high-energy electrons ($\underline{e}^-$), usually of 70 eV energy;

the result is the generation of $[M]^{+\bullet}$ ions which are usually radical cations, and thermal energy free electrons (e^-) (eqn 1).

$$M(g) + \underline{e}^- \rightarrow M^{+\bullet}(g) + 2e^- \tag{1}$$

In many cases the molecular ions, $[M]^{+\bullet}$, are unstable and fragment to generate more stable products (eqn 2).

$$M^{+\bullet}(g) \rightarrow A^+(g) + B^{\bullet}(g) \tag{2}$$

Fragmentation upon EI can be seen as both advantageous and disadvantageous. On the plus side, the fragmentation pattern resulting from decomposition of a molecular ion can provide structural information, allowing its identification. On the negative side, however, fragmentation may be so extensive that the molecular ion may not be observed, and thus the molecular weight of the compound of interest not determined. EI can be used to generate either positive ions $[M]^{+\bullet}$ or negative ions $[M]^{-\bullet}$. Negative ions are generated via an electron capture event, which involves the capture of secondary low-energy electrons generated by ionisation of a bath gas (*e.g.* Ar, N_2) (eqn 3).

$$Ar(g) + \underline{e}^- \rightarrow Ar^{+\bullet}(g) + 2e^- \tag{3a}$$

$$M(g) + e^- \rightarrow M^{-\bullet}(g) \tag{3b}$$

A prerequisite of EI is that the sample to be ionised must be in the gas phase; this is also true for GC and has led to the extensive development of derivatisation chemistry to allow the vaporisation of many small biomolecules without their decomposition.[1-9]

CI is a close relative of EI. It differs in that analyte ionisation is achieved via proton attachment rather than electron ejection (positive ion).[10] In CI the ion source contains a reagent gas, often methane, which becomes ionised by EI and acts as a proton donor to the analyte (eqn 4).

$$CH_4(g) + \underline{e}^- \rightarrow CH_4^{+\bullet}(g) + 2e^- \tag{4a}$$

$$CH_4^{+\bullet}(g) + CH_4(g) \rightarrow CH_5^+(g) + CH_3^{\bullet}(g) \tag{4b}$$

$$CH_5^+(g) + M(g) \rightarrow MH^+(g) + CH_4(g) \tag{4c}$$

The resulting ion, $[M+H]^+$, is an even-electron protonated molecule, which is more stable than the equivalent odd-electron molecular ion, $[M]^{+\bullet}$, formed by EI, and thus fragments to only a minor extent.

Electron-capture negative ionisation (ECNI), also called electron-capture negative chemical ionisation (EC-NCI), exploits the electron capturing properties of groups with high electron affinities (eqn 5).[11] The method often utilises fluorinated agents in the preparation of volatile derivatives with high electron affinities. For example, trifluoroacetic, pentafluoropropionic or heptafluorobutyric anhydrides can be used to prepare acyl derivatives of amines and hydroxyl groups, perfluorinated alcohols can be used to generate esters of carboxylic acids, while carbonyl groups can be converted to oximes which can then be converted to pentafluorobenzyl oximes[12] or pentafluorobenzylcarboxymethoximes,[13] for example (Scheme 1.1). Ionisation proceeds with the capture of a secondary low-energy electron generated under CI conditions, by the high-electron affinity

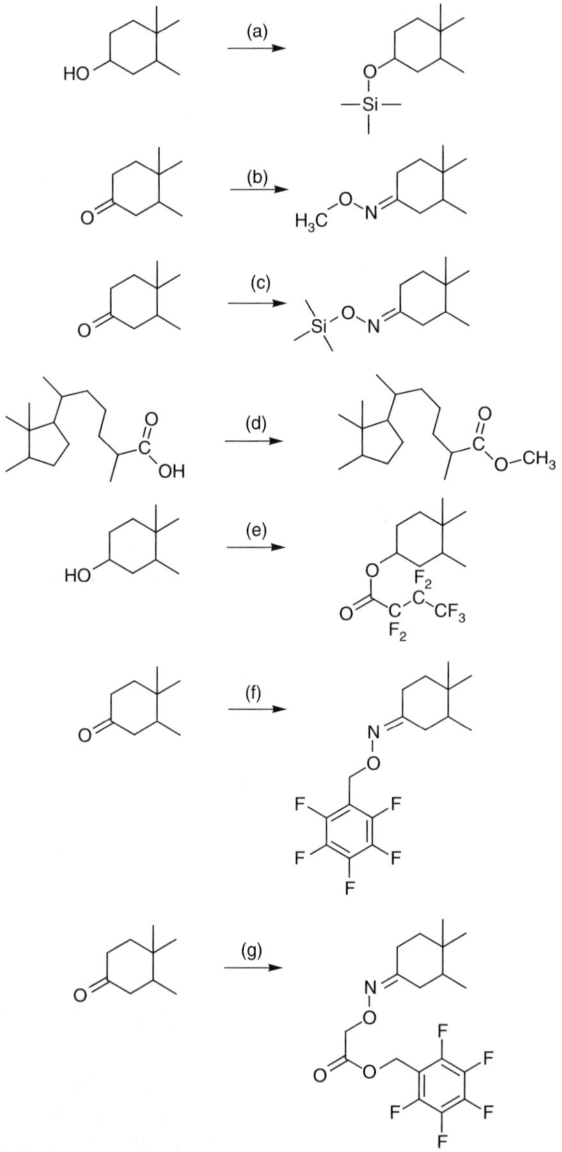

Scheme 1.1 Common derivatisation reactions exploited in metabolite analysis by GC-EI-MS and GC-EC-NCI. Reactions (a)–(c) and (e)–(g) are illustrated on the A-ring of a steroid.

fluorinated groups. Ionisation may lead to the formation of stable $[M]^{-\bullet}$ ions, or it may be dissociative,[12,13] depending on the analyte and the derivative used. The major advantage of EC-NCI is that ionisation is specific to compounds containing the electron capturing tag and provides excellent sensitivity in terms of signal to noise ratio when either the stable $[M]^{-\bullet}$ ion or a negatively charged fragment ion is monitored.

$$M(g) + e^- \rightarrow M^{-\bullet}(g) + e^- \qquad (5a)$$

$$M^{-\bullet}(g) \rightarrow A^-(g) + B^{\bullet}(g) \qquad (5b)$$

$$M^{-\bullet}(g) \rightarrow A^{-\bullet}(g) + B(g) \qquad (5c)$$

It is of historical interest that HPLC has been combined with EI and CI, but as both of these ionisation modes require high vacuum, the necessary removal of HPLC solvent has made this combination difficult. The natural marriage for HPLC is with atmospheric ionisation (API) methods discussed later.

1.2.2.2 Liquid Secondary Ion Mass Spectrometry (LSIMS) and Fast Atom Bombardment (FAB) Ionisation

It can be argued that the introduction of the FAB method of ionisation by Barber and colleagues in 1981 initiated the revolution in biological mass spectrometry.[14] Although rarely used today,[15] FAB was widely used for metabolite analysis throughout the 1980's and into the early 1990's, for example in bile acid[16-20] and steroid[21-23] analysis, and protocols developed for metabolite analysis during this era are easily incorporated into analytical procedures using API methods.[24-26]

FAB is most suitable for the ionisation of polar or ionic biomolecules. FAB ionisation is achieved by the generation of a fast atom beam of neutral atoms (6–8 keV kinetic energy, usually Ar or Xe atoms) in the FAB gun by a process of ionisation, acceleration and neutralisation, which impinges on a viscous solution of sample dissolved in a matrix, usually of glycerol. In the positive-ion mode proton transfer reactions result in the formation of protonated molecules, $[M + H]^+$, while in the negative-ion mode deprotonated molecules are formed, $[M - H]^-$. Usually, both protonated and deprotonated molecules are stable, and little fragmentation occurs in the ion source. LSIMS is very similar to FAB; however, a beam of Cs^+ ions (20–30 keV), rather than a beam of neutral atoms, is used to bombard the matrix. LSIMS spectra are essentially identical to those generated by FAB, and in this chapter, for simplicity, both ionisation modes will be referred to as FAB.

Negative-ion FAB was found to be particularly suitable for the ionisation of bile acids and steroid sulphates, alleviating the need for hydrolysis, solvolysis and derivatisation reactions necessary for GC-MS analysis.[16,17,21-23] Urine and plasma samples can be analysed by FAB following a simple C_{18} solid-phase extraction step, allowing the rapid diagnosis of certain liver diseases, *e.g.* cerebrotendinous xanthomatosis (CTX),[18] and 7α-hydroxylase deficiency.[15]

FAB is a vacuum ionisation method and is not suitable for combination with regular HPLC. However, capillary column HPLC (250 μm i.d., 2 μL/min flow-rate) has been successfully interfaced with FAB.[26]

1.2.2.3   *Atmospheric Pressure Ionisation (API)*

(a) Electrospray (ES).   Like FAB, ES is suitable for the analysis of polar and ionic biomolecules. ES occurs at atmospheric pressure and is readily

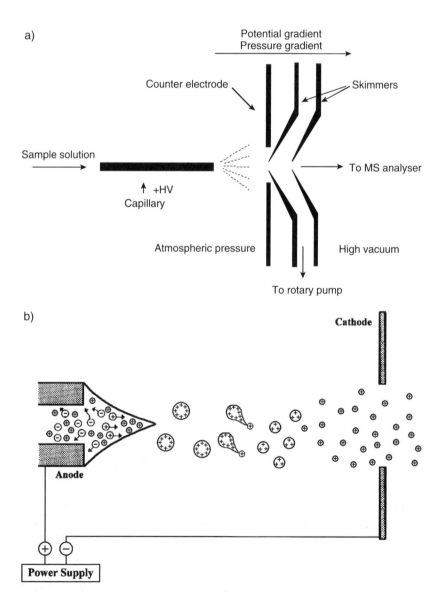

coupled with HPLC. Fenn and colleagues were the first practitioners of ES mass spectrometry[27–29] and by the turn of the last century, ES had almost completely replaced FAB for the analysis of polar and ionic metabolites.

In ES, the analyte is dissolved in a solvent (very often methanol, ethanol, aqueous methanol or ethanol, a mixture of acetonitrile and water, or of chloroform and alcohol), and sprayed from a metal or fused silica capillary (needle) of 20–100 μm i.d. at a flow rate of 1–500 μL/min (Figures 1.2 and 1.3). An electrospray is achieved by raising the potential on the spray capillary to ~4 kV (+4 kV in the positive-ion mode, and −4 kV in the negative-ion mode) and applying a back pressure to the contents of the capillary (*e.g.* via a syringe pump or HPLC pump). The resulting spray of charged droplets is directed toward a counter electrode which is at a lower electrical potential (Figure 1.2). As the spray of fine droplets travels towards the counter electrode, the droplets lose solvent, shrink and break up into smaller droplets. The small offspring droplets are derived from the surface of their predecessors, which contain the highest concentration of charge, and hence the offspring droplets are generated with an enhanced charge-to-mass ratio. Eventually, the droplets become so small that the charge density on the droplets exceeds the surface tension and gas-phase ions are desorbed (ion evaporation model),[30,31] or alternatively very small droplets containing a single charged species completely lose solvent leaving the residual charged species free (charge residue model).[32] Surface active compounds tend to be enhanced in the small droplets, and hence are preferentially brought into the gas phase. The counter electrode contains a circular orifice through which ions are transmitted into the vacuum chamber of

---

Figure 1.2 (a) Features of the ES interface and (b) schematic representation of the ES process. In the positive-ion mode a high positive potential is applied to the capillary (anode), causing positive ions in solution to drift towards the meniscus. Destabilisation of the meniscus occurs, leading to the formation of a cone[43] and a fine jet, emitting droplets with excess positive charge. Gas phase ions are formed from charged droplets in a series of solvent evaporation–Coulomb fission cycles. With the continual emission of positively charged droplets from the capillary, to maintain charge balance oxidation occurs within the capillary.[44] If the capillary is metal, oxidation of the metal may occur at the liquid/metal interface:

$$M(s) \rightarrow M^{n+}(aq) + ne^-(\text{in metal})$$

Alternatively, negative ions may be removed from solution by electrochemical oxidation:

$$4OH^-(aq) \rightarrow O_2(g) + 2H_2O(l) + 4e^-(\text{in metal})$$

When the interface is operated in the negative-ion mode, potentials are reversed, and to maintain charge balance cations or neutral molecules in solution may become reduced at the walls of the capillary.

$$C^{n+}(aq) + ne^- \rightarrow C(g/s)$$

$$H_2O(l) + 2e^- \rightarrow H_2(g) + O^{2-}(aq)$$

(Reproduced from Griffiths *et al.*,[45] with permission.)

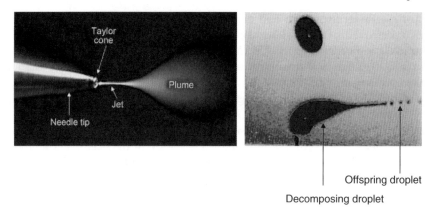

Offspring droplet

Decomposing droplet

Figure 1.3 The ES process viewed through a high-powered microscope (left panel). In the positive-ion mode a high positive potential is applied to the capillary (anode), and as the liquid exits the capillary it becomes charged and assumes a conical shape, known as the Taylor cone.[43] At the tip of the cone, the liquid is drawn into a filament, which then becomes unstable, breaking up into a mist of charged droplets. As the droplets are charged they repel each other and fly apart, becoming dispersed. Gas-phase ions are formed from charged droplets in a series of solvent evaporation–Coulomb fission cycles. Photograph of a decomposing droplet in an ES source (right panel). The droplet distorts, creating a miniature Taylor cone, and when the parent droplet reaches the Rayleigh limit, fission occurs leading to a string of multiple, small, highly charged offspring droplets. The critical point at which the surface tension of the droplet is overcome by the electrostatic repulsion of the surface charges is known as the Rayleigh stability limit. (Left panel reproduced from www.newobjective.com/electrospray/ with permission. Right panel from ref. 46, with permission.)

the mass spectrometer. By traversing differentially pumped regions via skimmer lenses, the ions are transmitted to the high vacuum region of the mass spectrometer for subsequent analysis (Figure 1.2). Early ES experiments were performed at flow rates of 5–50 µL/min, compatible with microbore HPLC columns,[33] although today ES interfaces are compatible with 4.2 mm columns operating at 500 µL/min flow rates. It should, however, be remembered that ES is a concentration rather than a mass dependent process, which effectively means that maximum sensitivity is achieved with high concentration, low volume samples, analysed at low flow rates (*e.g.* 1 pmol/µL, 1 µL, 1 µL/min) rather with dilute, high volume samples, analysed at high flow rates (*cf.* 0.02 pmol/µL, 50 µL, 50 µL/min). So, theoretically, a combination of low flow rate HPLC with ES should provide better sensitivity than conventional flow HPLC with ES. This has been particularly exploited by the proteomics community, who combine low flow rate HPLC with micro-ES or nano-ES. For a more detailed discussion on the electrospray process the reader is directed to the excellent volume edited by Cole,[34] and series of articles published in the *Journal of Mass Spectrometry*[35–40] and *Mass Spectrometry Reviews*.[41,42]

Today, the terms "micro-ES" and "nano-ES" are interchangeable, although the terms were originally coined to discriminate between two slightly different forms of ES. The term micro-ES was initially coined by Emmett and Caprioli[47] and used to refer to a miniaturised form of pressure-driven ES (*i.e.* pumped flow) operated at sub μL/min flow rate. Alternatively, nano-ES was invented by Wilm and Mann[48] and differs from micro-ES in that it is a pure form of ES, where sample flow is at nL/min rates and initiated by the electrical potential between the capillary tip and counter electrode (Figure 1.2), rather than being pressure driven. We prefer the term low flow rate ES to either micro-ES or nano-ES, and use it here to refer to ES operated at flow rates of 1 μL/min and below.

(b) Thermospray (TS). TS mass spectrometry predates ES.[49] TS, like ES, is a technique which involves the spraying of analyte dissolved in solvent from a capillary into an ion source at atmospheric pressure. But TS differs from ES in that vaporisation and ionisation are a result of thermally heating the spray, as opposed to raising the sprayer to a high potential. TS performs best with buffered aqueous mobile phase (0.1–0.01 M ammonium acetate), and droplets become charged by an uneven distribution of cations and anions between droplets, with the result that gaseous $[M+H]^+$, $[M+NH_4]^+$ or $[M+Na]^+$ ions are formed in the positive-ion mode in an ion evaporation process similar to that occurring in ES (ion evaporation). In the negative-ion mode $[M-H]^-$ ions are formed, and this is the ionisation mode of choice for the analysis of acidic metabolites. TS can be operated in two modes: "filament-on" mode or "filament-off" mode (direct ion evaporation). By the incorporation of a filament in the ion source, analytes which are not readily ionised by direct ion evaporation can be ionised via a chemical ionisation process in the "filament-on" mode. Spectra of biomolecules, *e.g.* steroids and bile acids, generated by ES and TS are similar; however, dehydration of the protonated or deprotonated molecule tends to occur to a much greater extent in TS than in ES.[50]

(c) Atmospheric Pressure Chemical Ionisation (APCI). APCI is a technique that has become popular for the ionisation of neutral biomolecules. It is very similar to TS described above.[51,52] Analyte dissolved in solvent is sprayed into an atmospheric pressure ion source. Vaporisation of sample and solvent is achieved by the application of heat, and ionisation of analyte is achieved by a chemical ionisation event. The APCI source differs from the ES source in that it additionally contains a corona discharge needle. Analyte ionisation can be achieved by two processes.

(i) The first is a primary CI process. The nebulised spray results in small droplets of differing charge formed as a result of statistical random sampling of buffer ions (*cf.* TS). The charged droplets will shrink, as in the ES process, with the eventual formation of gas phase buffer ions, analyte molecules (M) and solvent molecules (Sv). If ammonium acetate (NH_4^+ $CH_3CO_2^-$) is the buffer gas phase, $[NH_4]^+$ and $[CH_3CO_2]^-$ ions will be generated. The buffer gas ions will be free to react with analyte molecules in a CI event to generate analyte ions (eqns 6 and 7).

$$NH_4^+(g) + M(g) \rightarrow NH_3(g) + MH^+(g) \tag{6a}$$

$$NH_4^+(g) + M(g) \rightarrow [M + NH_4]^+(g) \tag{6b}$$

$$NH_4^+(g) + Sv(g) \rightarrow NH_3(g) + SvH^+(g) \tag{6c}$$

$$SvH^+(g) + M(g) \rightarrow [M + SvH]^+(g) \tag{6d}$$

In the positive-ion mode the exact products of the CI event will depend on the gas-phase basicity of the analyte, solvent and buffer. Adduct ions can also be formed (*e.g.* $[M + NH_4]^+$ and $[M + CH_3CN + H]^+$ in aqueous acetonitrile buffered with ammonium acetate). When the ion source is operated in the negative-ion mode, the products of the CI event will depend on the gas-phase acidity of the sprayed components.

$$CH_3CO_2^-(g) + M(g) \rightarrow CH_3CO_2H(g) + [M - H]^-(g) \tag{7a}$$

$$CH_3CO_2^-(g) + M(g) \rightarrow [M + CH_3CO_2]^-(g) \tag{7b}$$

(ii)   APCI can also be achieved in a secondary process, in which electrons from the corona discharge ionise nitrogen gas in the APCI source, leading to the eventual CI of the analyte. In the positive-ion mode, again the eventual products depend on the proton affinity of the components (eqn 8), while in the negative-ion mode the gas phase acidity of the components will define which deprotonated molecules are generated.

$$N_2^{+\bullet}(g) + Sv(g) \rightarrow Sv^{+\bullet}(g) + N_2(g) \tag{8a}$$

$$Sv^{+\bullet}(g) + Sv(g) \rightarrow SvH^+(g) + [Sv - H]^\bullet(g) \tag{8b}$$

$$SvH^+(g) + M(g) \rightarrow MH^+(g) + Sv(g) \tag{8c}$$

$$SvH^+(g) + M(g) \rightarrow [M + SvH]^+(g) \tag{8d}$$

Both primary and secondary processes can operate simultaneously, although by turning the filament off only the primary processes proceed.

APCI has been extensively used for the ionisation of neutral metabolites, *e.g.* steroids,[53–64] oxysterols,[65] and bile acid acyl glucosides,[66] and also ionic metabolites, *e.g.* bile acid glucuronides,[67] free bile acids[68] and steroid glucuronides.[64] It is compatibility with HPLC operated at high flow rates (100–500 µL/min), and this makes it a favoured mode of ionisation for high-throughput analysis.[69]

(d)   Electron Capture Atmospheric Pressure Chemical Ionisation (ECAPCI).   ECAPCI is suitable for the ionisation of molecules which contain a group with high electron affinity, and it can be regarded as the atmospheric

pressure equivalent of EC-NCI. For steroid analysis, for example, it is necessary to derivatise the analyte so as to introduce an electron-capturing group, *e.g.* pentafluorobenzyl moiety[70] or 2-nitro-4-trifluoromethylphenylhydrazone moiety[71] (Scheme 1.2). The resultant derivative is then introduced into a conventional APCI source operated in the negative-ion mode with the corona discharge needle "on". Low-energy secondary electrons are generated as a side-product of N_2 ionisation (eqn 9), and are captured by the electron capturing group tagged to the analyte to generate $[M]^{-\bullet}$ ions. When the pentafluorobenzyl derivatising group is employed, dissociative electron capture occurs with the loss of the pentafluorobenzyl radical and formation of a negative ion, $[A]^-$, which reflects the structure of the original analyte. Alternatively, 2-nitro-4-trifluoromethylphenylhydrazones give stable $[M]^{-\bullet}$ ions which can be fragmented in a tandem mass spectrometry (MS/MS) experiment (see below) to give product ions $[A]^-$, and neutral fragments $[B]^{\bullet}$, which reflect the structure of the analyte or are characteristic of the derivatising agent.

$$N_2(g) + \underline{e}^- \rightarrow N_2^{+\bullet}(g) + 2e^- \qquad (9a)$$

$$e^- + M(g) \rightarrow M^{-\bullet}(g) \qquad (9b)$$

$$M^{-\bullet} \rightarrow A^-(g) + B^{\bullet}(g) \qquad (9c)$$

ECAPCI is specific to molecules containing an electron-capturing group, and hence provides high specificity, and has been shown to provide enhancement in sensitivity compared to APCI of 20-fold[72] to 200-fold.[70]

(e) Atmospheric Pressure Photoionisation (APPI). APPI was introduced in 2000 by Robb, Covey and Bruins.[73] Photons (10 or 10.6 eV) are provided by a krypton lamp, and dopant molecules, *e.g.* toluene BzH, introduced into the ion source in combination with analyte (M) and mobile phase,

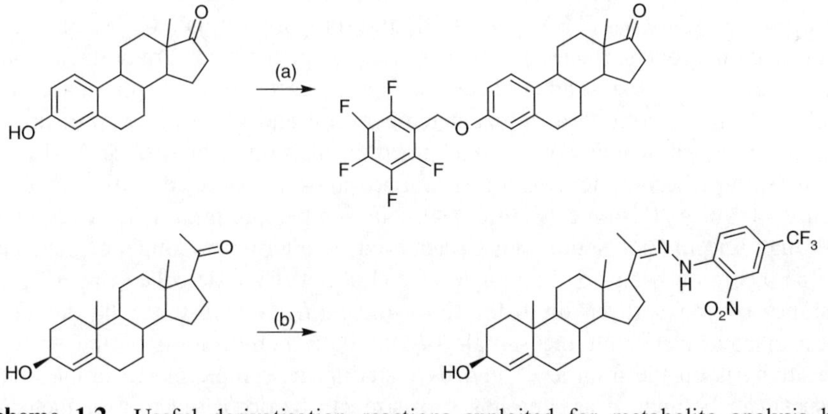

Scheme 1.2 Useful derivatisation reactions exploited for metabolite analysis by ECAPCI-MS. Reactions are illustrated on the A-ring of estrone and the side-chain of pregnenolone.

are vaporised and photoionised. This results in the formation of radical cations, *e.g.* $[BzH]^{+\bullet}$ (eqn 10).

$$BzH(g) + hv\,(10 \text{ or } 10.6 \text{ eV}) \rightarrow BzH^{+\bullet}(g) \tag{10a}$$

$$BzH^{+\bullet}(g) + Sv(g) \rightarrow Bz^{\bullet}(g) + SvH^{+}(g) \tag{10b}$$

$$M(g) + SvH^{+}(g) \rightarrow MH^{+}(g) + Sv(g) \tag{10c}$$

$$BzH^{+\bullet}(g) + M(g) \rightarrow M^{+\bullet}(g) + BzH(g) \tag{10d}$$

The radical cations, *e.g.* $[BzH]^{+\bullet}$, react with solvent molecules (Sv) producing protonated solvent molecules $[SvH]^{+}$ if the proton affinity of the solvent molecule is higher than that of the benzyl radical $(Bz)^{\bullet}$. The analyte is then ionised by proton transfer if the proton affinity of the analyte is higher than that of the solvent molecule. Alternatively, if the proton affinity of the solvent molecule is lower than that of the benzyl radical $(Bz)^{\bullet}$, a charge exchange reaction between the radical cation $[BzH]^{+\bullet}$ and the analyte can take place if the ionisation energy of the analyte is lower than that of the radical cation; the result is the formation of an analyte radical cation $[M]^{+\bullet}$.[63] APPI has been used for the analysis of biomolecules including sterols,[74,75] steroids[63,76–78] and other neutral lipids (see Chapter 6). A disadvantage of the technique, as illustrated in the study of Greig *et al.* on corticosteroids, is the formation of overlapping $[M+1]^{+\bullet}$ (due to the ^{13}C isotope) and $[M+H]^{+}$ ions.[76] This can cause confusion in molecular weight determination.

(f) Desorption Electrospray Ionisation Mass Spectrometry (DESI).   While the discovery in 1981 of FAB ionisation by Barber and colleagues[14] started the revolution in biomedical mass spectrometry, the next breakthrough that has the potential to lift mass spectrometry into a new dimension is DESI, introduced in 2004 by Cooks and colleagues.[79–83] DESI bears features of both classical desorption ionisation (*i.e.* FAB, MALDI) and spray ionisation (*i.e.* ES). Sample is deposited on a solid surface, *e.g.* urine or blood on paper,[80] solid tumour slice on a glass plate, or is the solid surface itself, *e.g.* heart tissue or pharmaceutical tablet,[81,82] and is ionised at ambient temperature and atmospheric pressure by electrospraying a nebulised beam of charged droplets onto the surface. Molecules at the surface become ionised and vaporised and are collected at the sampling orifice of an API source (Figure 1.4). No sample pre-treatment is required, although selective ionisation can be achieved by altering the angle of incidence of the incoming beam (α), the angle of collection of the desorbed ions (β), the distance of the sprayer tip from the sample (d_1), the distance of the mass spectrometer inlet from the sample (d_2) and the composition of the solvent. Sensitivity is on the fmol level, and by rastering the sample plate, samples (*e.g.* tissues) can be imaged as sampling spot size can be decreased to a diameter of 50 µm by using nano-ES. Methanol/water/acetic acid provides a good spraying solvent, but can also include reactive reagents (*e.g.* HCl, trifluoroacetic acid) to

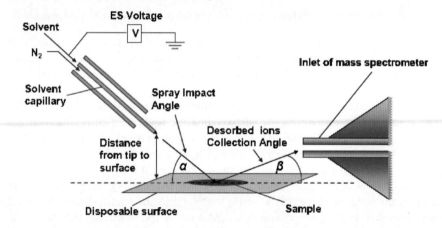

Figure 1.4 Schematic diagram of DESI source, from www.prosolia.com/ with permission.

provide selective ionisation. Alternatively, conductive organic solvents can be used. Flow rates are of the order of 0.05–$5\,\mu L/min$, and in the positive ion mode $[M+H]^+$, $[M+nH]^{n+}$ or $[M]^{+\bullet}$ ions are formed. The ionisation source can also be operated in the negative-ion mode and $[M-H]^-$, $[M+Cl]^-$ or $[M]^{-\bullet}$ ions may be observed depending on the analyte and operating conditions. For metabolite analysis this new ionisation is particularly exciting as ionisation can be achieved off paper, and one can easily imagine that by altering the properties of the paper, selective ionisation can be achieved. The optimal instrument settings for the analysis of neutral biomolecules are rather different from those of charged analytes and this can allow selective ionisation. Ion formation by DESI has been suggested to occur via three different mechanisms.[79] The first is thought to occur via charge transfer between gaseous projectile ions, $[PH]^+(g)$, in the electrospray beam and the analyte compound, M(sur), on the surface (eqn 11). The second mechanism is the "droplet pickup mechanism" and involves impact of electro-sprayed droplets, $[PH_n]^{n+}(g)$, on the surface, dissolution of the analyte, M(sur), at the surface onto or into the droplets and then release of "electrospray-like" droplets from the surface. Subsequent solvent evaporation and Coulomb fission generates ions in a process similar to that thought to occur in ES. The third mechanism of ion formation involves volatilisation/desorption of neutral species from the surface, $M(sur \rightarrow g)$, followed by gas phase ionisation through proton transfer/electron transfer or other ion molecule reactions at atmospheric pressure. This third mechanism is very similar to APCI and has been called DAPCI.

$$PH^+(g) + M(sur) \rightarrow MH^+(g) + P(g \text{ or sur}) \qquad (11a)$$

$$\begin{aligned}[PH_n]^{n+}(g \rightarrow sur) + M(sur) \rightarrow [MH_m]^{m+} \\ (sur \rightarrow g) + [PH_{n-m}]^{n-m+}(sur \text{ or } g)\end{aligned} \qquad (11b)$$

$$PH^+(g) + M(sur \rightarrow g) \rightarrow MH^+(g) + P(g) \qquad (11c)$$

1.2.2.4   Matrix Assisted Laser Desorption/Ionisation (MALDI)

MALDI is extensively used in imaging (see Chapter 9). It is also popular in lipidomics,[84–86] particularly for phospholipids analysis,[87–90] and to a lesser extent steroid and bile acid analysis.[91–95] MALDI is usually combined with time-of-flight (TOF) analysers, and its major advantages are robustness and ease of use. Sample is mixed with a solution of matrix, often α-cyano-4-hydroxycin-namic acid (or 3,5-dimethoxy-4-hydroxycinnamic acid or 2,5-dihydroxybenzoic acid) in aqueous acetonitrile containing 0.1% trifluoroacetic acid, and spotted on a stainless steel target plate and allowed to co-crystallise in air. The MALDI plate is then admitted to the high vacuum system of a mass spectrometer, and irradiated with laser light, usually of 337 nm from a N_2 laser. The matrix absorbs the light energy with the result that sample and matrix become ionised and vaporised. In the positive-ion mode $[M + H]^+$ ions are usually formed, while $[M - H]^-$ ions are formed in the negative-ion mode. The exact mechanism of ionisation is not fully understood and will not be speculated on here.[96] For steroid analysis, the most promising MALDI studies have been performed on steroids which have been derivatised so as to posess a preformed charge.[97–99]

A relative of MALDI is DIOS, an acronym for desorption ionisation on silicon, and as the name suggests, a silicon support is used as an alternative to matrix. Suizdak and colleagues have demonstrated the use of DIOS for the analysis of steroids in plasma.[100]

1.2.3   Mass Analysers

Once a sample has been ionised, it is transported from the ion source to the mass analyser (Figure 1.1). Trapping mass analysers can be an exception in that ionisation and mass analysis can be achieved in the trap itself (*e.g.* EI). Mass analysers operate by separating ions according to their m/z. Historically, early mass analysers used in metabolite research were based on magnetic sector fields,[101–103] but such analysers are less widely used for biological applications today. However, ion cyclotron resonance (ICR) analysers also use magnetic fields and are currently gaining popularity in biological mass spectrometry. Rather than using a magnetic field to separate ions according to their m/z most of today's mass analysers separate ions by their behaviour in electrical fields.

1.2.3.1   Magnetic Sector Analysers

Like all other mass analysers, magnetic sector analysers separate ions according to their m/z. Usually, the magnetic sector (B) is arranged in series with an electric sector (E) and the combination (either EB or BE) can give resolutions in excess of 100 000 (10% valley definition) for ions in the usual metabolite mass range.[104] Modern magnetic sector instruments are most commonly interfaced with an EI source, with or without a GC inlet, although in the past they have been coupled to TS, FAB, MALDI and ES ion sources. Magnetic sectors are still used today for accurate (exact) mass measurements (<5 ppm) at high resolution ($>10\,000$, 10% valley), and these instruments offer excellent dynamic range, important in isotope abundance measurements. Magnetic sector analysers operate with keV ion beams. This can be seen as a disadvantage, as the ion source must be raised to high potential relative to the rest of the instrument, and was initially problematic with API sources. However, a major advantage of keV ion beams is seen when the instrument is operated as a tandem mass spectrometer, allowing collision-induced dissociation (CID) reactions to occur at high collision energy (keV).

The figures of merit of typical magnetic sector instruments can be viewed as follows, but naturally, the exact characteristics vary depending on model and design.[105,106]

- Mass resolving power: 10^3–10^5 (10% valley)
- Mass accuracy: 1–5 ppm
- m/z range: 10^4
- Linear dynamic range: 10^9
- Scan speed: $\sim$ s
- Efficiency (transmission × duty cycle): $<$1% (scanning)
- Compatible with ionisation: API/vacuum (continuous)
- MS/MS: keV CID
- Cost: moderate to high
- Size/weight/laboratory requirements: MS laboratory

1.2.3.2 Linear Quadrupole Mass Filters

The quadrupole mass filter provides a much smaller and lower cost analyser than the magnetic sector. It consists of four parallel rods arranged equidistant from a central axis (Figure 1.5). By the application of a combination of radio frequency (rf) alternating current (ac) and direct current (dc) voltage components to the rods, ions of one particular m/z can be transmitted along the central axis between the rods, and conveyed to the detector. Others are

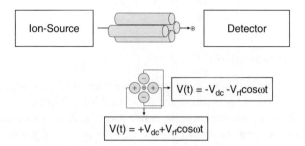

Figure 1.5   Schematic diagram of a quadrupole mass filter, showing the direction of ion travel and the equations for the potentials applied to the rods. $V(t)$ = voltage at time t; V_{dc} = direct current voltage component; $V_{rf} \cos \omega t$ = radio frequency voltage component.

deflected from the central axis and are not transmitted. By scanning the voltages applied to the rods an m/z range can be scanned (usually up to 2000 or 4000 m/z). For metabolite analysis, quadrupole mass filters are usually operated at a unit mass resolution (*i.e.* 0.7 Da, full width at half maximum height, FWHM), they are most often interfaced with EI or API sources, and offer the advantages of fast scanning ($\sim$ s) and stability. Quadrupole mass filters are additionally compatible with both GC and LC interfaces. Quadrupole mass filters can be used for accurate mass measurements,[107] but are rarely used for such applications. Some of the performance characteristics of quadrupole mass filters can be enhanced at the expense of others. For example, Smith and colleagues extended the mass range of their quadrupole to 45 000 m/z by reducing the operating frequency from 1.2–1.5 MHz to 262 Hz, but this was at the expense of resolving power and sensitivity.[108] Collings and Douglas were able to obtain a resolving power of 8500 (FWHM) at m/z 5000 by incorporating a high-pressure collision quadrupole between the ion source and quadrupole mass filter so as to collisionally focus the ions into a narrow beam for injection into the mass filter;[109] this gave relatively little loss in ion transmission as operating frequency was reduced to extend the mass range. Quadrupole mass filters can be arranged in series to give tandem quadrupole instruments,[110] and have also been coupled in series with magnetic sector[111] and orthogonally arranged TOF analysers[112–114] to give hybrid tandem mass spectrometers. The figures of merit of quadrupole mass filters can be summarised as follows.

- Mass resolving power: 10^2–10^4 (FWMH)
- Mass accuracy: 100 ppm
- m/z range: 10^4
- Linear dynamic range: 10^7
- Scan speed: $\sim$ s
- Efficiency (transmission $\times$ duty cycle): $<1\%$ (scanning) to 95%
- Compatible with ionisation: API/vacuum (continuous)
- MS/MS: eV CID
- Cost: low to moderate (MS/MS)
- Size/weight/laboratory requirements: bench top

1.2.3.3   Quadrupole Ion Trap

A cylindrical, or three dimensional quadrupole ion trap can be imagined as a quadrupole bent around on itself to form a closed loop. The inner rod is reduced to a point at the centre of the trap; the outer rod is a circular ring electrode, and the top and bottom rods become two end cap electrodes (Figure 1.6). Ions can be formed by EI within the trap or can be introduced from an external source (*e.g.* ES, MALDI). In the case of EI, electrons are admitted through a small central hole in one of the end caps; alternatively, ions formed in an external source can be transported to the trap and similarly admitted through the end

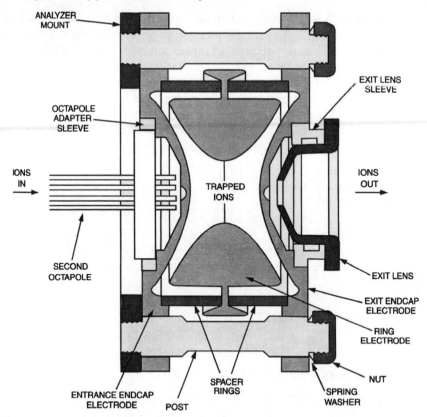

Figure 1.6   Schematic diagram of the LCQ ion trap manufactured by Thermo Electron.

cap. Initially ions of all m/z values are confined in the trap, and are expelled and detected according to their m/z, by ramping linearly the amplitude of the radio frequency potential applied to the ring electrode. Each ion species is ejected from the potential well at a specific radio frequency amplitude and, because the initial amplitude and ramping rate are known, the m/z can be determined for each ion species upon ejection. This method for measuring the m/z of confined ions was developed by Stafford *et al.* and is known as the "mass-selective axial instability scan-mode".[115]

Commercial ion traps have m/z ranges up to 6000 depending on their application; they are fast scanning (1000 m/z units/s); and are capable of "zoom scans" or enhanced resolution scans offering resolutions of 10 000 (FWHM) by scanning a short m/z range slowly. Unfortunately, as a result of space charging within the trap, mass accuracy is considerably lower than that achievable on beam instruments. Like the quadrupole mass filter, the ion trap can be used as a tandem mass spectrometer, but has the additional capability of multiple stages of fragmentation or MS^n. The cylindrical ion trap is now being replaced by a new generation of linear ion traps, where ions are trapped within a quadrupole mass analyser by potentials applied to end plates. This allows better mass accuracy as a result of reduced space charging provided by the greater cell volume. Linear ion traps (LITs) are combined with quadrupole (*e.g.* QTrap, Applied Biosystems), Fourier Transform (FT)ICR (*e.g.* LTQFT, Thermo Electron) and Orbitrap (*e.g.* LTQ-Orbitrap, Thermo Electron) mass analysers to give tandem instruments. Many excellent books and reviews have been written describing the principles and application of ion-trap mass spectrometers and the interested reader is directed to these.[116,117] Some of the figures of merit of ion-trap instruments are as follows.

- Mass resolving power: 10^3–10^4 (FWMH)
- Mass accuracy: 100 ppm
- *m/z* range: 10^4
- Linear dynamic range: 10^2–10^5
- Scan speed: $\sim$ s
- Efficiency (transmission × duty cycle): <1% (scanning) to 95%
- Compatible with ionisation: API/vacuum (continuous/pulsed)
- MS/MS: eV CID, MS^n
- Cost: low
- Size/weight/laboratory requirements: bench top

1.2.3.4   Time-of-Flight (TOF) Analysers

The first generation of TOF mass analysers were coupled to EI sources,[118] but it was not until the advent of MALDI that the modern era of TOF mass spectrometry was initiated. The pulsed nature of the MALDI source complements the necessity for time measurements in TOF. Simply, in the TOF mass spectrometer, ionisation occurs in the ion source and ions are pulsed into the TOF drift tube, and the time taken for the ions to traverse the drift tube to the detector is a measure of their *m/z* (Figure 1.7).

The TOF analyser is based on the following: ions formed in the ion source are accelerated through a potential *V*, and will gain a kinetic energy $mv^2/2$, so that:

$$mv^2/2 = zeV \tag{12}$$

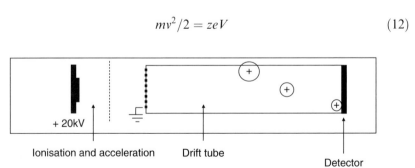

Figure 1.7   Main components of a linear TOF mass spectrometer.

where m is the mass of the ion, z the number of charges on the ion, and e the charge of an electron.

Then:

$$m/z = 2eV/v^2 \tag{13}$$

$$(m/z)^{1/2} = (2eV)^{1/2}.t/d \tag{14}$$

where t is the time the ion takes to travel down the drift tube of length d and reach the detector. For a given instrument operated at constant accelerating potential, d and V are constants, then:

$$t \propto (m/z)^{1/2} \tag{15}$$

TOF analysers can also be interfaced to API or EI sources,[113] in which case the source is arranged at right angles to the TOF drift tube and ions are pulsed orthogonally into the TOF analyser. The combination of API and orthogonal TOF is currently gaining popularity in metabolomic research,[119,120] as TOF analysers are fast "scanning", theoretically have unlimited mass range, and when combined with delayed extraction and a reflectron or ion mirror offer resolutions of up to 20 000 (FWHM). TOF analysers can be coupled to quadrupole mass filters to give Q-TOF tandem instruments[112,114] (Figure 1.8), ion traps to give QIT-TOF,[121] or arranged in series to give TOF-TOF instruments.[122,123] The figures of merit for TOF analysers are as follows.

- Mass resolving power: 10^4 (FWMH)
- Mass accuracy: 2–50 ppm
- m/z range: $>10^5$
- Linear dynamic range: 10^2–10^6
- Scan speed: ms
- Efficiency (transmission × duty cycle): 1–95%
- Compatible with ionisation: API/vacuum (continuous/pulsed)
- MS/MS: eV or keV CID
- Cost: low to high
- Size/weight/laboratory requirements: bench top or floor standing

1.2.3.5 Fourier Transform Ion Cyclotron Resonance (FTICR) Mass Spectrometers

FTICR mass spectrometers[124] are ion trapping instruments. However, unlike the quadrupole ion trap, the trapping field is magnetic rather than electrostatic. Ionisation occurs in an external source and ions are transported into the high vacuum ICR cell. Within the cell ions move with cyclotron motion governed by their cyclotron frequency. Ions of differing m/z values have different cyclotron frequencies, which are detected as an induced- or image-current as ions pass

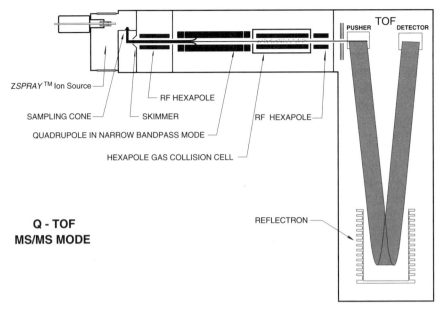

Figure 1.8   Layout of Waters Q-TOF MS/MS system. Ionisation is achieved by API.
The quadrupole can be operated in the narrow bandpass mode to transmit
a defined precursor-ion for fragmentation in the collision cell and the
recording of an MS/MS spectrum, or in the wide bandpass mode for the
recording of a first dimension mass spectrum in the TOF. The ion optics
make the beam "more parallel" before it enters the orthogonal accelerator
or pusher. The beam "fills" the first stage of the orthogonal accelerator
until a pushout pulse is applied. A packet of ions is thus sampled and
accelerated through grids to enter the drift region of the TOF. Reflecting
TOF optics are used to bring the ions to a space-time focus on the
detector. In this case the ion mirror shown has one stage (V-mode),
though two stages can also be used (W-optics). During the time that the
ions are in the drift-region (and ion mirror) the orthogonal accelerator is
"refilled" with new beam.

receiver plates, and can be converted into m/z values by application of Fourier
transform (FT). FTICR mass spectrometry is unlike most other forms of mass
spectrometry in that it is non-destructive and signal enhancement can be
achieved by signal averaging many ion cycles (this is also true of the Orbitrap
mass analyser). FTICR provides exceptionally high resolution in the mass
range of metabolites (>100 000 FWHM), and additionally provides high
accuracy of mass measurement (<2 ppm).[76] Like quadrupole ion traps, FTICR
instruments can be used as tandem mass spectrometers, and have MS[n] capability. The cost of the high performance of FTICR instruments is in time, as
FTICR is based on frequency measurements. This can have significant implications when FTICR instruments are interfaced to LC.[125] Many excellent
reviews have been published on the technology and application of FTICR and

the interested reader is referred publications by Amster,[126] Page[127] and Heeren.[128] Some of the performance merits of the FTICR are as follows.

- Mass resolving power: 10^4–10^6 (FWMH)
- Mass accuracy: 1–5 ppm
- *m/z* range: $>10^4$
- Linear dynamic range: 10^2–10^5
- Scan speed: ~s
- Efficiency (transmission × duty cycle): 1–95%
- Compatible with ionisation: API/vacuum (continuous/pulsed)
- MS/MS: eV, MS^n
- Cost: high
- Size/weight/laboratory requirements: floor standing

1.2.3.6 Linear Ion Trap – Orbitrap Mass Spectrometer (LIT-Orbitrap)

A new type of electrostatic ion trap is the Orbitrap analyser manufactured by Thermo Electron as part of a LIT-Orbitrap hybrid MS/MS instrument. This instrument provides resolution of up to 100 000 (FWHM) and mass accuracy of better than 5 ppm. MS/MS can be performed in the LIT and fragment ions formed analysed by the Orbitrap (Figure 1.9). The instrument offers the sensitivity advantages of a LIT, with high resolution and exact mass analysis of the Orbitrap. Ions from the ion source are initially stored in the LIT and analysed in either MS or MS^n modes. Ions can be detected at the LIT detector, or ejected axially and trapped in an intermediate C-trap from which they are "squeezed"

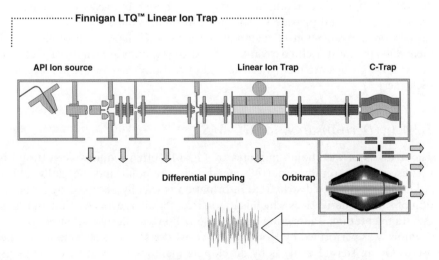

Figure 1.9 Schematic representation of the Thermo Electron LIT-Orbitrap.

into the Orbitrap. Trapped ions in the Orbitrap assume circular trajectories around the central electrode and perform axial oscillation. The oscillating ions induce an image current into the two halves of the Orbitrap, which can be detected. The axial oscillation frequency of an ion (ω) is proportional to the square root of the inverse of m/z [$\omega = k/(m/z)^{1/2}$], and the frequencies of complex signals derived from many ions can be determined using a Fourier transformation.[129,130] This instrument has already gained popularity for metabolite analysis,[131] in lipidomics[132] and for steroid and sterol analysis.[133–135] Some of the performance merits of the LIT-Orbitrap are as follows.

- Mass resolving power: 10^4–10^5 (FWMH)
- Mass accuracy: 1–5 ppm
- m/z range: 4000
- Scan speed: $\sim$ s
- Efficiency (transmission $\times$ duty cycle): 1–95%
- Compatible with ionisation: API
- MS/MS: eV, MS^n
- Cost: moderate to high
- Size/weight/laboratory requirements: floor standing

1.2.4   Gas Chromatography/Mass Spectrometry (GC-MS)

The earliest GC analyses of biological molecules were performed in the 1960's. VandenHeuvel *et al.* were the first practitioner in steroid analysis,[136] and it was the combination of GC with mass spectrometry that opened the door to the analysis of complex mixtures of metabolites. GC is ideally compatible with EI and CI as both rely upon gas-phase samples. A pre-requisite of GC-MS is sample vaporisation, and to achieve this without thermal decomposition, sample derivatisation is often required. Derivatives should be easy to prepare, be thermally stable and provide structurally informative mass spectra. Numerous reviews on derivatisation of metabolites for GC-MS have been published.[3–9] Generally, the most useful derivatives of hydroxyl groups are trialkylsilyl ethers, of carboxyl groups are methyl esters, and of carbonyl groups are alkyl oximes (Scheme 1.3).

1.2.4.1   Derivatisation for GC-MS

The analysis of metabolite mixtures by GC-MS often requires derivatisation because many naturally occurring metabolites are polar and thermally labile. Additionally, increased structural information is usually obtained by preparation of suitable derivatives which help to direct the charge localisation and thus the fragmentation. Pioneering work on methods to derivatise steroids, for example, was performed by Horning and co-workers[102] who also pioneered the use of GC in steroid analysis by developing methods for coating of supports with thin layers of stationary phase.[136]

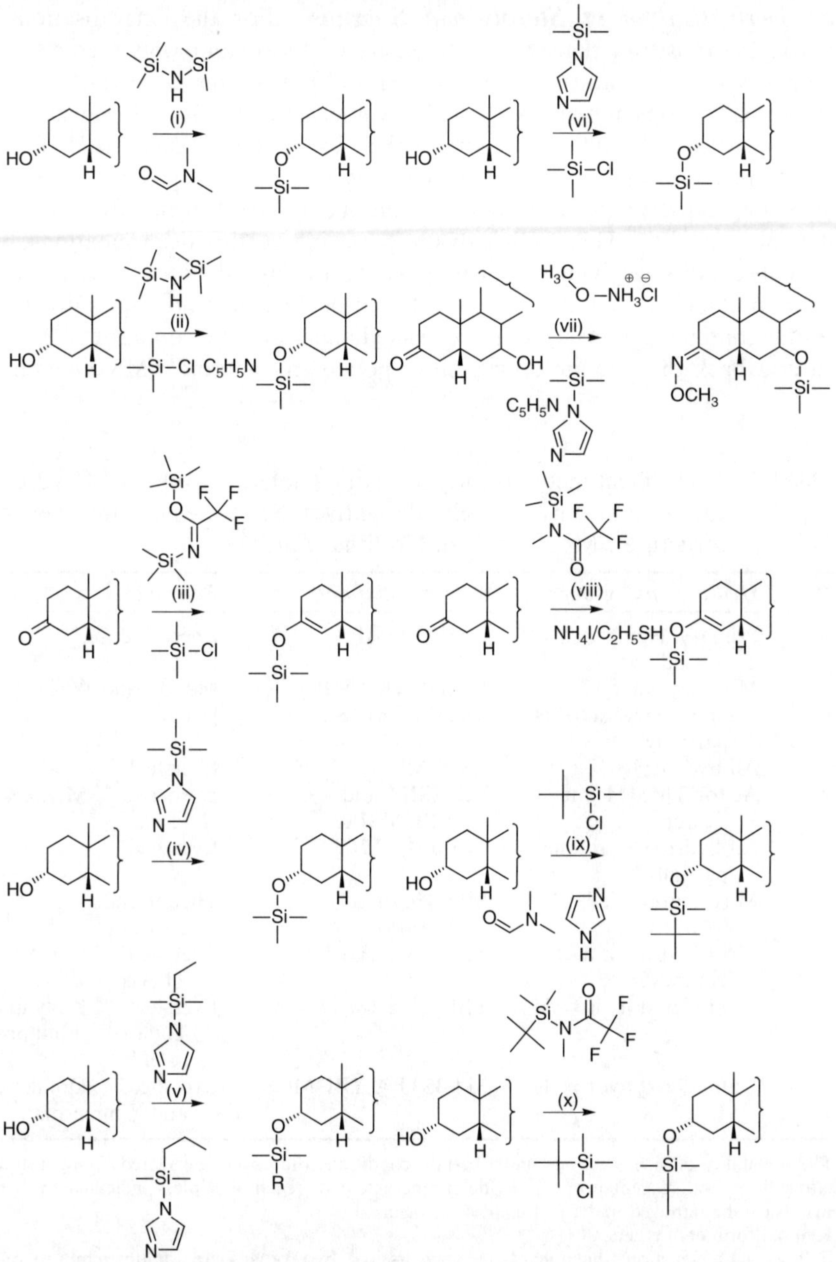

Scheme 1.3 Some derivatisation reactions useful for GC-MS analysis of steroids.

As the range of metabolite structures present in biology (almost) exceeds the imagination, in the following section we will exemplify chemical derivatisation for GC-MS by taking examples from just one class of biomolecules, *i.e.* sterols and steroids.

(a) Derivatisation of Sterols and Steroids. For the derivatisation of hydroxyl groups trimethylsilyl (TMS) ethers are most commonly used because of their ease of preparation (Scheme 1.3) and the commercial availability of deuterated versions of many reagents.[137] Numerous methods and reagents have been described since the first report of their use in steroid analysis.[138] References to reaction conditions useful in analyses of steroids are given in Table 1.1. Depending on their position and orientation, hydroxyl groups will react at different rates.[139,140] This can be advantageous in structure determinations, but normally conditions giving complete silylation are desired. Side reactions giving rise to artefacts are not common but should be kept in mind, especially when steroids containing an oxo group are reacted under forceful conditions.[141–143] It is a drawback that the molecular ion of polyhydroxysteroid TMS ethers may

Table 1.1 Convenient methods for partial or complete conversion of hydroxyl and oxo groups into silyl derivatives. See Scheme 3 for chemical structures. Modified from Griffiths *et al.*[137]

	Hydroxyl/oxo groups	*Reagent mixture*[a]	*Reference*
i	Primary, equatorial secondary	HMDS/DMF	Eneroth *et al.*[146]
ii	Most secondary	HMDS/TMCS/Pyr	Makita and Wells[147]
iii	Some tertiary, ketones partially[b]	BSTFA/TMCS	Poole[148]
iv	All hydroxyls	TMSIM[c]	Evershed[149]
v	As for TMSIM but slower	EDMSIM and n-PDMSIM[c]	Evershed,[149] Miyazaki *et al.*[150]
vi	All hydroxyls, ketones partially[b]	TMSIM/TMCS[c]	Evershed[149]
vii	Most ketones, all hydroxyls	MO·HCl/Pyr/ TMSIM[cd]	Thenot and Horning[151,152]
viii	Most hydroxyls and ketones[b]	MSTFA/NH₄I/ET	Evershed,[149] Donike,[153] Thevis *et al.*[154]
ix	Unhindered hydroxyls	TBDMCS/IM/DMF[c]	Evershed,[149] Kelly and Taylor,[155] Phillipou *et al.*[156]
x	Unhindered hydroxyls	MTBSTFA/TBDMCS	Evershed,[149] Donike and Zimmerman[157]

[a] The possibility of side reactions under certain conditions should be considered. Temperatures, heating times and reagent proportions determine extent of reaction. Unless indicated by [c], the reagents can be removed under a stream of nitrogen.
[b] Ketones form enol ethers.
[c] Following the reaction, the nonvolatile reagents and by-products are rapidly removed on a minicolumn of Lipidex 5000 in hexane.[158]
[d] The oxo groups are first converted into methyl oximes.
Abbreviations: HMDS, hexamethyldisilazane; TMCS, trimethylchlorosilane; BSTFA, *N,O*-bis(trimethylsilyl)trifluoroacetamide; TMSIM, trimethylsilylimidazole; EDMSIM, ethyldimethylsilylimidazole; n-PDMSIM, n-propyldimethylsilylimidazole; MO·HCl, methoxyammonium chloride; MSTFA, *N*-methyl-*N*-trimethylsilyltrifluoroacetamide; TBDMCS, t-butyldimethylchlorosilane; MTBSTFA, *N*-methyl-*N*-t-butyldimethylsilyltrifluoroacetamide; IM, imadazole; ET, ethanethiol; DMF, dimethylformamide; Pyr, pyridine.

not be seen because of extensive loss of trimethylsilanol to give $[M - 90]^{+\bullet}$ ions. Steric hindrance may prevent reaction of bulky alkylsilyl reagents with hydroxyl groups in many positions. This can be used to an advantage for preparation of mixed TMS-alkyldimethylsilyl derivatives as an aid in the interpretation of spectra and in quantitative analytical work. Sterically crowded alkylsilyl ethers are more stable than TMS ethers and can be purified by conventional chromatographic methods without undergoing hydrolysis.[144,145]

Depending on reagent and conditions, oxo groups may react to form enol ethers. This has found special applications since molecular ions of enol-TMS ethers are usually abundant and fragmentation may be characteristic.[159] However, it is often difficult to obtain single derivatives quantitatively, which complicates analyses of mixtures. Oxo groups are therefore usually converted into methyl oximes (MO) when biological mixtures are analysed, as first described by Fales and Luukkainen.[160] Thenot and Horning performed detailed studies of the preparation of MO-TMS derivatives of natural and synthetic steroid hormone metabolites.[151,152] Except for the formation of *E-Z* isomers, which may separate in GC-MS analyses depending on the stationary phase, their methods usually provide single derivatives.

For GC-MS analyses, conversion of carboxyl groups into methyl esters is most commonly used (Scheme 1.1). Many methods have been described, but side reactions depending on the steroid structure are often neglected. The mildest methylation is achieved with diazomethane. By using the commercially available trimethylsilyldiazomethane reagent, the preparation of diazomethane from toxic precursors can be avoided. Artefactual formation of ethyl and/or trimethylsilyl esters is sometimes seen due to transesterification or incomplete methylation, respectively. Using these methods of derivatisation it is also possible to analyse steroid conjugates with neutral and acidic sugars by GC-MS as first shown by Hornings group.[161]

Derivatisation can also be a means of introducing a functional group with special mass spectrometric properties into molecules, *e.g.* suitable for detection by ECNI (EC-NCI). Hydroxyl groups may be converted into perfluoroacyl esters and carbonyl groups into perfluorobenzyl oximes (Scheme 1.1).[12] These have found use in quantitative GC-MS analysis but are not suitable for polyfunctional steroids because of partial reactions.

In addition to derivatisation (complete or partial), microchemical or enzymatic reactions can aid in structure determinations, or be used to give desirable mass spectrometric properties. Simple reactions include conversion of vicinal hydroxyl groups into acetonides (isopropylidene derivatives)[162] or boronates,[163,164] periodate oxidation of vicinal hydroxyl groups, oxidation with sodium bismuthate, selective or complete oxidation of hydroxyl groups with chromic acid in acetone,[146] reduction of carbonyl groups with sodium borohydride or lithium aluminium hydride. One advantage with the cyclic alkaneboronates is that they give prominent molecular ions, this is usually not the case for the TMS ether derivatives of the same steroids. Cholesterol oxidase oxidises most 3β-hydroxy-5α- and 3β-hydroxy-Δ^5-steroids[165] and alcohol dehydrogenases are available that have defined substrate specificities.

1.2.5   Tandem Mass Spectrometry (MS/MS)

While most biomolecules fragment in the ion source upon EI to give structurally informative fragment ions, in-source fragmentation with MALDI or API is
usually only minor. To obtain detailed fragmentation information on molecules ionised by MALDI or API (or CI methods) it is usual to perform MS/MS.
MS/MS is usually performed by incorporating a CID step, although other
methods to induce fragmentation exist. CID is regarded to occur within two
different collision energy regimes, *i.e.* high collision energy ($> 1000 \, eV$) or at
low collision energy ($< 200 \, eV$). The spectra recorded under these two regimes
may be very different in appearance. For example, cholesterol sulphate
$[M - H]^-$ ions fragment under high collision energy conditions by cleavages
within the steroid ring system, while under low collision energy conditions the
$[HSO_4]^-$ ion at m/z 97 is the only significant fragment ion observed.[166] When
spectra are recorded at intermediate collision energy (*e.g.* 400–800 eV) the
nature of the collision gas dictates whether the spectra appear more like high or
low collision energy spectra. Heavy collision gas atoms (*e.g.* Xe) promote high
collision energy like CID, while light gas atoms (*e.g.* He) promote low collision
energy like CID.

   High collision energy spectra are only recorded on magnetic sector instruments or on TOF/TOF instruments, while all other tandem mass spectrometers
give low-energy CID spectra. Low-energy CID spectra recorded on beam
instruments (*e.g.* tandem quadrupole, Q-TOF instruments) often differ in
appearance to those recorded on trapping instruments (quadrupole ion trap,
FTICR, LIT-Orbitrap). With ion-beam instruments, fragmentation occurs in a
multiple collision process and a broad distribution of energy is imparted into
the fragmenting ion. Initially formed fragment ions may decompose further,
resulting in a mix of different fragment ions eventually reaching the detector.
With an ion trap, the ion of interest is selected while all others are expelled from
the trap. In the conventional quadrupole ion trap collisional activation involves
a competition between ion excitation and ion ejection. Poor efficiencies are
obtained at low trapping levels due to precursor ion ejection. This is avoided by
using higher trapping levels during excitation, but this can result in the loss of
information if some of the product ions fall below the low mass cut-off. The low
mass cut-off corresponds to the bottom third of the MS/MS spectrum recorded
on an ion trap where ions are not detected, and is perhaps the biggest
disadvantage of the ion trap. With a quadrupole ion trap, the product ion
current is usually concentrated into a single or just a few fragmentation
channels, *i.e.* those with the lowest activation energy. This results in the
formation of only few different fragment ions, but those that are formed are
of high abundance. The MS/MS spectrum is obtained by sequentially ejecting
the ions out of the trap towards the detector. Alternatively, one fragment ion
can be maintained in the trap while the others are ejected, this fragment can
then be activated to give a MS^3 spectrum. This process can be repeated to give
up to MS^6. The advantages provided by MS^n scans for metabolite identification
is illustrated by examples given in Chapters 3 and 6.

FTICR instruments can also be used for MS/MS experiments; again the ion of interest is trapped while all others are expelled, the ion is then activated and the fragment ions detected. Many different methods of ion activation can be employed in an FTICR instrument, *e.g.* sustained off-resonance irradiation (SORI),[126,167] infrared multiphoton dissociation (IRMPD),[168] black body infrared dissociation (BIRD),[169] each giving collision energies in the low-energy regime. The FTICR instruments offers the highest performance characteristics of all MS/MS instruments (except a capacity to perform high-energy CID). Precursor ions can be selected at high resolution, and fragment ions measured with high mass accuracies and at high resolution. Additionally, the non-destructive nature of the FTICR ion detection system allows ion re-measurement with the accompanied gain in signal to noise ratio. The non-destructive nature of the FTICR also allows MS^n on a single population of ions, with out the necessity of re-populating the trap between MS^n steps. The down side to the high performance of FTICR is the cost in time, which introduces severe limitations when the mass spectrometer is combined with a chromatography inlet. To counteract this problem, many of the current generation of FTICR instruments are quadrupole ion trap–FTICR hybrids, where the FTICR can be used as a high resolution, high mass accuracy analyser and detector, and the quadrupole ion trap can perform the MS/MS and MS^n process.[170] The Orbitrap which is also a FT-based mass analyser is also used as a high resolution, high mass accuracy analyser and detector when combined with a quadrupole ion trap in the LIT-Orbitrap hybrid instrument.[129,130]

Despite being unable to perform MS^n, beam instruments offer a major advantage over trapping instruments in that they are able to perform multiple different types of scan. In addition to product ion scans, magnetic sector, tandem quadrupole, and to a lesser extent Q-TOF type instruments can perform precursor ion and neutral loss scans (Figure 1.10). Tandem quadrupole instruments can also perform single reaction monitoring (SRM) or multiple reaction monitoring (MRM) "scans", where the precursor ion and fragment ion(s) are predefined, alleviating the need to scan the quadrupoles, thus achieving maximum ion transmission and sensitivity.

1.2.6 Liquid Chromatography–Mass Spectrometry (LC-MS)

The decision on what form of LC to couple with mass spectrometry depends on the desired application, and it may in fact be preferable to negate an LC separation step all together (see Chapters 4 and 5). Conventional LC encompasses both normal bore (3–4.6 mm i.d., 0.5–3 mL/min) and narrow bore (1–2 mm i.d., 20–300 μL/min) columns, while capillary LC includes micro bore (150–800 μm i.d., 2–20 μL/min) and nano bore columns (20–100 μm i.d., 100–1000 nL/min). The earliest generation of ES interface (which is still in use today) is well matched with flow rates of the order of 5–100 μL/min, with maximum sensitivity being achieved at the lower end of this flow rate range, and is most compatible with narrow bore or micro-bore columns. It is possible

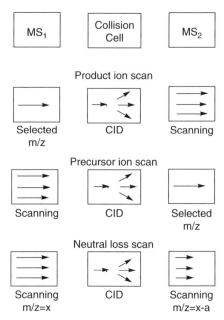

Figure 1.10 MS/MS experiments on spatially separated mass analysers.

to interface normal bore LC columns to such ES interfaces, but with the requirement of a post-column split. Pneumatically assisted ES, sometimes called ion spray or turbo ion spray, has been developed to allow the direct coupling of normal bore columns with the ES interface which is modified to receive flow rates of up to 1 mL/min. APCI interfaces are also capable of operating at this flow rate and receiving eluate from normal bore columns.

Theoretically, a reduction in column diameter produces a higher concentration of sample in an eluting peak.[171] As ES is also a concentration-dependent process, this dictates that maximum sensitivity can be achieved by using miniaturised LC, and has led to the increasing popularity of capillary LC-MS and capillary LC-MS/MS in biological mass spectrometry. A new generation of micro-ES interface has been developed which perform optimally at low flow rate ($< 1 \mu L$/min) and thereby provide maximum sensitivity when coupled with capillary column LC. Despite providing maximum sensitivity, capillary LC performed at low flow rate has its limitations. Although the concentration of sample in an eluting peak is dependent on the reciprocal square of column i.d., the column loading capacity and optimum injection volume also follow a similar relation (with respect to column i.d). This creates a problems in terms of sample injection and column overloading, particularly for columns of i.d. $< 300 \mu m$ where optimum injection volumes are less than $1 \mu L$.[172] The problem of low injection volume can be simply overcome by on-line sample pre-concentration on a trap column arranged in series with the analytical column. Sample pre-concentration is performed at μL/min flow rate on the trap column, which is then flushed, and sample separated on the analytical column.[173] However, the problem of column loading capacity still exists. The best solution to column overloading is to include a group separation step prior to capillary LC. This is illustrated in work performed by Yang *et al.*,[26] in which a urine extract from a child with cholestatic liver disease was separated into four fractions according to acidity on an anion exchange column. Each fraction was analysed by capillary-LC-ES-MS and MS/MS in a 1 hour run, allowing the partial characterisation of over 150 bile acids and conjugated bile alcohols. The remaining drawback with capillary column LC is one of analysis time. For example in the study performed by Yang *et al.*,[26] each LC run took 60 min precluding the possibility of high throughput analysis. However, with the development of ultra high pressure (performance) liquid chromatography (UPLC, 15 000 psi) using smaller particles ($<2\,\mu m$), or alternatively monolithic columns operated at lower pressure, it is likely that the time constraint associated with capillary chromatography will be overcome.

While capillary column LC combined with micro-ES will provide the maximum sensitivity for metabolite analysis, many screening and quantitative studies require high throughput as their main priority, in which case narrowbore chromatography combined with ES or APCI is the method of choice. High speed separations using UPLC offer an attractive alternative to HPLC, and the additional high reproducibility of retention time provided by UPLC marks this as a technique for the future.[119,120,174,175]

1.3 Future Perspectives

Mass spectrometry, along with NMR, is the key analytical techniques in metabolomics, metabonomics and metabolite profiling. Over the last 50 years mass spectrometry has been continually used for metabolite identification and metabolite profiling,[101,102] but with the advent of the "omic" revolution has become metamorphosised from an analytical technique mainly practised by chemists into one widely exploited in mainstream biology! Even as recently as the early 1980's it would not be unfair to quote Curt Brunnée's description of a mass spectrometer as "an instrument that only just doesn't quite work"; however, the concurrent development in API, TOF, FTMS, fast and reliable electronics and computer power, has, as we write (January 2007), made mass spectrometry a reliable tool routinely capable of unattended acquisition of high quality data over many days.

The future is bright in many directions of metabolite research. GC-MS instruments range from low-cost units where a quadrupole filter operated at unit mass resolution acts as a mass sensitive detector, through moderately expensive instruments employing a double-focusing magnetic sector or orthogonal TOF as the mass analyser, to true high resolution (60 000–100 000 10% valley) instruments. The GC-TOF and small GC-double focussing instruments are currently gaining popularity as they offer moderately high resolutions (~ 5000, 10% valley; $\sim 10\,000$, FWHM), exact mass measurements ($<5\,ppm$), fast scan times (1–0.1 s) and occupy a small footprint.

The exact mass capability of small moderately priced instruments has revived the use of this measurement in the structural determination of metabolites by GC-MS. As far back as 1954, Beynon in his classic paper in the journal *Nature*[176] reported the use of exact mass measurements to determine the elemental composition of ions. The fact that masses of all the isotopes of the elements have non-integer values (with the exception of ^{12}C), results in polyatomic ions having non-integer masses. In the mass range below 500 Da unique elemental compositions are separated by at least 1 ppm, and in many cases determination of mass to an accuracy of 5 ppm allows elemental compositions to be determined when valence rules and complementary information are considered.[177]

The ability of the modern generation of API–orthogonal TOF instruments to record mass spectra at moderate to high resolution (5000–20 000 FWHM) is now also being exploited for exact mass measurements in metabolomics.[178] As discussed above, the combination of UPLC with API–orthogonal TOF instruments offers fast chromatography at high performance with reproducibility in retention time and exact mass measurement. This combination will be of great value in the comparison of a metabolome under different conditions.

UPLC will offer further advantages when interfaced with tandem mass spectrometers. One exciting development is the combination of UPLC with the so-called MS(E) scan,[179] where full mass spectra are acquired on a Q-TOF instrument operated at low (5 eV) and "high" collision energy (25 eV) in consecutive scans. At low collision energy, the spectrum recorded is essentially a "normal" mass spectrum, while at "high" collision energy all precursor ions are simultaneously fragmented and the mass spectrum recorded is effectively a composite CID spectrum of all precursor components in the previous low energy scan. By taking advantage of exact mass and chromatographic peak shape for each component, the "high" energy scan can be de-convoluted to allow identification of the precursor ions. Conceptually the "high" energy scan and its de-convolution is very reminiscent of GC-MS performed at 70 eV. Note the rather loose usage of the term "high" collision energy to refer to $\sim$25 eV, which is distinct from the accepted meaning of high collision energy which is > 1 keV.

While UPLC is performed at very high pressure (15 000 psi) on columns packed with 1.7 μm particles,[119] and HPLC is usually performed at pressures below 6000 psi on columns packed with 3–5 μm particles, high speed chromatography can also be performed at HPLC pressures using particles of intermediate size (*i.e.* 1.9 μm).[135] To take full advantage of high speed chromatography requires sensitive and fast scanning mass spectrometers. Hybrid ion trap–FTMS instruments offer such an example, whether in the FTICR or FT-Orbitap format. Using these instruments the FTMS can be used to record a mass spectrum at high resolution with exact mass, while in parallel the ion trap can be used to perform MS^n on ions identified in the previous FTMS scan.

Although for global metabolite analysis fast chromatography is advantageous, for deep mining experiments on low abundance analytes low flow rate chromatography is the way forward (as demonstrated by the proteomics community) and the introduction of nano-UPLC is expected to expedite this process.

Acknowledgements

This work was supported by funding from UK Biotechnology and Biological Sciences Research Council (BBSRC grant no. BB/C515771/1 and BB/C511356/1), and The School of Pharmacy.

References

1. K. Blau and G. King, Eds, *Handbook of derivatives for chromatography*, Heyden & Son Ltd, London, 1977.
2. K. Blau and J. M. Halket, Eds, *Handbook of derivatives for chromatography*, 2nd edition, Wiley, Chichester, 1993.
3. J. M. Halket and V. G. Zaikin, Derivatization in mass spectrometry-1. Silylation, *Eur. J. Mass Spectrom.*, 2003, **9**, 1–21.
4. V. G. Zaikin and J. M. Halket, Derivatization in mass spectrometry-2. Acylation, *Eur. J. Mass Spectrom.*, 2003, **9**, 421–434.
5. J. M. Halket and V. G. Zaikin, Derivatization in mass spectrometry-3. Alkylation (arylation), *Eur. J. Mass Spectrom.*, 2004, **10**, 1–19.
6. V. G. Zaikin and J. M. Halket, Derivatization in mass spectrometry-4. Formation of cyclic derivatives, *Eur. J. Mass Spectrom.*, 2004, **10**, 555–568.
7. J. M. Halket, D. Waterman, A. M. Przyborowska, R. K. Patel, P. D. Fraser and P. M. Bramley, Chemical derivatization and mass spectral libraries in metabolic profiling by GC/MS and LC/MS/MS, *J. Exp. Bot.*, 2005, **56**, 219–243.
8. J. M. Halket and V. G. Zaikin, Derivatization in mass spectrometry-5. Specific derivatization of monofunctional compounds, *Eur. J. Mass Spectrom.*, 2005, **11**, 127–160.
9. V. G. Zaikin and J. M. Halket, Derivatization in mass spectrometry-6. Formation of mixed derivatives of polyfunctional compounds, *Eur. J. Mass Spectrom.*, 2005, **11**, 611–636.
10. A. G. Harrison, *Chemical ionization mass spectrometry*, CRC Press, Boca Raton, FL, 1992.
11. D. F. Hunt, G. C. Stafford, F. W. Crow and J. W. Russell, Pulsed positive negative ion chemical ionization mass spectrometry, *Anal. Chem.*, 1976, **48**, 2098–2104.
12. M. Vallée, J. D. Rivera, G. F. Koob, R. H. Purdy and R. L. Fitzgerald, Quantification of neurosteroids in rat plasma and brain following swim stress and allopregnanolone administration using negative chemical ionization gas chromatography/mass spectrometry, *Anal. Biochem.*, 2000, **287**, 153–166.
13. Y. S. Kim, H. Zhang and H. Y. Kim, Profiling neurosteroids in cerebrospinal fluids and plasma by gas chromatography/electron capture negative chemical ionization mass spectrometry, *Anal. Biochem.*, 2000, **277**, 187–195.

14. M. Barber, R. S. Bordoli, R. D. Sedgwick and A. N. Tyler, Fast atom bombardment of solids (FAB): a new ion source for mass spectrometry, *J. Chem. Soc. Chem. Commun.*, 1981, 325–327.
15. K. D. R. Setchell, M. Schwarz, N. C. O'Connell, E. G. Lund, D. L. Davis, R. Lathe, H. R. Thompson, R. W. Tyson, R. J. Sokol and D. W. Russell, Identification of a new inborn error in bile acid synthesis: mutation of the oxysterol 7α-hydroxylase gene causes severe neonatal liver disease, *J. Clin. Invest.*, 1998, **102**, 1690–1703.
16. J. Whitney, S. Lewis, K. M. Straub, M. M. Thaler and A. L. Burlingame, Analysis of conjugated bile salts in human duodenal bile using fast atom bombardment and field desorption mass spectrometry, *Koenshu-Iyo Masu Kenkyukai*, 1981, **6**, 33–44.
17. A. M. Ballatore, C. F. Beckner, R. M. Caprioli, N. E. Hoffman and J. G. Liehr, Synthesis and spectroscopic analysis of modified bile salts, *Steroids*, 1983, **41**, 197–206.
18. B. Egestad, P. Pettersson, S. Skrede and J. Sjövall, Fast atom bombardment mass spectrometry in the diagnosis of cerebrotendinous xanthomatosis, *Scand. J. Lab. Invest.*, 1985, **45**, 443–446.
19. P. T. Clayton, J. V. Leonard, A. M. Lawson, K. D. R. Setchell, S. Andersson, B. Egestad and J. Sjövall, Familial giant cell hepatitis associated with synthesis of 3β,7α-dihydroxy- and 3β,7α,12α-trihydroxy-5-cholenoic acids, *J. Clin. Invest.*, 1987, **79**, 1031–1038.
20. E. Wahlén, B. Egestad, B. Strandvik and J. Sjövall, Ketonic bile acids in urine of infants during the neonatal period, *J. Lipid Res.*, 1989, **30**, 1847–1857.
21. C. H. Shackleton and K. M. Straub, Direct analysis of steroid conjugates: the use of secondary ion mass spectrometry, *Steroids*, 1982, **40**, 35–51.
22. C. H. Shackleton, V. R. Mattox and J. W. Honour, Analysis of intact steroid conjugates by secondary ion mass spectrometry (including FABMS) and by gas chromatography, *J. Steroid Biochem.*, 1983, **19**, 209–217.
23. C. H. Shackleton, Inborn errors of steroid biosynthesis: detection by a new mass-spectrometric method, *Clin. Chem.*, 1983, **29**, 246–249.
24. L. J. Meng, W. J. Griffiths and J. Sjövall, The identification of novel steroid N-acetylglucosaminides in the urine of pregnant women, *J. Steroid Biochem. Mol. Biol.*, 1996, **58**, 585–598.
25. L. J. Meng, W. J. Griffiths, H. Nazer, Y. Yang and J. Sjövall, High levels of (24S)-24-hydroxycholesterol 3-sulfate 24-glucuronide in the serum and urine of children with severe cholestatic liver disease, *J. Lipid Res.*, 1997, **38**, 926–934.
26. Y. Yang, W. J. Griffiths, H. Nazer and J. Sjövall, Analysis of bile acids and bile alcohols in urine by capillary column liquid chromatography-mass spectrometry using fast atom bombardment or electrospray ionisation and collision induced dissociation, *Biomed. Chromatogr.*, 1997, **11**, 240–255.
27. M. Yamashita and J. B. Fenn, Electrospray ion source. Another variation on the free-jet theme, *J. Phys. Chem.*, 1984, **88**, 4451–4459.

28. M. Yamashita and J. B. Fenn, Negative ion production with the electrospray ion source, *J. Phys. Chem.*, 1984, **88**, 4671–4675.
29. J. B. Fenn, M. Mann, C. K. Meng, S. F. Wong and C. M. Whitehouse, Electrospray ionization for mass spectrometry of large biomolecules, *Science*, 1989, **246**, 64–71.
30. J. V. Iribarne and B. A. Thomson, On the evaporation of charged ions from small droplets, *J. Chem. Phys.*, 1976, **64**, 2287–2294.
31. J. V. Iribarne, P. J. Dziedzic and B. A. Thomson, Atmospheric pressure ion evaporation-mass spectrometry, *Int. J. Mass Spectrom. Ion Phys.*, 1983, **50**, 331–347.
32. M. Dole, L. L. Mack, R. L. Hines, R. C. Mobley, L. D. Ferguson and M. B. Alice, Molecular beams of macro ions, *J. Chem. Phys.*, 1968, **49**, 2240–2249.
33. L. O. G. Weidolf, E. D. Lee and J. D. Henion, Determination of boldenone sulfoconjugate and related steroid sulfates in equine urine by high-performance liquid chromatography/tandem mass spectrometry, *Biomed. Environ. Mass Spectrom.*, 1988, **15**, 283–289.
34. R. B. Cole, Ed., *Electrospray Ionization Mass Spectrometry, Fundamentals, Instrumentation and Applications*, John Wiley and Sons, New York, 1997.
35. P. Kebarle, A brief overview of the present status of the mechanisms involved in electrospray mass spectrometry, *J. Mass Spectrom.*, 2000, **35**, 804–817.
36. R. B. Cole, Some tenets pertaining to electrospray ionization mass spectrometry, *J. Mass Spectrom.*, 2000, **35**, 763–772.
37. G. J. Van Berkel, Electrolytic deposition of metals on to the high-voltage contact in an electrospray emitter: implications for gas-phase ion formation, *J. Mass Spectrom.*, 2000, **35**, 773–783.
38. M. H. Amad, N. B. Cech, G. S. Jackson and C. G. Enke, Importance of gas-phase proton affinities in determining the electrospray ionization response for analytes and solvents, *J. Mass Spectrom.*, 2000, **35**, 784–789.
39. M. Gamero-Castanño and J. Fernandez de la Mora, Kinetics of small ion evaporation from the charge and mass distribution of multiply charged clusters in electrosprays, *J. Mass Spectrom.*, 2000, **35**, 790–803.
40. S. J. Gaskell, Electrospray: principles and practice, *J. Mass Spectrom.*, 1997, **32**, 677–688.
41. A. P. Bruins, Mass spectrometry with ion sources operating at atmospheric pressure, *Mass Spectrom. Rev.*, 1991, **10**, 53–77.
42. N. B. Cech and C. G. Enke, Practical implications of some recent studies in electrospray ionisation fundamentals, *Mass Spectrom. Rev.*, 2001, **20**, 362–387.
43. G. I. Taylor, Disintegration of water drops in an electric Field, *Proc. R. Soc. London Ser. A.*, 1964, **280**, 383–397.
44. P. Kebarle and Y. Ho, On the mechanism of electrospray mass spectrometry, In *Electrospray Ionization Mass Spectrometry: Fundamentals,*

Instrumentation and Applications, R. B. Cole Ed., John Wiley and Sons, New York, 1997, p. 3.

45. W. J. Griffiths, A. P. Jonsson, S. Liu, D. K. Rai and Y. Wang, Electrospray and tandem mass spectrometry in biochemistry, *Biochem. J.*, 2001, **355**, 545–561.

46. A. Gomez and K. Tang, Charge and fission of droplets in electrostatic sprays, *Phys. Fluids*, 1994, **6**, 404–414.

47. M. R. Emmett and R. M. Caprioli, Micro-electrospray mass spectrometry: ultra-high sensitivity analysis of peptides and proteins, *J. Am. Soc. Mass Spectrom.*, 1994, **5**, 605–613.

48. M. S. Wilm and M. Mann, Electrospray and Taylor-cone theory: Dole's beam of macromolecules at last? *Int. J. Mass Spectrom. Ion Processes*, 1994, **136**, 167–180.

49. C. R. Blakley, M. J. McAdams and M. L. Vestal, Corossed beam liquid chromatography–mass spectrometry combination, *J. Chromatogr.*, 1978, **158**, 261–276.

50. C. Eckers, P. B. East and N. J. Haskins, The use of negative ion thermospray liquid chromatography/tandem mass spectrometry for the determination of bile acids and their glycine conjugates, *Biol. Mass Spectrom.*, 1991, **20**, 731–739.

51. T. R. Covey, E. D. Lee, A. P. Bruins and J. D. Henion, Liquid chromatography/mass spectrometry, *Anal. Chem.*, 1986, **58**, 1451A–1461A.

52. M. L. Vestal, Methods of Ion Generation, *Chem. Rev.*, 2001, **101**, 361–375.

53. S. Cristoni, D. Cuccato, M. Sciannamblo, L. R. Bernardi, I. Biunno, P. Gerthoux, G. Russo, G. Weber and S. Mora, Analysis of 21-deoxycortisol, a marker of congenital adrenal hyperplasia, in blood by atmospheric pressure chemical ionization and electrospray ionization using multiple reaction monitoring, *Rapid Commun. Mass Spectrom.*, 2004, **18**, 77–82.

54. K. Mitamura, T. Nakagawa, K. Shimada, M. Namiki, E. Koh, A. Mizokami and S. Honma, Identification of dehydroepiandrosterone metabolites formed from human prostate homogenate using liquid chromatography-mass spectrometry and gas chromatography-mass spectrometry, *J. Chromatogr. A*, 2002, **961**, 97–105.

55. Y.-C. Ma and H.-Y. Kim, Determination of steroids by liquid chromatography/mass spectrometry, *J. Am. Soc. Mass Spectrom.*, 1997, **8**, 1010–1020.

56. A. Lagana, A. Bacaloni, G. Fago and A. Marino, Trace analysis of estrogenic chemicals in sewage effluent using liquid chromatography combined with tandem mass spectrometry, *Rapid Commun. Mass Spectrom.*, 2000, **14**, 401–407.

57. G. Rule and J. Henion, High-throughput sample preparation and analysis using 96-well membrane solid-phase extraction and liquid chromatography-tandem mass spectrometry for the determination of steroids in human urine, *J. Am. Soc. Mass Spectrom.*, 1999, **10**, 1322–1327.

58. B. Starcevic, E. DiStefano, C. Wang and D. H. Catlin, Liquid chromatography-tandem mass spectrometry assay for human serum testosterone

and trideuterated testosterone, *J. Chromatogr. B Analyt. Technol. Biomed. Life Sci.*, 2003, **792**, 197–204.

59. R. Draisci, L. Palleschi, E. Ferretti, L. Lucentini and P. Cammarata, Quantitation of anabolic hormones and their metabolites in bovine serum and urine by liquid chromatography-tandem mass spectrometry, *J. Chromatogr. A.*, 2000, **870**, 511–522.

60. A. E. Nassar, N. Varshney, T. Getek and L. Cheng, Quantitative analysis of hydrocortisone in human urine using a high-performance liquid chromatographic-tandem mass spectrometric-atmospheric-pressure chemical ionization method, *J. Chromatogr. Sci.*, 2001, **39**, 59–64.

61. K. Shimada and Y. Mukai, Studies on neurosteroids. VII. Determination of pregnenolone and its 3-stearate in rat brains using high-performance liquid chromatography-atmospheric pressure chemical ionization mass spectrometry, *J. Chromatogr. B Biomed. Sci. Appl.*, 1998, **714**, 153–160.

62. C. Wang, D. H. Catlin, B. Starcevic, A. Leung, E. DiStefano, G. Lucas, L. Hull and R. S. Swerdloff, Testosterone metabolic clearance and production rates determined by stable isotope dilution/tandem mass spectrometry in normal men: influence of ethnicity and age, *J. Clin. Endocrinol. Metab.*, 2004, **89**, 2936–2941.

63. A. Leinonen, T. Kuuranne and R. Kostiainen, Liquid chromatography/mass spectrometry in anabolic steroid analysis – optimization and comparison of three ionization techniques: electrospray ionization, atmospheric pressure chemical ionization and atmospheric pressure photoionization, *J. Mass Spectrom.*, 2002, **37**, 693–698.

64. T. Kuuranne, M. Vahermo, A. Leinonen and R. Kostiainen, Electrospray and atmospheric pressure ionization tandem mass spectrometric behavior of eight anabolic steroid glucuronides, *J. Am. Soc. Mass Spectrom.*, 2000, **11**, 722–730.

65. I. Burkard, K. M. Rentsch and A. von Eckardstein, Determination of 24S- and 27-hydroxycholesterol in plasma by high-performance liquid chromatography-mass spectrometry, *J. Lipid Res.*, 2004, **45**, 776–781.

66. T. Goto, A. Shibata, T. Iida, N. Mano and J. Goto, Sensitive mass spectrometric detection of neutral bile acid metabolites. Formation of adduct ions with an organic anion in atmospheric pressure chemical ionization, *Rapid Commun. Mass Spectrom.*, 2004, **18**, 2360–2364.

67. J. Goto, N. Murao, C. Nakada, T. Motoyama, J. Oohashi, T. Yanagihara, T. Niwa and S. Ikegawa, Separation and characterization of carboxyl-linked glucuronides of bile acids in incubation mixture of rat liver microsomes, *Steroids*, 1998, **63**, 186–192.

68. S. Ikegawa, T. Goto, H. Watanabe and J. Goto, Stereoisomeric inversion of (25R)- and (25S)-3 alpha,7 alpha,12 alpha-trihydroxy-5 beta-cholestanoic acids in rat liver peroxisome, *Biol. Pharm. Bull.*, 1995, **18**, 1027–1029.

69. N. Clarke, M. Goldman, Clinical Applications of HTLC-MS/MS in the Very High Throughput Diagnostic Environment: LC-MS/MS on *Steroids. Proc. 53rd ASMS Conference on Mass Spectrometry and Allied Topics*, San Antonio, TX, 2005.

70. G. Singh, A. Gutierrez, K. Xu and I. A. Blair, Liquid chromatography/ electron capture atmospheric pressure chemical ionization/mass spectrometry: analysis of pentafluorobenzyl derivatives of biomolecules and drugs in the attomole range, *Anal. Chem.*, 2000, **72**, 3007–3013.

71. T. Higashi, N. Takido, A. Yamauchi and K. Shimada, Electron-capturing derivatization of neutral steroids for increasing sensitivity in liquid chromatography/negative atmospheric pressure chemical ionization-mass spectrometry, *Anal. Sci.*, 2002, **18**, 1301–1307.

72. T. Higashi, N. Takido and K. Shimada, Studies on neurosteroids XVII. Analysis of stress-induced changes in neurosteroid levels in rat brains using liquid chromatography-electron capture atmospheric pressure chemical ionization-mass spectrometry, *Steroids*, 2005, **70**, 1–11.

73. D. B. Robb, T. R. Covey and A. P. Bruins, Atmospheric pressure photoionization: an ionization method for liquid chromatography-mass spectrometry, *Anal. Chem.*, 2000, **72**, 3653–3659.

74. E. R. Trösken, E. Straube, W. K. Lutz, W. Völkel and P. Patten, Quantitation of Lanosterol and Its Major Metabolite FF-MAS in an Inhibition Assay of CYP51 by Azoles with Atmospheric Pressure Photoionization Based LC-MS/MS, *J. Am. Soc. Mass Spectrom.*, 2004, **15**, 1216–1221.

75. J. Lembcke, U. Ceglarek, G. M. Fiedler, S. Baumann, A. Leichtle and J. Thiery, Rapid quantification of free and esterified phytosterols in human serum using APPI-LC-MS/MS, *J.Lipid Res.*, 2005, **46**, 21–26.

76. M. J. Greig, B. Bolaños, T. Quenzer and J. M. R. Bylund, Fourier transform ion cyclotron resonance mass spectrometry using atmospheric pressure photoionization for high-resolution analyses of corticosteroids, *Rapid Commun. Mass Spectrom.*, 2003, **17**, 2763–2768.

77. W. Z. Shou, X. Jiang and W. Naidong, Development and validation of a high-sensitivity liquid chromatography/tandem mass spectrometry (LC/MS/MS) method with chemical derivatization for the determination of ethinyl estradiol in human plasma, *Biomed. Chromatogr.*, 2004, **18**, 414–421.

78. T. Guo, M. Chan and S. J. Soldin, Steroid profiles using liquid chromatography-tandem mass spectrometry with atmospheric pressure photoionization source, *Arch. Pathol. Lab. Med.*, 2004, **128**, 469–475.

79. Z. Takats, J. M. Wiseman, B. Gologan and R. G. Cooks, Mass spectrometry sampling under ambient conditions with desorption electrospray ionization, *Science*, 2004, **306**, 471–473.

80. Z. Takats, J. M. Wiseman and R. G. Cooks, Ambient mass spectrometry using desorption electrospray ionization (DESI): instrumentation, mechanisms and applications in forensics, chemistry, and biology, *J. Mass Spectrom.*, 2005, **40**, 1261–1275.

81. J. M. Wiseman, S. M. Puolitaival, Z. Takats, R. G. Cooks and R. M. Caprioli, Mass spectrometric profiling of intact biological tissue by using desorption electrospray ionization, *Angew Chem. Int. Ed. Engl.*, 2005, **44**, 7094–7097.

82. H. Chen, N. N. Talaty, Z. Takats and R. G. Cooks, Desorption electrospray ionization mass spectrometry for high-throughput analysis of pharmaceutical samples in the ambient environment, *Anal. Chem.*, 2005, **77**, 6915–6927.

83. I. Cotte-Rodriguez, Z. Takats, N. Talaty, H. Chen and R. G. Cooks, Desorption electrospray ionization of explosives on surfaces: sensitivity and selectivity enhancement by reactive desorption electrospray ionization, *Anal. Chem.*, 2005, **77**, 6755–6764.

84. J. Schiller, R. Suss, B. Fuchs, M. Muller, O. Zschornig and K. Arnold, MALDI-TOF MS in lipidomics, *Front Biosci.*, 2007, **12**, 2568–2579.

85. A. S. Woods and S. N. Jackson, Brain tissue lipidomics: direct probing using matrix-assisted laser desorption/ionization mass spectrometry, *AAPS J.*, 2006, **8**, E391–E395.

86. R. M. Adibhatla, J. F. Hatcher and R. J. Dempsey, Lipids and lipidomics in brain injury and diseases, *AAPS J.*, 2006, **8**, E314–E321.

87. D. J. Harvey, Matrix-assisted laser desorption/ionization mass spectrometry of phospholipids, *J. Mass Spectrom.*, 1995, **3**, 1333–1346.

88. S.N. Jackson, H.Y. Wang and A.S. Woods, In situ Structural Characterization of Glycerophospholipids and Sulfatides in Brain Tissue Using MALDI-MS/MS, *J. Am. Soc. Mass Spectrom.*, 2007, **18**, 1176–1182.

89. J. Schiller, R. Suss, J. Arnhold, B. Fuchs, J. Lessig, M. Muller, M. Petkovic, H. Spalteholz, O. Zschornig and K. Arnold, Matrix-assisted laser desorption and ionization time-of-flight (MALDI-TOF) mass spectrometry in lipid and phospholipid research, *Prog. Lipid Res.*, 2004, **43**, 449–488.

90. K. A. Al-Saad, V. Zabrouskov, W. F. Siems, N. R. Knowles, R. M. Hannan and H. H. Hill, Matrix-assisted laser desorption/ionization time-of-flight mass spectrometry of lipids: Ionization and prompt fragmentation patterns, *Rapid Commun. Mass Spectrom.*, 2003, **17**, 87–96.

91. J. Schiller, S. Hammerschmidt, H. Wirtz and K. Arnhold, Lipid analysis of bronchoalveolar lavage fluid (BAL) by MALDI-TOF mass spectrometry and ^{31}P NMR spectroscopy, *Chem. Phys. Lipids*, 2001, **112**, 67–79.

92. J. Schiller, J. Arnhold, H. J. Glander and K. Arnold, Lipid Analysis of human spermatozoa and seminal plasma by MALDI-TOF mass spectrometry and NMR spectroscopy: effects of freezing and thawing, *Chem. Phys. Lipids*, 2000, **106**, 145–156.

93. M. Rujoi, J. Jin, D. Borchman, D. Tang and M. C. Yappert, *Invest. Ophthalmol. Vis. Sci.*, 2003, **44**, 1634–1642.

94. D. Mims and D. Hercules, Quantification of bile acids directly from urine by MALDI-TOF-MS, *Anal. Bioanal. Chem.*, 2003, **375**, 609–616.

95. D. Mims and D. Hercules, Quantification of bile acids directly from plasma by MALDI-TOF-MS, *Anal. Bioanal. Chem.*, 2004, **378**, 1322–1326.

96. M. Karas, M. Glückmann and J. Schäfer, Ionization in matrix-assisted laser desorption/ionization: singly charged molecular ions are the lucky survivors, *J. Mass Spectrom.*, 2000, **35**, 1–12.

97. W. J. Griffiths, S. Liu, G. Alvelius and J. Sjövall, Derivatisation for the characterisation of neutral oxosteroids by electrospray and matrix-assisted laser desorption/ionisation tandem mass spectrometry: the Girard P derivative, *Rapid Commun. Mass Spectrom.*, 2003, **17**, 924–935.

98. M. A. Khan, Y. Wang, S. Heidelberger, G. Alvelius, S. Liu, J. Sjövall and W. J. Griffiths, Analysis of derivatised steroids by matrix-assisted laser desorption/ionisation and post-source decay mass spectrometry, *Steroids*, 2006, **71**, 42–53.

99. Y. Wang, M. Hornshaw, G. Alvelius, K. Bodin, S. Liu, J. Sjövall and W. J. Griffiths, Matrix-Assisted Laser Desorption/Ionization High-Energy Collision-Induced Dissociation of Steroids: Analysis of Oxysterols in Rat Brain, *Anal. Chem.*, 2006, **78**, 164–173.

100. Z. Shen, J. J. Thomas, C. Averbuj, K. M. Broo, M. Engelhard, J. E. Crowell, M. G. Finn and G. Siuzdak, Porous silicon as a versatile platform for laser desorption/ionization mass spectrometry, *Anal. Chem.*, 2001, **73**, 612–619.

101. J. H. Beynon, *Mass Spectrometry and its Applications to Organic Chemistry*, Elsevier, Amsterdam, 1960.

102. E. C. Horning, Gas phase analytical methods for the study of steroid hormones and their metabolites, In *Gas Phase Chromatography of Steroids*, K. B. Eik-Nes, E. C. Horning, Eds, Springer Verlag, Berlin, 1968, p. 1.

103. R. Ryhage, The mass spectrometry laboratory at the Karolinska Institute 1944–1987, *Mass Spectrom. Rev.*, 1993, **12**, 1–49.

104. R. P. Morgan, J. H. Beynon, R. H. Bateman and B. N. Green, The MMZAB-2F double focusing mass spectrometer and mike spectrometer, *Int. J. Mass Spectrom. Ion Phys.*, 1978, **28**, 171–191.

105. S. A. McLuckey and J. M. Wells, Mass Analysis at the Advent of the 21st Century, *Chem. Rev.*, 2001, **101**, 571–606.

106. G. Siuzdak, *The Expanding Role of Mass Spectrometry in Biotechnology*, 2nd edition, MCC Press, San Diego, 2006.

107. A. N. Tyler, E. Clayton and B. N. Green, Exact mass measurements of polar organic molecules at low resolution using electrospray ionization and a quadrupole mass spectrometer, *Anal. Chem.*, 1996, **68**, 3561–3569.

108. B. E. Winger, K. J. Light-Wahl, R. R. O. Loo, H. R. Udseth and R. D. Smith, Observation and implications of high mass-to-charge ratio ions from electrospray ionization mass spectrometry, *J. Am. Soc. Mass Spectrom.*, 1993, **4**, 536.

109. B. A. Collings and D. J. Douglas, An extended mass range quadrupole for electrospray mass spectrometry, *Int. J. Mass Spectrom. Ion Processes*, 1997, **162**, 121–127.

110. R. A. Yost and C. G. Enke, Selected ion fragmentation with a tandem quadrupole mass spectrometer, *J. Am. Chem. Soc.*, 1978, **100**, 2274–2275.

111. R. A. Yost and R. K. Boyd, Tandem mass spectrometry: quadrupole and hybrid instruments, *Methods Enzymol.*, 1990, **193**, 154–200.

112. H. R. Morris, T. Paxton, A. Dell, B. Langhorn, M. Berg, R. S. Bordoli, J. Hoyes and R. H. Bateman, High sensitivity collisionally-activated decomposition tandem mass spectrometry on a novel quadrupole/ orthogonal-acceleration time-of-flight mass spectrometer, *Rapid Commun. Mass Spectrom.*, 1996, **10**, 889–896.

113. M. Guilhaus, D. Selby and V. Mlynski, Orthogonal acceleration time-of-flight mass spectrometry, *Mass Spec. Rev.*, 2000, **19**, 65–107.

114. I. V. Chernushevich, A. V. Loboda and B. A. Thomson, An introduction to quadrupole–time-of-flight mass spectrometry, *J. Mass Spectrom.*, 2001, **36**, 849–865.

115. G. C. Stafford Jr, P. E. Kelley, J. E. P. Syka, W. E. Reynolds and J. F. J. Todd, Recent improvements in and analytical applications of advanced ion trap technology, *Int. J. Mass Spectrom. Ion Processes*, 1984, **60**, 85–98.

116. R. E. March, An introduction to quadrupole ion trap mass spectrometry, *J. Mass Spectrom.*, 1997, **32**, 351–369.

117. R. E. March and J. F. Todd, *Quadrupole Ion Trap Mass Spectrometry*, John Wiley & Sons, Chichester, 2005.

118. W. C. Wiley and I. H. McLaren, Time-of-Flight Mass Spectrometer with Improved Resolution, *Rev. Sci. Instrumen.*, 1955, **26**, 1150.

119. R. S. Plumb, K. A. Johnson, P. Rainville, J. P. Shockcor, R. Williams, J. H. Granger and I. D. Wilson, The detection of phenotypic differences in the metabolic plasma profile of three strains of Zucker rats at 20 weeks of age using ultra-performance liquid chromatography/orthogonal acceleration time-of-flight mass spectrometry, *Rapid Commun. Mass Spectrom.*, 2006, **20**, 2800–2806.

120. R. S. Plumb, J. H. Granger, C. L. Stumpf, K. A. Johnson, B. W. Smith, S. Gaulitz, I. D. Wilson and J. Castro-Perez, A rapid screening approach to metabonomics using UPLC and oa-TOF mass spectrometry: application to age, gender and diurnal variation in normal/Zucker obese rats and black, white and nude mice, *Analyst*, 2005, **130**, 844–849.

121. N. Ojima, K. Masuda, K. Tanaka and O. Nishimura, Analysis of neutral oligosaccharides for structural characterization by matrix-assisted laser desorption/ionization quadrupole ion trap time-of-flight mass spectrometry, *J. Mass Spectrom.*, 2005, **40**, 380–388.

122. K. F. Medzihradszky, J. M. Campbell, M. A. Baldwin, A. M. Falick, P. Juhasz, M. L. Vestal and A. L. Burlingame, The characteristics of peptide collision-induced dissociation using a high-performance MALDI-TOF/ TOF tandem mass spectrometer, *Anal. Chem.*, 2000, **72**, 552–558.

123. A. L. Yergey, J. R. Coorssen, P. S. Backlund Jr, P. S. Blank, G. A. Humphrey, J. Zimmerberg, J. M. Campbell and M. L. Vestal, De novo sequencing of peptides using MALDI/TOF-TOF, *J. Am. Soc. Mass Spectrom.*, 2002, **13**, 784–791.

124. M. B. Comisarow and A. G. Marshall, Fourier transform ion cyclotron resonance spectroscopy, *Chem. Phys. Lett.*, 1974, **25**, 282–283.

125. W. Schrader and H. W. Klein, Liquid chromatography/Fourier transform ion cyclotron resonance mass spectrometry (LC-FTICR MS): an early overview, *Anal. Bioanal. Chem.*, 2004, **379**, 1013–1024.

126. I. J. Amster, Fourier transform MS, *J. Mass Spectrom.*, 1996, **31**, 1325–1337.

127. J. S. Page, C. D. Masselon and R. D. Smith, FTICR mass spectrometry for qualitative and quantitative bioanalyses, *Curr. Opin. Biotechnol.*, 2004, **15**, 3–11.

128. R. M. Heeren, A. J. Kleinnijenhuis, L. A. McDonnell and T. H. Mize, A mini-review of mass spectrometry using high-performance FTICR-MS methods, *Anal. Bioanal. Chem.*, 2004, **378**, 1048–1058.

129. A. Makarov, E. Denisov, A. Kholomeev, W. Balschun, O. Lange, K. Strupat and S. Horning, Performance evaluation of a hybrid linear ion trap/orbitrap mass spectrometer, *Anal. Chem.*, 2006, **78**, 2113–2120.

130. M. Scigelova and A. Makarov, Orbitrap mass analyzer-overview and applications in proteomics, *Proteomics*, 2006, **6**(Suppl 2), 16–21.

131. M. Thevis, G. Sigmund, A. K. Schiffer and W. Schanzer, Determination of N-desmethyl- and N-bisdesmethyl metabolites of Sibutramine in doping control analysis using liquid chromatography-tandem mass spectrometry, *Eur. J. Mass Spectrom.*, 2006, **12**, 129–136.

132. C. S. Ejsing, T. Moehring, U. Bahr, E. Duchoslav, M. Karas, K. Simons and A. Shevchenko, Collision-induced dissociation pathways of yeast sphingolipids and their molecular profiling in total lipid extracts: a study by quadrupole TOF and linear ion trap-orbitrap mass spectrometry, *J. Mass Spectrom.*, 2006, **41**, 372–389.

133. M. Thevis, A. A. Makarov, S. Horning and W. Schanzer, Mass spectrometry of stanozolol and its analogues using electrospray ionization and collision-induced dissociation with quadrupole-linear ion trap and linear ion trap-orbitrap hybrid mass analyzers, *Rapid Commun. Mass Spectrom.*, 2005, **19**, 3369–3378.

134. Y. Wan, K. Karu and W.J. Griffiths, Analysis of neurosterols and neurosteroids by mass spectrometry, *Biochimie*, 2007, **89**, 182–191.

135. K. Karu, M. Hornshaw, G. Woffendin, K. Bodin, M. Hamberg, G. Alvelius, J. Sjövall, J. Turton, Y. Wang and W.J. Griffiths, Liquid Chromatography Combined with Mass Spectrometry Utilising High-Resolution, Exact Mass, and Multi-Stage Fragmentation for the Identification of Oxysterols in Rat Brain, *J. Lipid Res.*, 2007, **48**, 976–987.

136. W. J. A. VandenHeuvel, C. C. Sweeley and E. C. Horning, Separation of steroids by gas chromatography, *J. Am. Chem. Soc.*, 1960, **82**, 3481–3482.

137. W. J. Griffiths, C. Shackleton and J. Sjövall, Steroid Analysis, In *The Encylopedia of Mass Spectrometry, Vol. 3*, R. M. Capriolli, Ed., Elsevier, Amsterdam, 2005, p. 447.

138. T. Luukainen, W. J. A. VandenHeuvel, E. O. Haahti and E. C. Horning, Gas chromatographic behaviour of trimethylsilyl ethers of steroids, *Biochim. Biophys. Acta*, 1961, **52**, 599–601.

139. E. M. Chambaz and E. C. Horning, Conversion of steroids to trimethylsilyl derivatives for gas phase analytical studies; reactions of silylating reagents, *Anal. Biochem.*, 1969, **30**, 7–24.

140. T. Iida, M. Hikosaka, J. Goto and T. Nambara, Capillary gas chromatographic behaviour of tert-hydroxylated steroids by trialkylsilylation, *J. Chromatogr. A*, 2001, **937**, 97–105.

141. E. M. Chambaz, G. Maume, B. Maume and E. C. Horning, Silylation of steroids. Formation of trimethylsilyl ethers and oxysilylation products, *Anal. Lett.*, 1968, **1**, 749–761.

142. R. M. Thompson and E. C. Horning, Aromatization of the A-ring of norethynodrel, a steroidal oral contraceptive, during trimethylsilylation, *Steroids. Lipids Res.*, 1973, **4**, 135–142.

143. E. Schwartz, S. Abdel-Baky, P. W. Lequesne and P. Vouros, Aromatization and catecholization of the A-ring of nor-19-methyl steroidal 3-ketoepoxides by trimethylsilylation, *Int. J. Mass Spectrom. Ion Phys.*, 1983, **47**, 511–514.

144. J. Sjövall and M. Axelson, Newer approaches to the isolation, identification and quantitation of steroids in biological materials, *Vitam. Horm.*, 1982, **39**, 31–144.

145. C. H. L. Shackleton, J. Merdinck and A. M. Lawson, Steroid and bile acid analyses, In *Mass Spectrometry of Biological Materials*, C. N. McEwen, B. S. Larsen, Eds, Marcel Dekker Inc, New York, 1990, p. 297.

146. P. Eneroth, B. Gordon, R. Ryhage and J. Sjövall, Identification of mono- and dihydroxy bile acids in human feces by gas-liquid chromatography and mass spectrometry, *J. Lipid Res.*, 1966, **7**, 511–523.

147. M. Makita and W. W. Wells, Quantitative analysis of fecal bile acids by gas-liquid chromatography, *Anal. Biochem.*, 1963, **5**, 523.

148. C. F. Poole, Recent advances in the silylation of organic compounds for gas chromatography, In *Handbook of Derivatives for Chromatography*, K. Blau, G. King, Eds, Heyden & Son Ltd, London, 1977, p. 152.

149. R. P. Evershed, Advances in silylation, In *Handbook of Derivatives for Chromatography*, 2nd edition, K. Blau, J. M. Halket, Eds, Wiley, Chichester, 1993, p.51.

150. H. Miyazaki, M. Ishibashi, M. Itoh and T. Nambara, Use of new silylating agents for identification of hydroxylated steroids by gas chromatography and gas chromatography-mass spectrometry, *Biomed. Mass Spectrom.*, 1977, **4**, 23–25.

151. J.-P. Thenot and E. Horning, MO-TMS derivatives of human urinary steroids for GC and GC-MS studies, *Anal. Lett.*, 1972, **5**, 21–33.

152. J.-P. Thenot and E. Horning, GC-MS derivatization studies. The formation of dexamethasone MO-TMS, *Anal. Lett.*, 1972, **5**, 905–913.

153. M. Donike, *N*-Methyl-*N*-(trimethylsilyl)trifluoracetamide, a new silylation agent in the silylated amide series, *J. Chromatogr.*, 1969, **42**, 103–104.

154. M. Thevis, G. Opferman, H. Schmickler and W. Schänzer, Mass spectrometry of steroid glucuronide conjugates. I. Electron impact fragmentation of 5α/5β-androstan-3α-ol-17-one glucuronides, 5α-estran-3α-ol-17-one

glucuronide and deuterium-labelled analogues, *J. Mass Spectrom.*, 2001, **36**, 159–168.

155. R. W. Kelly and P. L. Taylor, tert-Butyl dimethylsilyl ethers as derivatives for qualitative analysis of steroids and prostaglandins by gas phase methods, *Anal. Chem.*, 1976, **48**, 465–467.

156. G. Phillipou, D. A. Bigham and R. F. Seamark, Steroid tert-butyldimethylsilyl ethers as derivatives for mass fragmentography, *Steroids*, 1975, **26**, 516–524.

157. M. Donike and J. Zimmerman, Preparation of trimethylsilyl-, triethylsilyl- and tert-butyldimethylsilyl enol ethers of oxo steroids for gas chromatographic and mass spectrometric studies, *J. Chromatogr.*, 1980, **202**, 483–486.

158. M. Axelson and J. Sjövall, Analysis of unconjugated steroids in plasma by liquid-gel chromatography and glass capillary gas chromatography-mass spectrometry, *J. Steroid. Biochemistry*, 1977, **8**, 683–692.

159. E. M. Chambaz, G. Defaye and C. Madani, Trimethylsilyl ether-enol-trimethylsilyl ether. New type of derivative for the gas phase study of hormonal steroids, *Anal. Chem.*, 1973, **45**, 1090–1098.

160. H. M. Fales and T. Luukkainen, O-Methyloximes as carbonyl derivatives in gas chromatography, mass spectrometry and nuclear magnetic resonance, *Anal. Chem.*, 1965, **37**, 955–957.

161. P. I. Jaakonmaki, K. A. Yarger and E. C. Horning, Gas-liquid chromatographic separation of human steroid glucuronides, *Biochim. Biophys. Acta*, 1967, **137**, 216–219.

162. J. Sjövall and K. Sjövall, Identification of 5α-pregnane-3α,20α,21-triol in human pregnancy plasma, *Steroids*, 1968, **12**, 359–366.

163. C. J. W. Brooks, B. S. Middleditch and D. J. Harvey, The mass spectra of some corticosteroid boronates, *Org. Mass Spectrom.*, 1971, **5**, 1429–1453.

164. C. J. W. Brooks, W. J. Cole, H. B. McIntyre and A. G. Smith, Selective reactions in the analysis and characterization of steroids by gas chromatography-mass spectrometry, *Lipids*, 1980, **15**, 745–755.

165. C. J. W. Brooks and A. G. Smith, Cholesterol oxidase. Further studies of substrate specificity in relation to the analytical characterisation of steroids, *J. Chromatogr.*, 1975, **112**, 499–511.

166. W. J. Griffiths, Tandem mass spectrometry in the study of fatty acids, bile acids, and steroids, *Mass Spectrom. Rev.*, 2003, **22**, 81–152.

167. J. W. Gauthier, T. R. Trautman and D. B. Jacobsen, Sustained off-resonance irradiation for collision-activated dissociation involving Fourier transform mass spectrometry. Collision-activated dissociation technique that emulates infrared multiphoton dissociation, *Anal. Chim. Acta*, 1991, **246**, 211–225.

168. D. P. Little, J. P. Speir, M. W. Senko, P. B. O'Connor and F. W. McLafferty, Infrared multiphoton dissociation of large multiply charged ions for biomolecule sequencing, *Anal. Chem.*, 1994, **66**, 2809–2815.

169. P. D. Schnier, W. D. Price, R. A. Jockush and E. R. Williams, Blackbody infrared radiative dissociation of bradykinin and its analogues, energetics,

dynamics, and evidence for salt-bridge structures in the gas phase, *J. Am. Chem. Soc.*, 1996, **118**, 7178–7189.

170. J. V. Olsen, S. E. Ong and M. Mann, Trypsin cleaves exclusively C-terminal to arginine and lysine residues, *Mol. Cell. Proteomics*, 2004, **3**, 608–614.

171. J. Abian, A. J. Oosterkamp and E. Gelpi, Comparison of conventional, narrow-bore, and capillary liquid chromatography/mass spectrometry for electrospray ionization mass spectrometry: Practical considerations, *J. Mass Spectrom.*, 1999, **34**, 244–254.

172. K. B. Tomer, M. A. Moseley, L. J. Deterding and C. E. Parker, Capillary liquid chromatography/mass spectrometry, *Mass Spectrom. Rev.*, 1994, **13**, 431–457.

173. S. Liu, J. Sjövall and W. J. Griffiths, Neurosteroids in rat brain: extraction, isolation, and analysis by nanoscale liquid chromatography-electrospray mass spectrometry, *Anal. Chem.*, 2003, **75**, 5835–5846.

174. P. Yin, X. Zhao, Q. Li, J. Wang, J. Li and G. Xu, Metabonomics study of intestinal fistulas based on ultraperformance liquid chromatography coupled with Q-TOF mass spectrometry (UPLC/Q-TOF MS), *J. Proteome Res.*, 2006, **5**, 2135–2143.

175. A. Nordström, G. O'Maille, C. Qin and G. Siuzdak, Nonlinear data alignment for UPLC-MS and HPLC-MS based metabolomics: quantitative analysis of endogenous and exogenous metabolites in human serum, *Anal. Chem.*, 2006, **78**, 3289–3295.

176. J. H. Beynon, Qualitative analysis of organic compounds by mass spectrometry, *Nature*, 1954, **174**, 735–737.

177. D. H. Russell and R. D. Edmondson, High-resolution mass spectrometry and accurate mass measurements with emphasis on the characterization of peptides and proteins by matrix-assisted laser desorption/ionization time-of-fight mass spectrometry, *J. Mass Spectrom.*, 1997, **32**, 263–276.

178. D. O'Connor and R. Mortishire-Smith, High-throughput bioanalysis with simultaneous acquisition of metabolic route data using ultra performance liquid chromatography coupled with time-of-flight mass spectrometry, *Anal. Bioanal. Chem.*, 2006, **385**, 114–121.

179. R. S. Plumb, K. A. Johnson, P. Rainville, B. W. Smith, I. D. Wilson, J. M. Castro-Perez and J. K. Nicholson, UPLC/MS(E); a new approach for generating molecular fragment information for biomarker structure elucidation, *Rapid Commun. Mass Spectrom.*, 2006, **20**, 1989–1994.

CHAPTER 2

1D and 2D NMR Spectroscopy: From Metabolic Fingerprinting to Profiling

MARK R. VIANT,[1] CHRISTIAN LUDWIG[2] AND ULRICH L. GÜNTHER[2]

[1] School of Biosciences, The University of Birmingham, Edgbaston, Birmingham B15 2TT, UK
[2] Cancer Research UK Institute for Cancer Studies, The University of Birmingham, Edgbaston, Birmingham B15 2TT, UK

2.1 Introduction

Metabolomics has in the past few years become a versatile approach which is broadly used by industry and academia in the medical, biological and environmental sciences. The analysis of the small molecule composition of a tissue or biofluid sample enables an unprejudiced investigation of changes in an organism's metabolic status as a consequence of disease, toxic insult, genetic manipulation or environmental stress, and as a response to clinical intervention with drugs or other therapies.[1-6] This wealth of metabolic information can be used to interpret hypothesis-driven research, as well as to support discovery-driven research and the detection of novel biomarkers of disease. NMR spectroscopy is one of the leading technologies used to measure metabolite levels and its use currently dominates the metabolomics scientific literature.[5,7-9] It is a versatile approach that can be applied to solid[10] or liquid samples that are either *in vivo* or *in vitro*. Here we focus on the application of ^{1}H NMR spectroscopy to the *in vitro* studies of biofluids and tissue extracts. Although the use of NMR to measure metabolite levels considerably pre-dates the term metabolomics,[11] a significant advance in the field occurred when spectral processing and multivariate methods were applied to NMR spectra, an approach pioneered by researchers at Imperial College.[3,4,7]. During the past

44

few years the field has exploded with further significant advances in NMR method development, spectral processing and multivariate analyses.

^{1}H NMR spectroscopy has several benefits for measuring metabolite levels, including:

- rapid, simple sample preparation;
- observation of all high abundance metabolites that contain non-exchangeable H-atoms;
- it does not require chromatographic separation or the ionisation of metabolites;
- potentially quantitative metabolic measurements with a high degree of reproducibility;
- capable of relatively high throughput and automated analyses (>100 samples/day);
- inexpensive on a consumables basis.

Undoubtedly the largest disadvantage of NMR spectroscopy is its relatively poor sensitivity, which limits the observation to an estimated hundred or so metabolites per sample. This number of metabolites most likely accounts for less than 10% of an organism's metabolome. The high degree of connectivity within metabolic networks, however, somewhat reduces this problem as changes in low concentration (invisible to NMR) metabolites may lead to indirect changes in higher concentration (visible) metabolites. A considerable challenge in NMR studies is the unambiguous identification and quantification of metabolites. Although one-dimensional (1D) ^{1}H NMR has formed the bedrock of metabolomics studies to date, this approach has a significant problem of severe spectral overlap of the resonances. This reduces our ability to identify and quantify metabolites, in particular those at low concentration. State-of-the-art two-dimensional (2D) NMR methods, however, potentially offer a solution to this problem.

Two related strategies have evolved for the analysis of NMR metabolic data. The traditional method, termed 'fingerprinting', is based upon the analysis of an intact NMR spectrum that can be considered a 'fingerprint' of unassigned signals arising from low molecular weight metabolites. This requires an initial processing of the NMR fingerprint to convert it into a format for multivariate analysis. Multivariate methods such as principal components analysis (PCA) or partial least squares regression (PLS) can be conducted to identify similarities and differences between the NMR fingerprints. A plethora of other algorithms exist for multivariate analyses,[12–14] but most have the common goals of constructing classification models that help to visualise variation between data sets. The second strategy for analysing NMR data, termed 'profiling', uses computational methods to deconvolute each NMR spectrum into a list of metabolites and their concentrations. Although this approach provides considerably more meaningful data from a biochemical perspective, the spectral processing that is required is extremely challenging, in particular for 1D NMR spectra with a large number of overlapping signals. Following spectral deconvolution, multivariate methods of analysis can be applied directly to the list of metabolites and their concentrations.

Here we describe traditional NMR fingerprinting approaches that are used in metabolite studies, including the application of 1D NMR pulse sequences, water suppression methods, spectral processing and multivariate analyses. Although such methods have proved relatively successful to date, we will highlight their limitations in terms of the difficulty of unambiguous metabolite identification and quantification. To address this problem we also describe the recent application of 2D NMR in metabolomics, including homonuclear *J*-resolved spectroscopy and Hadamard encoded TOCSY (total correlation spectroscopy). These methods have considerable potential to advance NMR spectroscopy from a fingerprinting to a profiling tool, which is critical if NMR is to achieve its full potential. Throughout the chapter we provide examples of the application of these NMR methods to disease diagnosis.

2.2   NMR Methods in Metabolomics

2.2.1   Sample Preparation

The majority of NMR metabolomics studies have investigated the metabolic fingerprints of biofluids, in particular plasma and urine.[7] Biofluids are easier to collect than tissue samples and facilitate time-course studies of metabolism. The metabolic information obtained from a biofluid has the advantage that it is effectively an integration of the individual metabolic changes occurring within different compartments of a multi-organ animal and therefore represents a useful sample for diagnostic screening. This can also be a disadvantage in that biofluid metabolite concentrations can vary so dramatically between individuals that significant differences between groups can be hard to achieve with realistic numbers of samples. This is less of a problem with plasma (versus urine) as it is under greater homeostatic regulation. Tissues are in general under even greater homeostatic regulation than plasma and can therefore provide highly consistent metabolic measurements between individuals, but have the disadvantage that they are more difficult to collect. Tissue-specific metabolic fingerprints are of interest when investigating certain diseases[2,15] or sites of toxicity.[16]

Since high resolution NMR spectroscopy requires samples in a liquid state (with the exception of magic angle spinning NMR),[10] biofluids require considerably less preparation than tissues. The preparation of a generic sample involves three steps.

- The first is to rapidly collect and freeze the sample to quench metabolism and preserve the metabolites. Samples are typically stored at −80°C to prevent any decay.[17]
- The second step involves de-proteinising the sample to permanently halt metabolism and to extract the metabolites of interest. Although this is mandatory for tissue samples (using a homogeniser), it is an optional procedure for preparing biofluids. Several extraction methods can be

used, the choice depends on the polarity of the metabolites that are required. Solvents to extract only polar metabolites include perchloric acid, methanol or acetonitrile. If both polar and lipophilic metabolites are desired then extraction using methanol and chloroform can be used to fractionate the metabolite classes.[18,19]

- The third step is to optimise the solution for high resolution NMR spectroscopy. This typically entails buffering the sample pH to minimise variation in the chemical shifts of the NMR resonances (*e.g.* 100 mM phosphate buffer, pH 7.0), adding D_2O to provide a frequency lock for the spectrometer, and adding an internal chemical shift (and intensity) standard such as sodium 3-(trimethylsilyl)proprionate-2,2,3,3-d_4 (TMSP).

2.2.2 1D NMR Methods

Until recently, ^{1}H NMR metabolomics has almost always employed one-dimensional (1D) NMR methods at spectrometer frequencies of either 500 or 600 MHz. The primary advantage of this approach is that a metabolic finger-print can be acquired rapidly in 3–15 min with relatively good sensitivity. The acquisition time will depend on the sample size, with some studies reporting 1D acquisition times of more than 2 hours per sample.[20] The 1D ^{1}H NMR spectrum of a complex biological sample will typically be composed of several hundred or even a thousand resonances with chemical shifts between 0–10 ppm which are estimated to arise from a hundred or more metabolites. This large number of signals will always result in a highly congested spectrum.

2.2.2.1 Standard 1D ^{1}H NMR Experiment

A standard 1D ^{1}H NMR experiment with water suppression (discussed below) is commonly used in metabolomics and comprises [recovery delay–excite–acquire]. 1D spectra are composed of highly congested signals and, depending on the origin of the sample, the baseline may contain several broad resonances from high molecular weight macromolecules. Typical acquisition parameters include a 60° hard pulse (enabling a shorter recovery delay than if a 90° pulse was used), 12 ppm spectral width, 2.5 s recovery delay, and 40–160 transients collected into 32k data points. The resulting free induction decay (FID) is typically processed by zero-filling to 64k data points, applying an exponential line broadening of 0.5 Hz, Fourier transformation, phasing of the spectrum, baseline correction and calibration of the spectrum by setting the TMSP signal to 0.0 ppm. Examples of 1D NMR spectra recorded at 600, 500 and 800 MHz are shown in Figures 2.1a, 2.2a and 2.2b, respectively. Figure 2.1a shows the metabolic fingerprint of a medaka fish egg and illustrates the high line density typical of 1D spectra. Figures 2.2a and 2.2b each show two narrow chemical shift ranges within the NMR spectra of dab liver extracts (a flatfish), which clearly highlights the extent of signal overlap in 1D spectra and also the increased spectral resolution at 800 MHz versus 500 MHz.

(a)

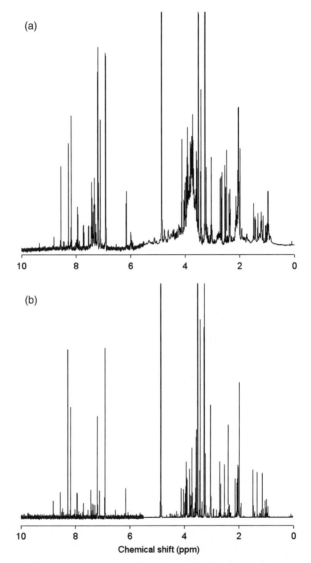

(b)

Chemical shift (ppm)

Figure 2.1   Representative 600 MHz NMR spectra of the polar metabolite extracts of 8-day-old medaka fish embryos: (a) 1D ^{1}H NMR spectrum and (b) 1D projection of a 2D ^{1}H,^{1}H *J*-resolved NMR spectrum (p-JRES). The lower line density in the projected JRES spectrum results in improved metabolite specificity. For both spectra the aromatic region has been vertically expanded by a factor of 7 to enable better comparison with the aliphatic region.

2.2.2.2   *1D ^{1}H CPMG Spin-Echo NMR Experiment*

The 1D ^{1}H Carr-Purcell-Meiboom-Gill (CPMG) spin-echo pulse sequence enables T_2-spectral editing and is often used to study plasma samples.[21] The

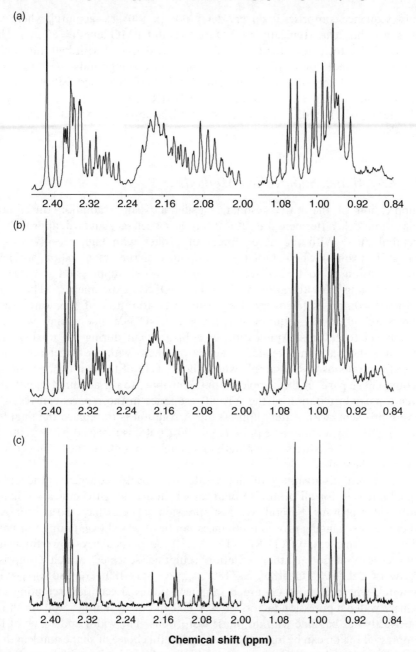

Figure 2.2   Sections of the NMR spectra of the polar metabolite extracts of flatfish liver: (a) 500 MHz 1D NMR spectrum, (b) 800 MHz 1D NMR spectrum and (c) 1D projection of an 800 MHz 2D ^{1}H,^{1}H *J*-resolved NMR spectrum. The improved spectral resolution of the 800 MHz spectra are readily apparent, as is the considerably reduced spectral congestion in the projected JRES spectrum.

pulse sequence comprises [recovery delay–90°–$(\tau$–180°–$\tau)_n$–acquire], where τ is the spin-echo delay (typically $\tau = 400$ µs, overall CPMG length $= 80$ ms). This method helps to remove the broad resonances associated with high molecular weight macromolecules and motionally constrained compounds, facilitating the observation of low molecular weight metabolites. The acquisition time is slightly longer than for the standard 1D NMR sequence and yields spectra with fewer baseline distortions but still with a large number of overlapping signals.

2.2.2.3   *Water Suppression Methods*

Suppression of the water resonance using a weak irradiation during the relaxation delay (termed pre-saturation) has been a standard approach in metabolomics. With the development of probes with high sensitivity (*e.g.* cryogenic probes) where radiation damping can present a major problem, pre-saturation is no longer sufficient. Better water suppression is achieved by combining pre-saturation with a 1D NOESY experiment.[22] The pulse sequence comprises [recovery delay with pre-saturation of the water reso-nance–90°–90°–τ–90°–acquire], where τ is the NOESY mixing time which is usually set to 100 ms. Application of a B_0-gradient during the mixing time improves the water suppression. Usually problems with the water resonance arise from regions of the sample where the B_1-field of the coil is not homoge-neous. Those parts of the sample do not contribute to the acquired signal. But even a partial excitation of the solvent resonance in those parts can result in severe problems with water suppression. A solution is to replace the final 90° proton pulse by a composite pulse (*e.g.* a 90°$_x$ pulse is replaced by 90°$_y$–90°$_{-x}$–90°$_{-y}$–90°$_x$). This reduces the strength of signals from outside the homogeneous centre of the coil.

An inherent disadvantage of all pre-saturation based sequences is the loss of signal intensity for all proton resonances which are in fast chemical exchange with water protons. Several water suppression schemes have been proposed where fast exchanging proton resonances can be observed (*e.g.* jump return and binomial sequences, WATERGATE,[23] *etc.*). A particularly powerful water suppression method is the excitation sculpting scheme[24] which comprises [recovery delay–90°–G_1–180°$_{x,soft}$–180°$_{-x,hard}$–G_1–G_2–180°$_{x,soft}$–180°$_{-x,hard}$–G_2–acquire], where G_1 and G_2 are B_0-gradient pulses. Excitation sculpting can achieve water suppression factors of $> 30\,000$ and the resulting spectrum shows a perfectly flat baseline with no phase distortions. The width and shape of the suppressed region can be manipulated by a careful choice of shape and length of the 180° soft pulses. Furthermore, multiple solvent lines can be suppressed simultaneously by using polychromatic pulses. This method can be easily incorporated into other pulse sequences (*e.g.* the 2D *J*-resolved experiment discussed below) and is therefore an excellent choice for metabolomics exper-iments. Typical acquisition parameters include a 3 ms soft pulse with Gaussian shape and gradient pulse lengths between 0.5 and 1 ms.

2.2.3 2D NMR Methods

Although 1D ^{1}H NMR has formed the bedrock of metabolomics studies to date, severe overlap of the resonances creates a serious limitation. This hinders the unambiguous identification and quantification of metabolites, especially for metabolites with small signals underlying intense resonances. Peak congestion is a particularly significant problem for metabolic profiling where concentrations of pre-selected metabolites need to be determined. The full potential of NMR in metabolomics, in particular its ability to quantify all metabolites in a spectrum using a *single* internal standard (which is a major advantage over mass spectrometry) will not be achieved until appropriate data collection methods have been developed. The use of 2D NMR spectroscopy to spread the overlapping resonances into a second dimension could solve this problem. To date, the limited implementation of 2D NMR in metabolomics is a consequence of the long acquisition times needed. For example, a typical 2D TOCSY spectrum of a biological sample[25] requires 16–20 hours to provide the same sensitivity as a comparable 1D spectrum obtained in only 15 minutes. This is clearly not a feasible approach when hundreds or thousands of samples must be analysed. State-of-the-art fast 2D NMR methods, however, have the potential to acquire a 2D spectrum of metabolites with little overlap of signals in 20 minutes or less. These methods have the potential to identify and quantify metabolites including small signals superimposed by larger resonances, and will significantly improve the scope of NMR metabolomics applications.

2.2.3.1 2D ^{1}H,^{1}H J-Resolved NMR Methods

The 2D ^{1}H,^{1}H *J*-resolved (JRES) experiment, first reported by Ernst and co-workers,[26] yields a spectrum that separates the chemical shift and spin-spin coupling data onto different axes (see Figure 2.3a). The projection of the 2D dataset onto the chemical shift axis yields a considerably less congested 'proton-decoupled' 1D ^{1}H spectrum (termed p-JRES; Figure 2.3b). Note that although the p-JRES spectrum contains far fewer resonances than a standard 1D NMR spectrum, there is no loss of metabolic information as the same number of metabolites are detected. The benefit of this spectral simplification for analysing a complex metabolite sample was reported in 1993,[27] but another 10 years passed before this approach was first used in a metabolomics investigation.[28] Since then, we and others have utilised this approach for unbiased biomarker discovery and the characterisation of metabolic phenotype.[29,30]

The 2D JRES pulse sequence comprises [recovery delay–90°–(t_1/2)–180°–(t_1/2)–acquire], where t_1 is an incremented time delay. Typical acquisition parameters for a 500 MHz spectrometer include 7 kHz spectral width in F2 (chemical shift axis) and 50 Hz in F1 (spin-spin coupling constant axis), 3.5 s recovery delay, and 8 transients per 32 increments collected into 8k or 16k data points. To save acquisition time the incremented dimension is usually incremented to obtain real rather than complex data points. The processing of 2D JRES spectra includes zero-filling to 128 points in F1 and to 16k or 32k points

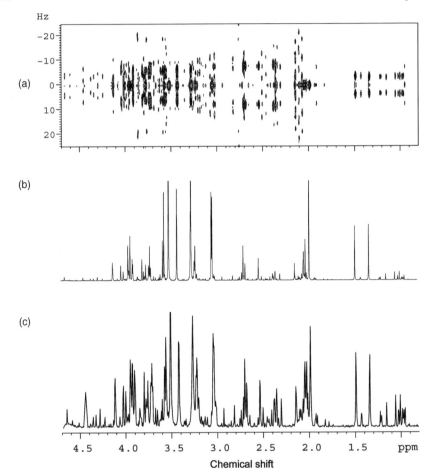

Figure 2.3   Representative 500 MHz ^{1}H,^{1}H *J*-resolved NMR spectra of extracts of
medaka fish embryos, including (a) original 2D ^{1}H,^{1}H *J*-resolved spectrum, (b) 1D projection of this 2D spectrum termed p-JRES and (c)
p-JRES spectrum after application of the generalised log transformation.
The most striking beneficial effect of the transformation includes an
increased weighting of the less intense signals.

in F2, multiplying both dimensions by sine-bell window functions, applying a
Fourier transformation, and calibration of the spectrum (TMSP, 0.0 ppm).
This results in a 2D JRES spectrum (Figure 2.3a) that can then be processed
further to obtain a simplified 1D p-JRES projection (Figure 2.3b), which
requires tilting the 2D spectrum by 45°, symmetrising about F1, and then
calculating the 1D projection of the 2D spectrum.

The reduced congestion in the p-JRES spectrum increases the likelihood that
a specific metabolite will appear as a well resolved and identifiable line, thereby
improving the extraction of metabolic information. This is demonstrated by
comparing the 1D and p-JRES spectra within Figures 2.1 and 2.2. Many weak

resonances that are partially or completely buried in the 1D spectra become fully resolved in the p-JRES spectra (Figure 2.2c). In terms of metabolite identification, the 2D JRES spectrum provides spin-spin coupling data that can greatly aid identification. Furthermore, due to the removal of spin-spin couplings in the 1D p-JRES dataset, the pattern of signals arising from a specific metabolite should be almost identical at all NMR spectrometer frequencies which will also aid metabolite identification (and reduce the number of metabolite libraries that need to be constructed for different spectrometer frequencies). The primary disadvantage of the 2D JRES approach is that it requires longer acquisition times than the corresponding 1D experiment, typically 20 minutes per sample. Also, strong coupling artefacts can arise in JRES spectra.[22,27] However, Keeler and co-workers have recently reported improvements in the 2D JRES pulse sequence to suppress these artifacts.[31]

2.2.3.2   2D Hadamard Encoded TOCSY

Whilst 2D TOCSY is a perfectly suitable method to reduce overlap between signals in different spin systems in spectra of biofluids or tissue extracts, its prohibitively long acquisition time is a major limitation. There have been several approaches to record NMR spectra in a more rapid way, including projection-reconstruction techniques,[32] covariance spectroscopy,[33] Hadamard encoding,[34,35] and Frydman's ultra-fast method to obtain 2D spectra.[36] Although all of these methods have the potential to acquire multi-dimensional NMR spectra in a relatively short time, they are also prone to introduce artefacts, suffer from low signal intensity, or are unsuitable to obtain homonuclear 2D spectra. Hadamard encoding has the additional limitation that the frequencies of the resonances must be known in advance. While this may seem to be a major limitation, this is not the case for most metabolomics applications where the frequencies are in fact known. Figure 2.4 shows TOCSY spectra recorded in 18 hours using conventional frequency encoding and in 20 minutes using Hadamard encoding for 93 signals which were picked from a 1D spectrum of the same sample.

The principle of Hadamard encoding for NMR spectroscopy dates back to 1962 when Ernst and Anderson built a multi-channel NMR spectrometer to excite multiple frequencies in an NMR spectrum. The electronic technology available at this time made this a rather cumbersome device which was soon superseded by pulse-excited Fourier transform spectroscopy. More recently, Hadamard encoded NMR methods were reintroduced by Kupce for applications in NMR spectroscopy of proteins.[37] On modern spectrometers polychromatic pulses can excite multiple frequencies over a broad frequency range at the same time and the width for individual bands can be adjusted to select individual signals or signal groups in a spectrum. The speed advantage arises from the fact that all of the selected frequencies are excited in every single sub-spectrum and therefore each sub-spectrum contributes to all signals. Therefore, for a limited number of increments, spectra can be recorded with a smaller number of scans compared to a conventional frequency encoded TOCSY.

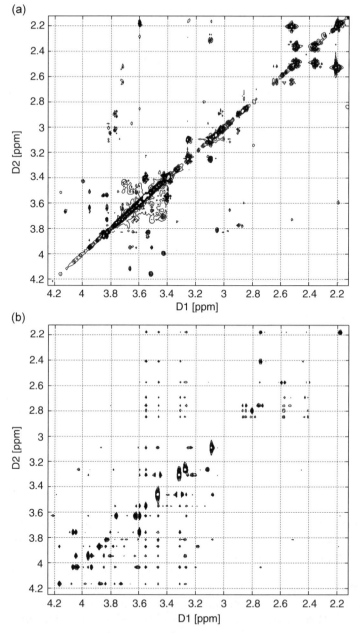

Figure 2.4   (a) TOCSY spectrum of a plasma sample recorded at 800 MHz proton frequency, requiring an 18-hour acquisition time. (b) Hadamard encoded TOCSY spectrum of the same sample with 93 signals selected and obtained in only 20 minutes.

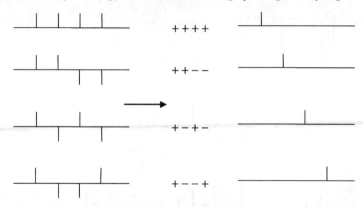

Figure 2.5 Principle of Hadamard encoding: For a spectrum of 4 lines Hadamard encoding is achieved by selectively inverting groups of lines in the spectrum using a polychromatic pulse. By adding (+) and subtracting (–) the four sub-spectra we obtain four linear combinations which each represent a spectrum with one single line.

The Hadamard principle is illustrated in Figure 2.5, which shows a schematic spectrum with four lines which are encoded by inverting individual signals in four different ways. Sub-spectra of individual lines can be obtained simply by adding and subtracting different sub-spectra. Hadamard-TOCSY excites all resonances of one spin system by selecting only one line for excitation, ideally a line which is isolated in the spectrum. The final spectrum after applying the Hadamard transformation will consist of a series of 1D sub-spectra each representing one spin system (Figure 2.6) which can be presented as a 2D TOCSY spectrum, as shown in Figure 2.4b. The advantage of selecting individual lines can be significant when the intensity of those lines is corrupted by overlapping signals. Figure 2.7a shows a typical example where the signal of lactate overlaps with that of LDL2 (red spectrum), whereas it is nicely separated in the Hadamard-TOCSY (blue spectrum). Since the concentrations of glucose and lactate are often altered in tumour patients due to an increased glycolytic energy metabolism (Warburg hypothesis) we would expect their signals to be anti-correlated. Figures 2.7b and 2.7c show the correlations between the NMR signals of glucose and lactate in a series of plasma samples from colon cancer patients. For the integrated peak intensities from Hadamard encoded TOCSY spectra, a clear anti-correlation (cyan) between the two signal sets is observed. The same result could not be achieved using 1D NMR spectra due to severe signal overlap.

Since all transients contribute to all signals there is also a risk of introducing artefacts by subtraction of signals. For this reason the quality of Hadamard encoded spectra requires a highly stable spectrometer, complete absence of vibrations and perfect temperature control. Another limitation is the size of memory for polychromatic pulses which is crucial to achieve artefact free excitation with relatively narrow excitation bands. Recent implementations of polychromatic pulses allow different widths of excitation profiles for individual flanks of the pulse. Since selection pulses can become very long (up to 40 ms)

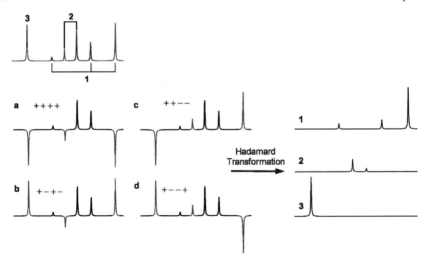

Figure 2.6 Hadamard encoded TOCSY for a spectrum consisting of three spin
systems (labelled 1–3, top left), two of which consist of more than one
line. The resonances displayed in red are selected for Hadamard encoding
in order to obtain four Hadamard encoded spectra (labelled a–d). After
Hadamard transformation three sub-spectra representing three separate
spin systems (*i.e.* three separate metabolites) are obtained.

relaxation optimised pulses may improve the sensitivity of the experiment.
Although some challenges still remain, Hadamard encoded TOCSY has the
potential to provide considerably more specific metabolic information than can
be achieved using 1D NMR, which would address one of the most major
limitations of the traditional NMR approach.

2.3   NMR Data Analysis

2.3.1   NMR Fingerprinting

Fingerprinting is the most commonly used method in NMR metabolomics and
is based upon the multivariate analysis of groups of whole NMR spectra, where
each spectrum can be considered a 'fingerprint' of unassigned signals. The
typical goal of such an analysis is to classify groups of samples according to
similarities and/or differences between their NMR fingerprints. If differences
between the spectra are discovered then further methods are used in an attempt
to assign those lines within the NMR fingerprint which are responsible for the
discrimination. NMR fingerprinting can be broken into three phases:

- spectral processing;
- multivariate analysis;
- identification of biomarkers.

Each of these is discussed below.

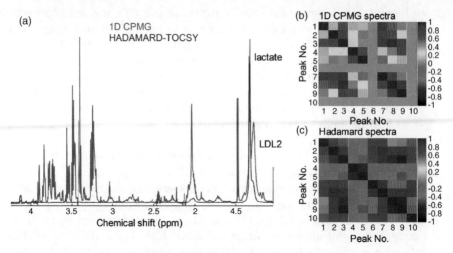

Figure 2.7 (a) Superimposed 1D CPMG spectrum of a plasma sample (red) and a Hadamard-TOCSY subspectrum of lactate from the same plasma sample (blue). Correlation plots are shown for the signals arising from glucose (peaks 1–3, 6–10) and lactate (peaks 4, 5) in plasma samples from colon cancer patients derived from (b) 1D spectra and (c) Hadamard-TOCSY spectra. The 1D data shows correlations between the glucose signals (red) but no clear correlations between the two lactate signals. The Hadamard-TOCSY data shows positive correlation between the glucose signals, positive correlation between the lactate signals and also anti-correlation between the lactate and glucose (cyan).

2.3.1.1 Spectral Processing

Several spectral processing algorithms are used to convert 1D NMR spectra (or 1D p-JRES projections of 2D JRES spectra) into a matrix format for multivariate analysis. A number of commercial packages (*e.g.* AMIX from Bruker BioSpin, ACD/1D NMR Processor from ACD, Inc.) and free software (*e.g.* ProMetab[28]) can be used for spectral processing. Although a range of algorithms have been reported,[12,38] many include the following four steps.

- First unwanted spectral features such as the residual water resonance are removed.
- Second, the intensities of the spectra are normalised to enable comparison across all spectra. This is often achieved by setting the total signal intensity of every spectrum to one (often referred to as total spectral area scaling), although other methods based upon normalising to sample mass or volume have been reported.[2,29]
- The third step typically incorporates a binning (or bucketing) algorithm that reduces the 32k or 64k data points comprising each spectrum into a much smaller number of segments each of a defined width (discussed further below). The spectral area within each bin is then integrated to yield a vector that contains intensity based descriptors of the original spectrum.

- Finally, binned spectra are scaled (or transformed) using one of a range of methods that include mean centring, auto-scaling, Pareto scaling[39] or a generalised logarithmic (glog) transformation.[40] The glog transformation has the effect of increasing the intensity of the weaker signals relative to the strong ones and of producing more similar variances (of the spectral intensities) across all the bins, as illustrated in Figure 2.3c. Overall, the input matrix for the multivariate analysis consists of N rows of samples with M columns of bin intensities.

The primary advantages of binning are: (i) the localisation of signals that show small variations in chemical shift (*e.g.* due to pH) into single bins, and (ii) an up to 200-fold decrease in the number of parameters that describe each NMR spectrum (enabling faster computation). The major disadvantages are the loss of spectral resolution, artefacts caused by signals on the border of bin boundaries, and compensation of intensities within individual bins. The majority of studies to date have employed a bin width of 0.04 ppm which reduces the 32k or 64k data points comprising the spectrum into only 256 segments. In this case each bin corresponds to typically 16 line widths and consequently many signals can be grouped within each bin, reducing the resolution of the subsequent multivariate analysis. Although NMR spectra are typically recorded at 500 or 600 MHz, the resolution of 0.04 ppm binned data is no better than could be obtained using a low field NMR spectrometer. This loss of resolution can be largely avoided with a 0.005 ppm bin that corresponds to only twice a typical line width.[28]

2.3.1.2   *Multivariate Analyses*

A series of tools are available for multivariate analyses,[12–14] including unsupervised methods that do not assign class identities to the samples (*e.g.* PCA) and supervised methods that do use class identifiers (*e.g.* PLS). In general, the primary goals of these methods are to identify differences between NMR fingerprints, to construct classification models, and/or to discover biomarkers. Although a detailed discussion of these multivariate analyses is beyond the scope of this chapter, we will introduce the most commonly used method of PCA. This method reduces the dimensionality of the binned data from typically 256–2000 bins down to just a few principal components (PCs) which account for a large amount of the total variance between all the NMR fingerprints. The results of a PCA are presented in terms of scores and loadings plots. A scores plot summarises the similarities and differences between each of the NMR fingerprints, where each data point corresponds to one sample and two closely spaced data points indicate a high degree of metabolic similarity between those two samples. Each PC is a weighted linear combination of the original chemical shift bins, and this information is shown in a loadings plot. If two groups of samples were shown to differ along the PC1 axis then the loadings plot for PC1 can be used to determine which signals in the NMR spectra are producing this metabolic difference.

Examples of PCA scores and loadings plots are shown in Figure 2.8, which is from a study of the biochemical changes that occur throughout the 8 days-post-fertilisation (dpf) of medaka fish embryos.[28] 2D JRES spectra were recorded for several replicates of embryos on each day of development, processed using *ProMetab*, and then subject to PCA. The scores plot (Figure 2.8a) shows groupings of the biological replicates for each dpf, indicating that on any single day of development all the metabolic fingerprints of the embryos were very similar. Considering all the samples, there is a clear trajectory from day 1 to day 8, indicating a continual change in metabolism as the embryos develop. This is termed a metabolic developmental trajectory that provides a succinct description of changes in the NMR-visible metabolome throughout embryogenesis, and has been used to examine the effects of toxicants on embryogenesis.[29] The metabolites that contribute to the changes along PC1 (in particular from 3–6

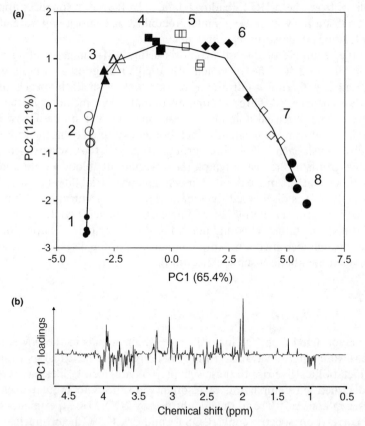

Figure 2.8 (a) Scores plot from a PCA of 1D projections of 2D JRES NMR spectra of several replicates of 1-, 2-, 3-, 4-, 5-, 6-, 7- and 8-day-old medaka fish embryos (age in days indicated on plot), illustrating the concept of a metabolic developmental trajectory. (b) Corresponding PC1 loadings plot which indicates some of the metabolic differences between the early and late stage embryos in terms of unidentified NMR peaks.

dpf) are summarised in the loadings plot (Figure 2.8b). Peaks with positive PC1 loadings correspond to metabolites that are at elevated concentration at 6 dpf (*i.e.* with positive PC1 scores) compared to 3 dpf, and vice versa.

The metabolic interpretation of differences identified between samples in a scores plot is typically of major importance. Only by identifying the metabolites that differentiate, for example, a diseased phenotype from the normal population, can useful biomarkers be determined. Attempts to identify these metabolites from a loadings plot again highlights the serious limitations of existing 1D NMR approaches. Figure 2.9 shows narrow regions of three loadings plots from three PCAs of NMR spectra of healthy and cancerous flatfish liver. The analyses were conducted using 1D NMR spectra recorded at 500 and 800 MHz and also using 2D JRES spectroscopy at 800 MHz. Comparison of the loadings plots from the 1D spectra (Figures 2.9a and 2.9b) and p-JRES spectra (Figure 2.9c) again demonstrates the vastly reduced congestion in the p-JRES data. The removal of *J*-couplings from the p-JRES loadings data, such that all resonances appear as singlets, substantially increases its interpretability. A complete interpretation of the 1D loadings data would be impossible.

A further example of disease diagnosis using NMR fingerprinting is illustrated in Figure 2.10. The approach was used to diagnose a muscle wasting disease called withering syndrome in red abalone, a shellfish species of commercial importance.[2,41] Three groups of animals were included in the study: healthy abalone, diseased animals and abalone showing stunted growth. The PCA scores plot from analysing the metabolites extracted from the foot muscle could distinguish each of the three groups (Figure 2.10a), whereas the metabolic fingerprints of the haemolymph (*i.e.* invertebrate 'blood') were under less homeostatic control and exhibited much greater variability (Figure 2.10b). Interestingly, it was discovered that changes in the concentration of a metabolite called homarine (1-methyl-2-pyridine carboxylic acid) could be associated with the disease status. This highlights the discovery driven nature of metabolomics, which identified a potential biomarker for the disease as a result of a non-targeted metabolic 'fishing expedition'.

2.3.1.3   Identification of Metabolic Biomarkers

Having determined from the loadings plot which signals in the NMR spectra appear to differentiate various groups of samples, the next task is to identify which metabolites give rise to these signals. To date, metabolites have typically been identified via comparison of resonance positions to tabulated chemical shifts (see references[42,43]) and then confirmed by 2D NMR experiments such as $^1H,^1H$ correlation spectroscopy (COSY) and/or $^1H,^{13}C$ heteronuclear single quantum coherence (HSQC). Although this process is time consuming, until the NMR signals are identified the interpretation lacks biochemical meaning and is therefore of limited value. The desire to achieve a more complete biochemical interpretation of NMR data has recently motivated a number of efforts to construct NMR spectral libraries of metabolites. These include

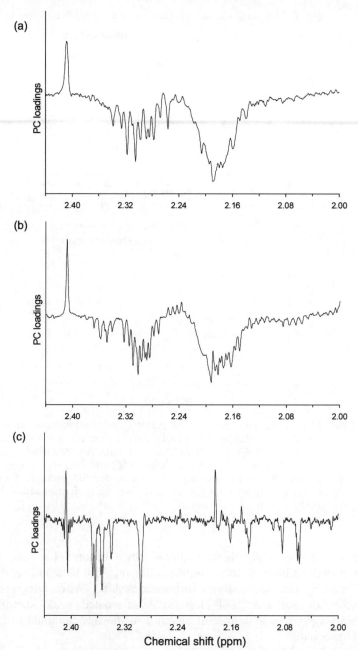

Figure 2.9 Sections of PCA loadings plots which indicate the metabolic differences between healthy and cancerous liver biopsies from dab, a marine flatfish. The NMR data used in the analyses were derived from (a) 500 MHz 1D spectra, (b) 800 MHz 1D spectra and (c) 1D projections of 800 MHz 2D *J*-resolved spectra, and clearly illustrate the reduced spectral complexity of the p-JRES approach.

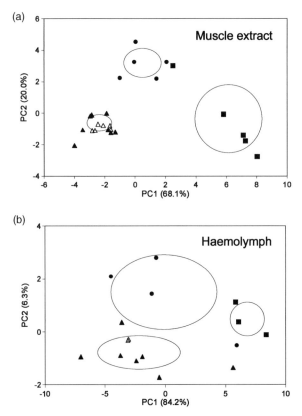

Figure 2.10 PCA scores plots from a 1D NMR fingerprinting study of withering syndrome in abalone using metabolites extracted from (a) foot muscle and (b) haemolymph. Three groups of animals were included, comprising healthy (▲) and diseased abalone (■) and animals showing stunted growth (●). The ellipses represent the mean ± SD (along PC1 and PC2) of each of these three groups and show clear discrimination between healthy, stunted and diseased.

(existing and planned) commercial libraries by Chenomx Inc. and Bruker BioSpin, as well as libraries in the public sector (*e.g.* the Human Metabolome Project[44] and by The University of Birmingham, UK). When integrated with the appropriate software tools for (semi-)automated peak identification these libraries promise a significant advance in the interpretability of NMR metabolomics data.

In summary, NMR fingerprinting is a proven approach for discovering metabolic changes in organisms associated with disease in humans[1,45–47] and wildlife,[2,48] toxicant exposure in mammals,[3,5,49,50] fish[29,51,52] and terrestrial invertebrates,[53,54] as well as genetic manipulation in yeast.[6] The advantages of the approach are that it uses established procedures, can identify overall similarities and differences between metabolite compositions of samples, and can determine which lines in the NMR spectra discriminate between these samples. Major disadvantages arise because much of the data typically remains as unidentified NMR signals. This locks the data into an instrument-dependent format (*i.e.* it cannot be compared to mass spectrometry data), and it limits the metabolic knowledge that is gained (*i.e.* this approach is of little value for characterising metabolic pathways). Below we consider methods that address the limitations of fingerprinting, and also present alternative analyses of multi-dimensional NMR data.

2.3.2 Multi-dimensional Analysis of 2D NMR Spectra

2D *J*-resolved spectra are usually processed by tilting by 45° and then 'folding', which takes the minimum signal value of both halves of the spectrum in the frequency dimension. A 1D projection of the 2D spectrum can then be generated that has narrow lines with no *J*-couplings. If PCA is employed for multivariate analysis it can either be applied to a series of 1D projected spectra or to an array of the concatenated 2D spectra. The latter method has the advantage that it preserves the full information of the 2D data set (in particular the *J*-couplings) but without the associated signal congestion that occurs in 1D spectra. In many cases better clustering is achieved in a PCA scores plot when the 2D spectra are used rather than 1D projections. This is illustrated in Figure 2.11, where PCA has been applied to extracts from KG1a cancer cells which were treated with two different inhibitors. Better clustering is achieved when PCA is applied to the 2D dataset (Figure 2.11b), in particular for the control data.

2.3.3 NMR Profiling

Profiling is a more challenging but ultimately more meaningful approach for analysing NMR spectra and is based on the deconvolution of the resonances in each NMR spectrum into a list of metabolites and their associated concentrations. This deconvolution is conducted on every NMR spectrum such that the input matrix for multivariate analysis consists of N rows of samples with M columns of identified and quantified metabolites. The same PCA and PLS multivariate methods of analysis as used in NMR fingerprinting can then be applied. The advantage of this approach is that the results of these analyses are considerably more meaningful to the biochemist as the variables that potentially discriminate two groups are reported as identified metabolites rather than unassigned lines. This data can be transformed into knowledge much more easily and can be used for biomarker discovery, characterising effects on specific metabolic pathways, and for instrument-independent data archiving. Although this approach addresses many of the disadvantages of NMR finger-printing, the spectral deconvolution that is required is extremely challenging, in particular considering the high chemical shift degeneracy of signals in 1D NMR spectra. In fact, since the chemical shift of a signal is subject to fluctuate somewhat, dependent upon the pH, temperature and other properties of the

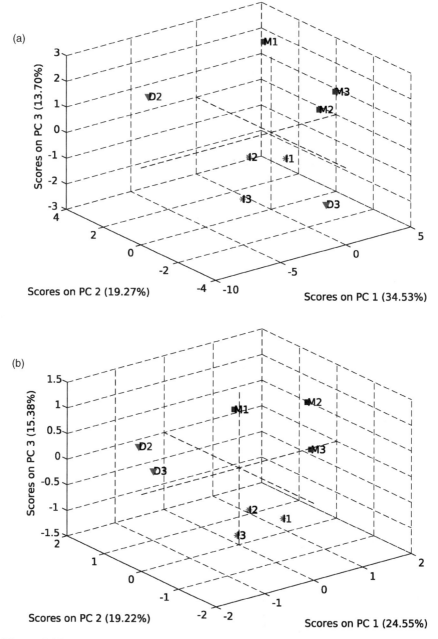

Figure 2.11   PCA of 2D JRES spectra obtained from KG1a cell extracts treated with different inhibitors (I and M, 3 samples each) compared to control spectra (D). (a) PCA was applied to a series of 1D projected spectra. (b) PCA was applied to a series of 2D JRES spectra. In both cases spectra were tilted and a symmetrisation using the minimum signal from both sides of the incremented dimension was applied.

sample matrix, it is currently infeasible to accurately identify and quantify the metabolites in a 1D NMR spectrum using fully automated methods. Considerable human intervention is required to guide the deconvolution process, and even then the most highly congested spectral regions are almost impossible to deconvolute accurately. One of the leading tools for NMR profiling using 1D NMR spectra is the Chenomx NMR Suite, which currently uses a library of 240 1D NMR spectra of pure metabolites to aid in both the identification and quantification of metabolites. In addition, Bruker BioSpin offers a module for AMIX which can identify individual metabolites by fitting spectra of known metabolites. The Chenomx NMR Suite software was recently applied to 800 MHz 1D NMR spectra of metabolites extracted from healthy and cancerous flatfish liver (unpublished). Over 30 metabolites were identified in every sample including acetate, adenine, adenosine, alanine, arginine, choline, creatine, formate, fumarate, glucose, glutamate, glutamine, glutarate, glycerol, histidine, isoleucine, lactate, leucine, lysine, malate, malonate, phosphocholine, oxalacetate, proline, propionate, serine, succinate, taurine, threonine, trimethylamine *N*-oxide, valine and myo-inositol. Statistical tests revealed that eight of these metabolites were significantly different between cancer and controls, which immediately provides metabolic information on the disease process. Identification of further compounds requires the construction of NMR spectral libraries containing metabolites that are found in tissues (as opposed to biofluids). Furthermore, NMR profiling is likely to take a significant step forward when fast 2D approaches such as JRES and Hadamard-TOCSY become more accepted and widely implemented, and in particular when NMR spectral libraries are developed for such considerably more metabolite-specific 2D methods.

2.3.4 Reporting Requirements for NMR Metabolomics

NMR metabolomic experiments are complex and large in size, and consequently the development of standardised 'reporting' requirements for data storage and data exchange within the scientific community presents an important (and urgent) challenge. The key components for a metabolomics reporting requirement includes the *biological metadata* (including a description of the biological sample, its origin, collection protocol and preparation), the *analytical instrument* (including acquisition parameters and the raw data), and *data analysis* (including data processing parameters, validation protocols and the processed data). Ideally this should re-use existing designs from related omics standardisation projects, both to avoid re-inventing the wheel and to maximise overlap between developing sets of standards.

To date, the most comprehensive description of *what* data to capture in experimental investigations of metabolism (not specifically *how* to capture it) has been developed and published by the Standard Metabolic Reporting Structure (SMRS) group.[55] In addition, a data model called ArMet (Architecture for Metabolomics) for the description of plant metabolomics experiments and their results was reported by Jenkins *et al.*[56] Neither of these publications describes the specific requirements for an NMR metabolomics data capture object model, which recently has been the subject of considerable effort. Recommended reporting requirements have now been prepared based upon the merging of the draft NMR data models from the universities of Cambridge and Aberystwyth as well as the instrument parameter and data processing parameter models developed at The University of Birmingham.[57] The Cambridge model is based on experience from both Cambridge and Imperial College London (based upon the SMRS policy document), whilst the Aberystwyth model is based on requirements for the UK Centre for Plant and Microbial Metabolomics based at Rothamsted Research. The new NMR data model has been developed to handle 1D and 2D NMR data as well as to be compatible with NMR fingerprinting and NMR profiling approaches.[57] Another important advance in data standardisation includes the effort by the Metabolomics Society to become an acknowledged authoritative body and a facilitator for the development of data standards for metabolic studies.

2.4 Conclusions and Future Developments

The use of NMR spectroscopy in metabolomics has grown rapidly in the past few years in a wide range of disciplines including medicine, basic biological research, and in the environmental sciences. The traditional NMR fingerprinting approach has been accepted as a useful tool for the classification of samples according to metabolic phenotype, but increasingly there is recognition that poor peak annotation is restraining the value of this NMR data. This recognition is emerging now since the tools are becoming more established and so they are moving from the hands of chemists towards those of biochemists and biologists who, not surprisingly, place a greater importance on the interpretation of the data than the art of the initial measurement. This transition, however, is a necessary step in the growth of NMR metabolomics and only when accepted as a valuable tool in the biochemists arsenal will the full benefits of metabolomics with its functional, phenotypically relevant measurements be revealed. We believe, however, that there are three crucial areas that the chemists and bioinformaticians must address as a priority.

- First is the further development of NMR as a profiling tool such that unambiguous identification and quantification of metabolites is readily achievable, and for which 2D NMR methods have considerable potential for high-throughout profiling analyses while maintaining minimal sample preparation and handling.
- Second is the construction of NMR metabolite libraries that are available in the public domain for identification of metabolites within these 1D and 2D NMR spectra.
- And finally the standards and reporting requirements for NMR-based metabolomic studies need to be accepted by the research community and

industry so that public domain databases can be set up to manage the large numbers of datasets being produced. This is a necessary route if measurements from NMR metabolomic studies are to be integrated with metabolic measurements by other tools such as mass spectrometry and furthermore if they are to be combined with genomic, transcriptomic and proteomic data, which is one of the goals of Systems Biology.

Acknowledgements

We thank Drs Stentiford, Lyons, and Feist from the Centre for the Environment, Fisheries and Aquaculture Science for the collection of the dab liver tissues, and thank Profs Johnson and Wakelam and Mr Ismail from The University of Birmingham for the colon cancer samples. The authors are indebted to Chenomx Inc., Canada for the use of the NMR Suite metabolomics data analysis software. MRV is grateful to the Natural Environment Research Council for an Advanced Fellowship in metabolomics (NER/J/S/2002/00618). Finally we thank several funding agencies for their support of the metabolomics research at The University of Birmingham, including the Natural Environment Research Council, the Biotechnology and Biological Sciences Research Council, the Wellcome Trust, and the European Union.

References

1. J. T. Brindle, H. Antti, E. Holmes, G. Tranter, J. K. Nicholson, H. W. L. Bethell, S. Clarke, P. M. Schofield, E. McKilligin, D. E. Mosedale and D. J. Grainger, *Nat. Med.*, 2002, **8**, 1439.
2. M. R. Viant, E. S. Rosenblum and R. S. Tjeerdema, *Environ. Sci. Technol.*, 2003, **37**, 4982.
3. J. K. Nicholson, J. Connelly, J. C. Lindon and E. Holmes, *Nat. Rev. Drug Discov.*, 2002, **1**, 153.
4. J. K. Nicholson and I. D. Wilson, *Nat. Rev. Drug. Discov.*, 2003, **2**, 668.
5. J. L. Griffin, *Curr. Opin. Chem. Biol.*, 2003, **7**, 648.
6. L. M. Raamsdonk, B. Teusink, D. Broadhurst, N. S. Zhang, A. Hayes, M. C. Walsh, J. A. Berden, K. M. Brindle, D. B. Kell, J. J. Rowland, H. V. Westerhoff, K. van Dam and S. G. Oliver, *Nat. Biotechnol.*, 2001, **19**, 45.
7. J. C. Lindon, J. K. Nicholson, E. Holmes and J. R. Everett, *Concept Magn. Reson.*, 2000, **12**, 289.
8. W. B. Dunn and D. I. Ellis, *Trends Anal. Chem.*, 2005, **24**, 285.
9. I. Pelczer, *Current Opinion in Drug Discovery & Development*, 2005, **8**, 127.
10. K. K. Millis, W. E. Maas, D. G. Cory and S. Singer, *Magn. Reson. Med.*, 1997, **38**, 399.

11. S. G. Oliver, M. K. Winson, D. B. Kell and F. Baganz, *Trends Biotechnol.*, 1998, **16**, 373.
12. J. C. Lindon, E. Holmes and J. K. Nicholson, *Prog. Nucl. Magn. Reson. Spec.*, 2001, **39**, 1.
13. L. Eriksson, E. Johansson, N. Kettaneh-Wold and S. Wold, Multi- and Megavariate Data Analysis-Principles and Applications, Umetrics, Umea, Sweden, 2001.
14. B. M. Wise, N. B. Gallagher, R. Bro, J. M. Shaver, W. Windig and R. S. Koch, PLS Toolbox Manual, Eigenvector Research, Manson, US, 2004.
15. J. L. Griffin, H. J. Williams, E. Sang, K. Clarke, C. Rae and J. K. Nicholson, *Anal. Biochem.*, 2001, **293**, 16.
16. N. J. Waters, E. Holmes, C. J. Waterfield, R. D. Farrant and J. K. Nicholson, *Biochem. Pharmacol.*, 2002, **64**, 67.
17. S. Deprez, B. C. Sweatman, S. C. Connor, J. N. Haselden and C. J. Waterfield, *J. Pharm. Biomed. Anal.*, 2002, **30**, 1297.
18. E. G. Bligh and W. J. Dyer, *Can. J. Biochem. Physiol.*, 1959, **37**, 911.
19. W. Weckwerth, K. Wenzel and O. Fiehn, *Proteomics*, 2004, **4**, 78.
20. J. L. Ward, C. Harris, J. Lewis and M. H. Beale, *Phytochemistry*, 2003, **62**, 949.
21. Q. N. Van, G. N. Chmurny and T. D. Veenstra, *Biochem. Biophys. Res. Comm.*, 2003, **301**, 952.
22. J. K. Nicholson, P. J. D. Foxall, M. Spraul, R. D. Farrant and J. C. Lindon, *Anal. Chem.*, 1995, **67**, 793.
23. M. L. Liu, X. A. Mao, C. H. Ye, H. Huang, J. K. Nicholson and J. C. Lindon, *J. Magn. Reson.*, 1998, **132**, 125.
24. T. L. Hwang and A. J. Shaka, *J. Magn. Reson. Series A*, 1995, **112**, 275.
25. H. R. Tang, Y. L. Wang, J. K. Nicholson and J. C. Lindon, *Anal. Biochem.*, 2004, **325**, 260.
26. W. P. Aue, J. Karhan and R. R. Ernst, *J. Chem. Phys.*, 1976, **64**, 4226.
27. P. J. D. Foxall, J. A. Parkinson, I. H. Sadler, J. C. Lindon and J. K. Nicholson, *J. Pharmaceut. Biomed.*, 1993, **11**, 21.
28. M. R. Viant, *Biochem. Biophys. Res. Comm.*, 2003, **310**, 943.
29. M. R. Viant, J. G. Bundy, C. A. Pincetich, J. S. de Ropp and R. S. Tjeerdema, *Metabolomics*, 2005, **1**, 149.
30. Y. L. Wang, M. E. Bollard, H. Keun, H. Antti, O. Beckonert, T. M. Ebbels, J. C. Lindon, E. Holmes, H. R. Tang and J. K. Nicholson, *Anal. Biochem.*, 2003, **323**, 26.
31. M. J. Thrippleton, R. A. E. Edden and J. Keeler, *J. Magn. Reson.*, 2005, **174**, 97.
32. E. Kupce and R. Freeman, *J. Am. Chem. Soc.*, 2003, **125**, 13–958.
33. R. Bruschweiler, *J. Chem. Phys.*, 2004, **121**, 409.
34. E. Kupce and R. Freeman, *J. Magn. Reson.*, 2003, **163**, 56.
35. E. Kupce and R. Freeman, *J. Magn. Reson.*, 2003, **162**, 300.
36. L. Frydman, A. Lupulescu and T. Scherf, *J. Am. Chem. Soc.*, 2003, **125**, 9204.

37. E. Kupce and R. Freeman, *J. Biomol. NMR*, 2003, **25**, 349.
38. T. M. Alam and M. K. Alam, *Annual Reports on NMR Spectroscopy*, 2005, **54**, 41.
39. H. C. Keun, T. M. D. Ebbels, H. Antti, M. E. Bollard, O. Beckonert, E. Holmes, J. C. Lindon and J. K. Nicholson, *Anal. Chim. Acta*, 2003, **490**, 265.
40. P. V. Purohit, D. M. Rocke, M. R. Viant and D. L. Woodruff, *OMICS*, 2004, **8**, 118.
41. E. S. Rosenblum, M. R. Viant, B. M. Braid, J. D. Moore, C. S. Friedman and R. S. Tjeerdema, *Metabolomics*, 2005, **1**, 199.
42. W. M. T. Fan, *Prog. Nucl. Mag. Res.*, 1996, **28**, 161.
43. V. Govindaraju, K. Young and A. A. Maudsley, *NMR Biomed.*, 2000, **13**, 129.
44. The Human Metabolome Project (http://www.metabolomics.ca).
45. K. Odunsi, R. M. Wollman, C. B. Ambrosone, A. Hutson, S. E. McCann, J. Tammela, J. P. Geisler, G. Miller, T. Sellers, W. Cliby, F. Qian, B. Keitz, M. Intengan, S. Lele and J. L. Alderfer, *Int. J. Cancer*, 2005, **113**, 782.
46. M. Coen, M. O'Sullivan, W. A. Bubb, P. W. Kuchel and T. Sorrell, *Clin. Infect. Dis.*, 2005, **41**, 1582.
47. M. A. Burns, W. L. He, C. L. Wu and L. L. Cheng, *Technol. Cancer Res. Treat.*, 2004, **3**, 591.
48. K. S. Solanky, I. W. Burton, S. L. MacKinnon, J. A. Walter and A. Dacanay, *Dis. Aquat. Org.*, 2005, **65**, 107.
49. J. C. Lindon, H. C. Keun, T. M. D. Ebbels, J. M. T. Pearce, E. Holmes and J. K. Nicholson, *Pharmacogenomics*, 2005, **6**, 691.
50. L. C. Robosky, D. G. Robertson, J. D. Baker, S. Rane and M. D. Reily, *Comb. Chem. High T. Scr.*, 2002, **5**, 651.
51. M. R. Viant, C. A. Pincetich, D. E. Hinton and R. S. Tjeerdema, *Aquat. Toxicol.*, 2006, **76**, 329.
52. M. R. Viant, C. A. Pincetich and R. S. Tjeerdema, *Aquat. Toxicol.*, 2006, **77**, 359–371.
53. J. G. Bundy, E. M. Lenz, N. J. Bailey, C. L. Gavaghan, C. Svendsen, D. Spurgeon, P. K. Hankard, D. Osborn, J. M. Weeks and S. A. Trauger, *Environ. Toxicol. Chem.*, 2002, **21**, 1966.
54. J. G. Bundy, D. J. Spurgeon, C. Svendsen, P. K. Hankard, J. M. Weeks, D. Osborn, J. C. Lindon and J. K. Nicholson, *Ecotoxicology*, 2004, **13**, 797.
55. J. C. Lindon, J. K. Nicholson, E. Holmes, H. C. Keun, A. Craig, J. T. M. Pearce, S. J. Bruce, N. Hardy, S. A. Sansone, H. Antti, P. Jonsson, C. Daykin, M. Navarange, R. D. Beger, E. R. Verheij, A. Amberg, D. Baunsgaard, G. H. Cantor, L. Lehman-McKeeman, M. Earll, S. Wold, E. Johansson, J. N. Haselden, K. Kramer, C. Thomas, J. Lindberg, I. Schuppe-Koistinen, I. D. Wilson, M. D. Reily, D. G. Robertson, H. Senn, A. Krotzky, S. Kochhar, J. Powell, F. van der Ouderaa, R. Plumb, H. Schaefer and M. Spraul, *Nat. Biotechnol.*, 2005, **23**, 833.
56. H. Jenkins, N. Hardy, M. Beckmann, J. Draper, A. R. Smith, J. Taylor, O. Fiehn, R. Goodacre, R. J. Bino, R. Hall, J. Kopka, G. A. Lane, B. M.

Lange, J. R. Liu, P. Mendes, B. J. Nikolau, S. G. Oliver, N. W. Paton, S. Rhee, U. Roessner-Tunali, K. Saito, J. Smedsgaard, L. W. Sumner, T. Wang, S. Walsh, E. S. Wurtele and D. B. Kell, *Nat. Biotechnol.*, 2004, **22**, 1601.

57. D. V. Rubtsov, H. Jenkins, C. Ludwig, J. Easton, M. R. Viant, U. Günther, J. L. Griffin and N. Hardy, *Metabolomics*, in press.

CHAPTER 3

Steroids, Sterols and the Nervous System

YUQIN WANG AND WILLIAM J. GRIFFITHS

The School of Pharmacy, University of London, 29–39 Brunswick Square, London WC1N 1AX, UK

3.1 Introduction

In this chapter we will concentrate on the analysis of steroids derived from cholesterol and found in mammalian systems; however, the interested reader is directed to a number of excellent articles discussing steroids in other systems.[1] Steroids are based on the cyclopentanoperhydrophenanthrene structure (see the appendix at the end of this chapter) and belong to the family of isoprenoids derived from mevalonic acid, and abnormalities in their synthesis and metabolism are related to a number of diseases (Schemes 3.1–3.4).[2] The classical steroid hormones are ligands to nuclear receptors, and recently a variety of steroid metabolites have been found to be ligands to "orphan" nuclear receptors, e.g. liver X receptor (LXR), farnesoid X receptor (FXR), steroid xenobiotic receptor (SXR), vitamin D receptor (VDR) and pregnane X receptor (PXR), and to exert important regulatory functions.[3] Mutations of a receptor may change its ligand specificity and allow the binding of other steroids, with resulting physiological changes.[4] Steroids can occur in the free form, or in covalent linkage to other molecules such as fatty acids, sulfuric acid, glucuronic acid, sugars and amino acids.

In humans, cholesterol (C^5-3β-ol) can be synthesised by most cells, but the main site of cholesterol synthesis is the liver. Cholesterol is also available from the diet, but one organ in which cholesterol is obtained almost exclusively by *de novo* synthesis is the brain, as the blood–brain barrier prevents the exchange of cholesterol between the circulation and the brain. The rate limiting step in *de novo* cholesterol synthesis is the conversion of 3-hydroxy-3-methylglutaryl-CoA (HMG-CoA) to mevalonate catalysed by the enzyme HMG-CoA reductase

71

a)

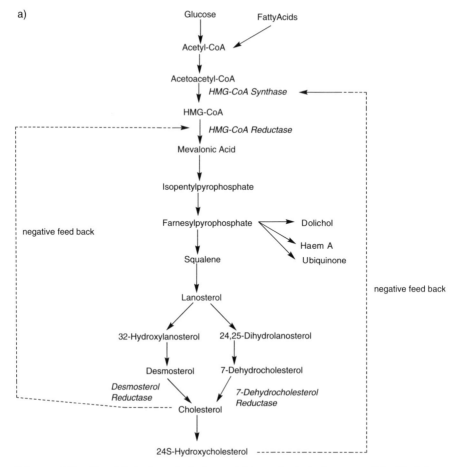

Scheme 3.1   Simplified view of cholesterol biosynthesis in humans. (Modified from Ref. 9.)

(Scheme 3.1). This step is responsive to a negative feedback mechanism, and is also the target of cholesterol-lowering drugs.[5] Defects in cholesterol biosynthesis lead to disease, for example mutations in the enzyme 7-dehydrocholesterol reductase can lead to its malfunction and a build up of 7- and 8-dehydrocholesterol ($C^{5,7}$-3β-ol, $C^{5,8}$-3β-ol) characteristic of Smith-Lemli-Opitz (SLO) syndrome, a devastating disorder characterised by severe physical anomaly and mental retardation.[6]

Cholesterol is removed from the body by oxidation to bile acids (Scheme 3.4). The primary bile acids are synthesised in the liver, amidated with either glycine or taurine, and excreted into the small intestine via the gallbladder. In the intestine bile acids play an important role in the absorption of lipids and are themselves efficiently reabsorbed (95%) and transported back to the liver as part of the enterohepatic circulation. Bile acid metabolism during the

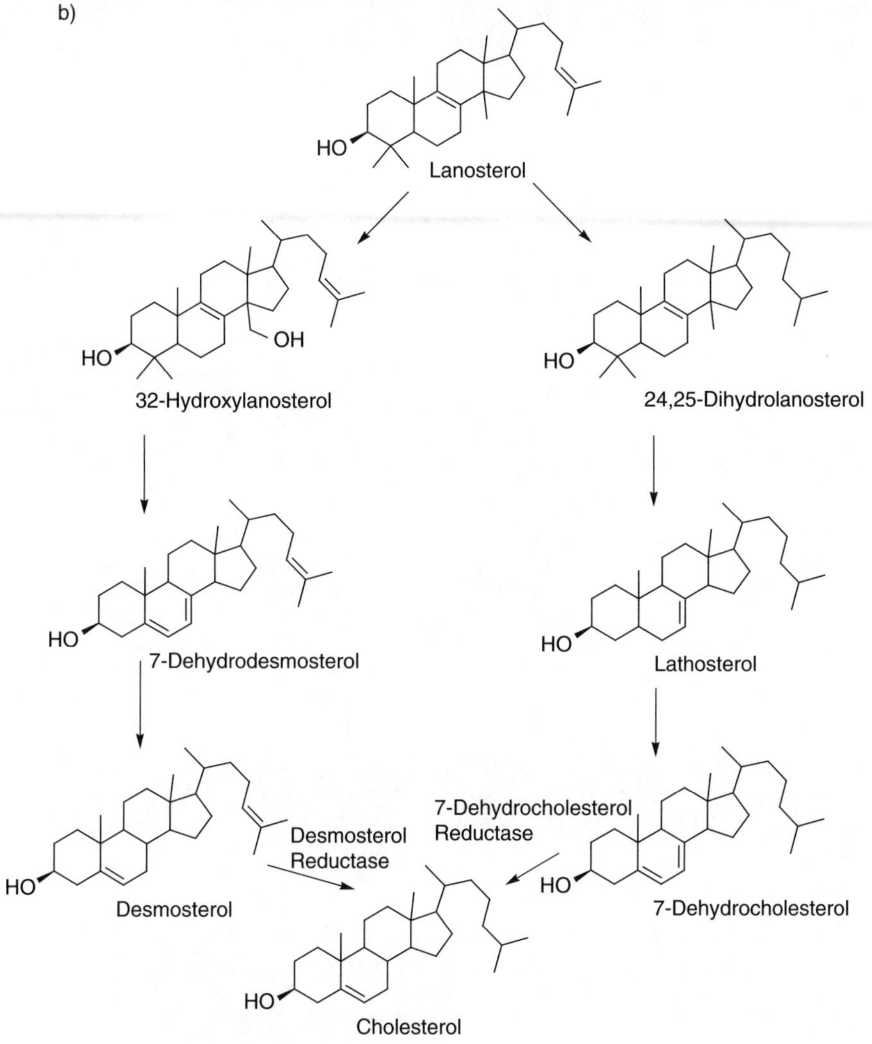

b)

Scheme 3.1 (*Continued*).

enterohepatic circulation is extensive and includes 7α-dehydroxylation [cholic acid (BA-3α,7α,12α-triol) → deoxycholic acid (BA-3α,12α-diol)], oxidoreductions, epimerisation and hydrolysis of conjugates. Under normal conditions only small amounts of bile acids are excreted in the urine; however, many liver diseases are characterised by the excretion of large quantities of bile acids.[7]

Defects in steroid biosynthesis and metabolism can be reflected by unusual profiles in body fluids, and as such provide a marker of disease. Mass spectrometry (MS) provides an ideal method for steroid profile analysis and has been extensively used in this role.[8]

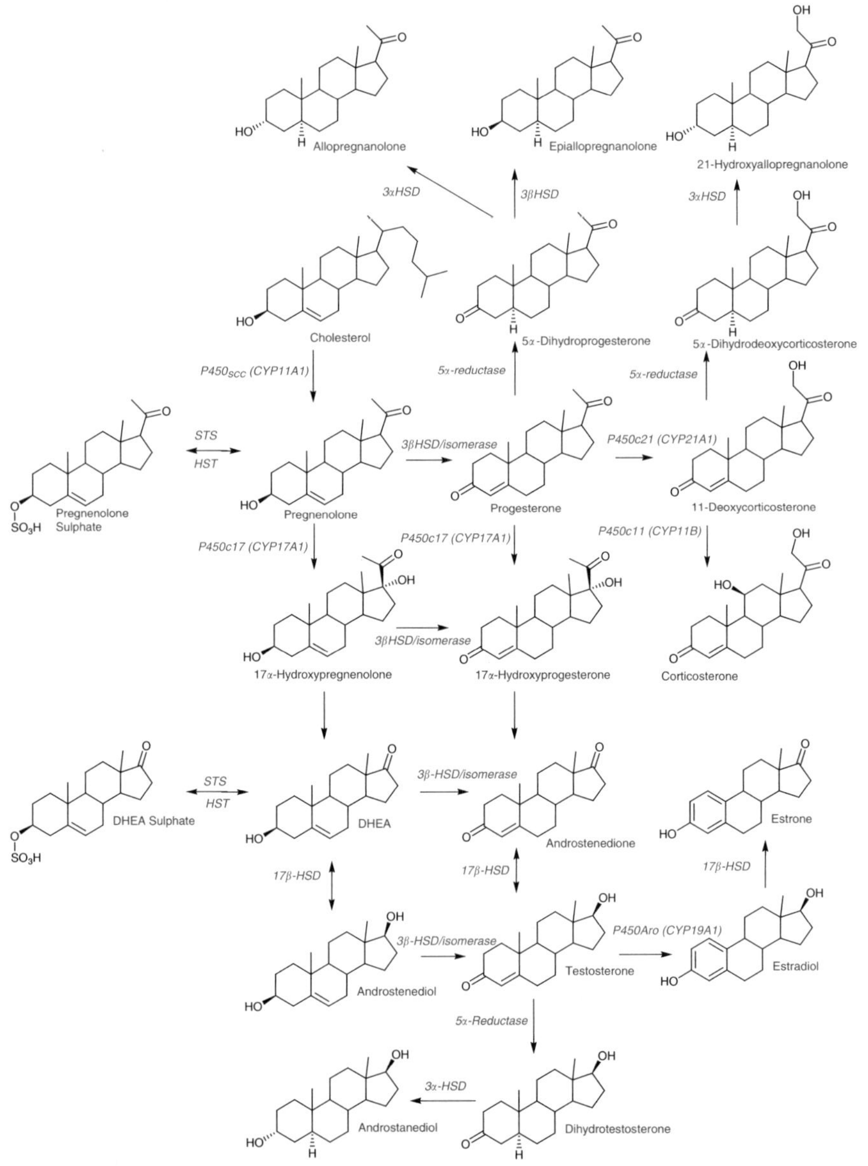

Allopregnanolone

Epiallopregnanolone

21-Hydroxyallopregnanolone

$3\alpha HSD$

$3\beta HSD$

$3\alpha HSD$

Cholesterol

5α-Dihydroprogesterone

5α-Dihydrodeoxycorticosterone

$P450_{scc}$ (CYP11A1)

5α-reductase

5α-reductase

Pregnenolone Sulphate

STS / HST

Pregnenolone

$3\beta HSD$/isomerase

Progesterone

$P450c21$ (CYP21A1)

11-Deoxycorticosterone

$P450c17$ (CYP17A1)

$P450c17$ (CYP17A1)

$P450c11$ (CYP11B)

17α-Hydroxypregnenolone

$3\beta HSD$/isomerase

17α-Hydroxyprogesterone

Corticosterone

DHEA Sulphate

STS / HST

DHEA

3β-HSD/isomerase

Androstenedione

Estrone

17β-HSD

17β-HSD

17β-HSD

Androstenediol

3β-HSD/isomerase

$P450Aro$ (CYP19A1)

Testosterone

Estradiol

5α-Reductase

Androstanediol

3α-HSD

Dihydrotestosterone

Scheme 3.2 Simplified view of neurosteroid biosynthesis in rat brain. Enzymes are shown in italics. $P450_{SCC}$, mitochondrial side-chain cleavage enzyme (CYP11A1); $P450c17$, mitochondrial 17-hydroxylase (CYP17A1); $P450aro$, aromatase (CYP19A1); $P450c21$, mitochondrial 21-hydroxylase (CYP21A1); $P450c11$, mitochondrial 11-hydroxylase (CYP11B); HST, hydroxysteroid sulfotransferase; STS, steroid sulfatase sulfohydrolase; HSD, hydroxysteroid dehydrogenase. (Modified from Ref. 14.)

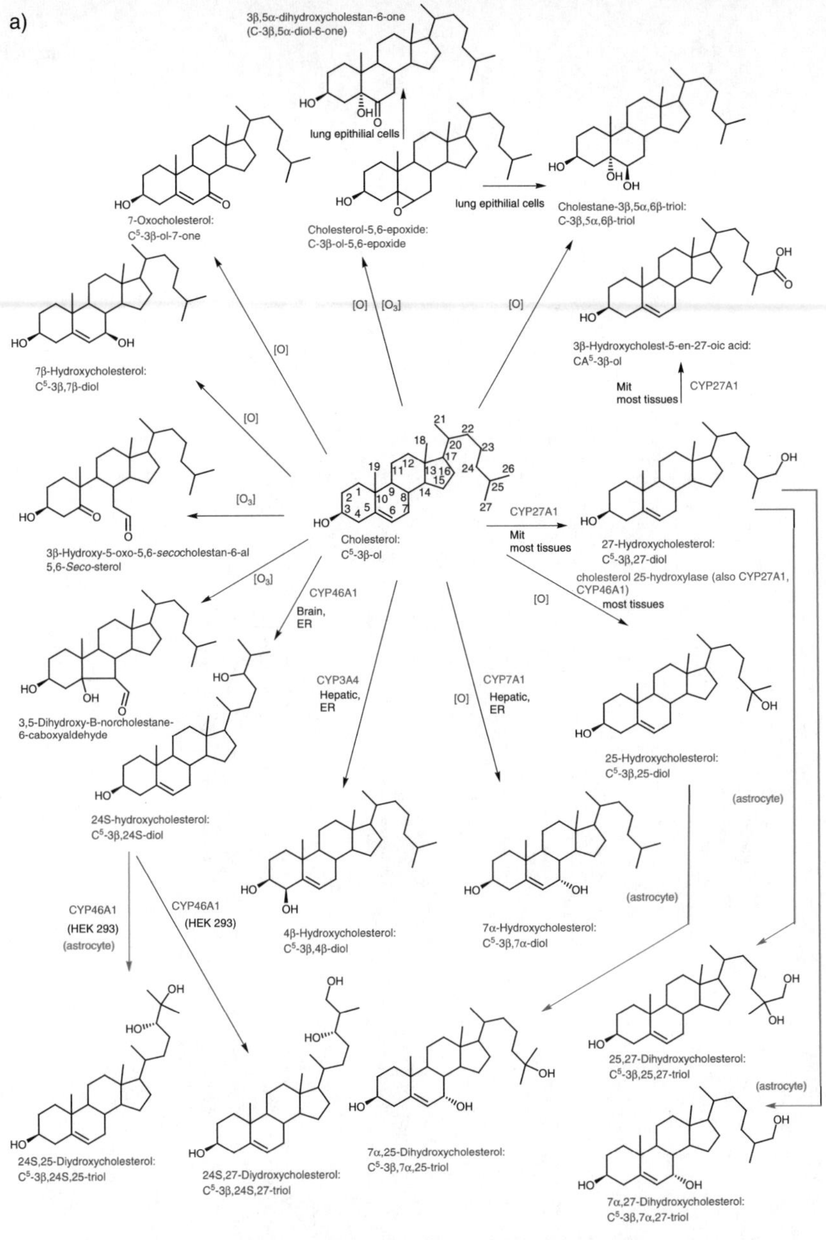

Scheme 3.3   Oxysterols found in human and rodent. (a) Enzymes and subcellular locations are shown. Autoxidation reactions indicated by [O], ozonolysis reactions indicated by [O$_3$]. Additional oxysterols found in cell cultures of HEK 293 cells transfected with human CYP46A1 cDNA, and cultures of rat astrocytes are also shown. (b) Further metabolites found in rat brain and cultured cells from rat brain. The metabolic pathway of cholesterol in astrocytes suggested by Zhang *et al.*[30] is shown by green arrows. Some confusion can arise concerning the nomenclature used to describe 27-hydroxycholesterol, which was previously denoted as 26-hydroxycholesterol. According to rules of priority of numbering the correct description is of 27-hydroxycholesterol is 25R,26-hydroxycholesterol, however, as the medical community uses the name 27-hydroxycholesterol, we will use this name in this chapter.

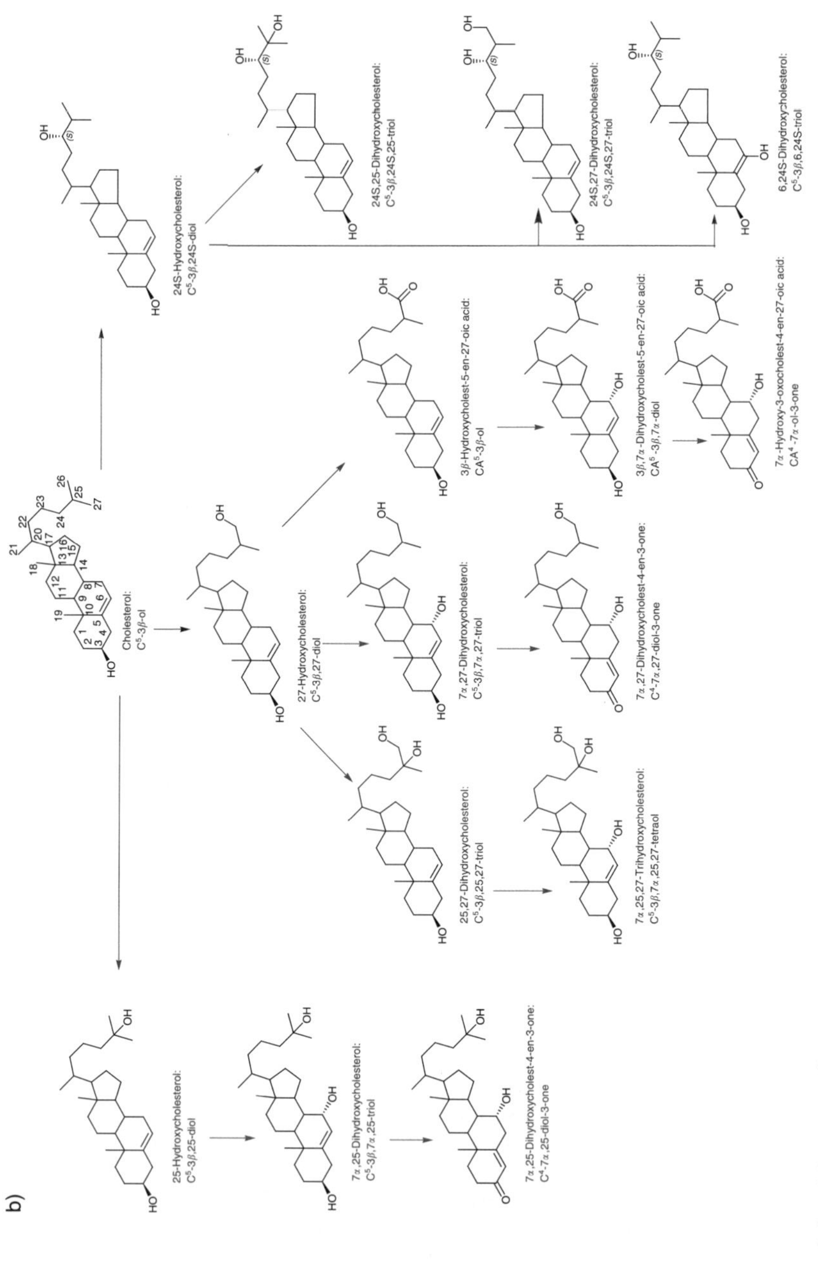

Scheme 3.3 (*Continued*).

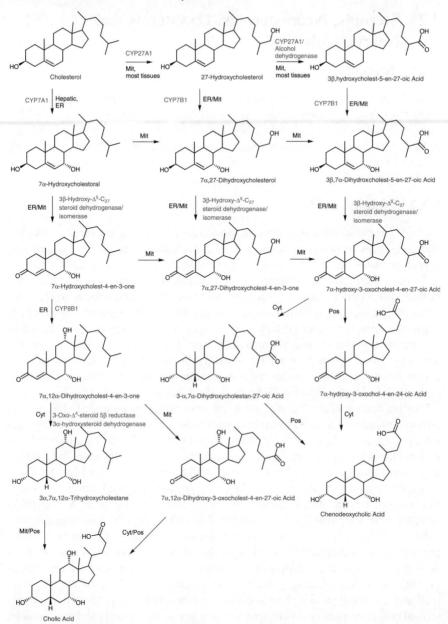

Scheme 3.4 The role of oxysterols in the biosynthesis of bile acids in man. Some enzymes, their distribution, and subcellular location are indicated. Endoplasmic reticulum, ER: Mitochondria, Mit: Peroxisome, Pos.

3.2 Steroids, Neurosteroids, Oxysterols and Neurosterols

3.2.1 Steroids

Steroid hormones are biosynthesised in the adrenal and gonadal glands from cholesterol. The first step involves the formation of pregnenolone (P^5-3β-ol-20-one) catalysed by the mitochondrial cholesterol side-chain cleavage enzyme $P450_{scc}$ (CYP11A1). Steroid hormones exert their influence in biological systems by binding to nuclear receptors which results in subsequent target gene transcription.

3.2.2 Neurosteroids

Steroids have also been shown to be formed in the nervous system.[10,11] Baulieu and colleagues showed that some steroids (*e.g.* allopregnanolone, 3α-hydroxy-5α-pregnane-20-one, 5α-P-3α-ol-20-one) could be detected in rodent brain up to 15 days after removal of steroidogenic glands, suggesting that steroids could be biosynthesised in brain.[12] Further, steroids have been shown to have an effect on the nervous system by acting on neurotransmitter-gated ion channels, hence the term neurosteroid has been coined. As neurosteroids synthesised in the nervous system are indistinguishable from those synthesised else where, the terms neurosteroid or neuroactive steroid are now used to describe steroids which have an effect in the nervous system. A neurosteroid biosynthesis pathway has been proposed (Scheme 3.2), although some of the routes have yet to be confirmed.[13,14] Early work on neurosteroids indicated that sulfate esters of pregnenolone and dehydroepiandrosterone (DHEA, A^5-3β-ol-17-one) were present in rat brain.[15,16] However, recent work by Baulieu's group in Paris and Sjövall's laboratory in Stockholm suggest that pregnenolone and DHEA are present in an unknown bound form in rat brain,[17] not as sulfate esters,[18] at concentrations much higher than those of the unconjugated steroids. Neurosteroids have been shown to act on the γ-aminobutyric acid A (GABA_A) receptor and the *N*-methyl-D-aspartate (NMDA) receptor. Allopregnanolone enhances GABAergic transmission and decreases NMDA transmission, while pregnenolone sulfate inhibits GABA-mediated currents but enhances NMDA-activated currents in rat hippocampal neurons.[19–21] Neurosteroids are present in brain at low levels (ng/g brain wet weight, Table 3.1), hence, sensitive methods are required for their analysis. Further, the observation that neurosteroid effects on neurotransmitter receptors are isomer specific defines the need for highly specific steroid analysis.

3.2.3 Oxysterols

Oxysterols are oxygenated derivatives of cholesterol, they are usually found in association with cholesterol but at far lower levels (10^{-3}–10^{-6}, Table 3.2). Oxysterols found in biological systems may be formed enzymatically, by

Table 3.1 Steroids present in the nervous system.

Neurosteroid	Chemical formula Mass (Da)	Human CSF	Human Plasma	Rat Brain	Rat Plasma
Allopregnanolone 5α-P-3α-ol-20-one	$C_{21}H_{34}O_2$ 318.26	0.05 ng/mL[1]	0.07 ng/mL[1]	≤2.5 ng/g[2] 0.5 ng/g[3] 2.1–11 ng/g[4] 0.6 ng/g[6]	0.12 ng/mL[1] 0.15 ng/mL[2]
Epipregnanolone 5β-P-3β-ol-20-one	$C_{21}H_{34}O_2$ 318.26			0.05–0.14 ng/g[3,5] 1.1–2.5 ng/g[4,5]	0.11 ng/mL[1]
Epiallopregnanolone 5α-P-3β-ol-20-one	$C_{21}H_{34}O_2$ 318.26			≤2.5 ng/g[2] 0.05–0.14 ng/g[3,5] 1.1–2.5 ng/g[4,5]	
Pregnanolone 5β-P-3α-ol-20-one	$C_{21}H_{34}O_2$ 318.26		0.11 ng/mL[1]	0.1–0.2 ng/g[3]	
5α-Pregnane-3,20-dione 5α-P-3,20-dione	$C_{21}H_{32}O_2$ 316.24			1.9 ng/g[7]	
Pregnenolone P⁵-3β-ol-20-one	$C_{21}H_{32}O_2$ 316.24	0.04 ng/mL[1]	4.18 ng/mL[1] 0.1–2.3 ng/mL[8]	0.8 ng/g[6] 8.9 ng/g[7] ≤2.5 ng/g[2] 7 ng/g[2,9] 0.6–1.2 ng/g[3] 2.7–3.9 ng/g[4] 2.8 ng/g[6] 7 ng/g[10] 61 ng/g[10,11]	1.2 ng/mL[7] 0.15 ng/mL[1]

Table 3.1 (*Continued*)

Neurosteroid	Chemical formula Mass (Da)	Human CSF	Human Plasma	Rat Brain	Rat Plasma
Pregnenolone sulphate P[5]-3β-ol-20-one 3-sulphate	$C_{21}H_{32}O_5S$ 396.20		87 ng/mL[12]	14.2 ng/g[7] <0.1 ng/g[13]	2.1 ng/mL[7]
Progesterone P[4]-3,20-dione	$C_{21}H_{30}O_2$ 314.22		0.1–2.5 ng/mL[8]	<0.3 ng/g[14] 0.5 ng/g[5] 0.05–0.5 ng/g[16] 2.2 ng/g[7] 1.0–3.4 ng/g[3] 4.4–21.0 ng/g[4] 1.9 ng/g[6] <0.5 ng/g[10] 13 ng/g[10,11] 0.2 ng/g[17]	1.9 ng/mL[7]
5α-androstane-3α,17β-diol 5α-A-3α,17β-diol	$C_{19}H_{32}O_2$ 292.24				
Androsterone 5α-A-3α-ol-17-one	$C_{19}H_{30}O_2$ 290.22	0.050 ng/mL[1]	35 ng/mL[1]		0.24 ng/mL[1]
DHEA A[5]-3β-ol-17-one	$C_{19}H_{28}O_2$ 288.21		1.3–12.5 ng/mL[8]	0.24 ng/g[7] ≤2.5 ng/g[2] 0.05–0.11 ng/g[3] 0.04–0.08 ng/g[4]	0.06 ng/mL[7]

DHEA sulphate A^5-3β-ol-17-one 3-sulphate	C$_{19}$H$_{28}$O$_5$S 368.17	0.20 ng/mL[1]	1500 ng/mL[12] 1000–2800 ng/mL[8]	1.7 ng/g[7] <0.1 ng/g[13]	0.20 ng/mL[7]
Testosterone A^4-17β-ol-3-one	C$_{19}$H$_{28}$O$_2$ 288.21		2.7 ng/mL[1] 1.2–11.1 ng/mL[8,18] 0.03–0.46 ng/mL[8,19]	<0.3 ng/g[14] ≤2.5 ng/g[2] 0.4–0.5 ng/g[3] 0.04–0.11 ng/g[4] 1.3 ng/g[17]	2.64 ng/mL[1] 2.5 ng/mL[2] 2.5 ng/mL[2]

[1] Data for male human and male rats, from Kim et al.[56]
[2] Data for male rat, frontal cortex, from Vallée et al.[57]
[3] Data for male rats, from Liu et al.[18]
[4] Data for female rats, from Liu et al.[18]
[5] Epipregnanolone or epiallopregnanolone.
[6] Data for adrenalectomized/castrated male rats, from Uzunov et al.[66]
[7] Data for male rats, from Baulieu[10,11] and Corpéchot et al.[12,15,16]
[8] Data from Shackleton.[70]
[9] Data for male rats, frontal cortex after swim stress, Vallée et al.[57]
[10] Data for male rats, from Higashi et al.[60]
[11] Data for male rats after fixation stress, Higashi et al.[60]
[12] Data for male adult, from Liu et al.[62]
[13] Data for male rats, from Liere et al.[17]
[14] Data for male and female rats, from Liu et al.[18]
[15] Data for male rats, from Mitamura et al.[59]
[16] Data for male rats, from Higashi et al.[64]
[17] Data for male rats, from Higashi et al.[61]
[18] Data for human adult male, from Shackleton.[70]
[19] Data for female adult female, from Shackleton.[70]

Table 3.2 Oxysterols present in the brain, CSF and plasma.

Sterol[1]	Chemical formula / Mass	Human plasma	Human CSF	Human brain	Rodent brain
Desmosterol[2] C[5,24]-3β-ol	$C_{27}H_{44}O$ 384.34				Rat (0.1% of total sterol)[3] Mouse (50 ng/mg)[4]
Cholesterol[2] C[5]-3β-ol	$C_{27}H_{46}O$ 386.35	(2 mg/mL)[5]		(7–8 µg/mg)[5]	Mouse (20 µg/mg)[4]
Cholesterol sulphate[6] C[5]-3β-ol 3-sulphate	$C_{27}H_{46}O_4S$ 466.31	(50–300 ng/mL)[7]			Rat (1.2 ng/mg)[6]
7-Oxocholesterol C[5]-3β-ol-7-one	$C_{27}H_{44}O_2$ 400.33		(0.9 ng/mL)[8]	(170–640 ng/mg)[9]	
24-Oxocholesterol C[5]-3β-ol-24-one	$C_{27}H_{44}O_2$ 400.33				Rat (~0.5 ng/mg)[10]
7α-Hydroxycholesterol C[5]-3β,7α-diol	$C_{27}H_{46}O_2$ 402.35	(40 ng/mL)[11]			
7β-Hydroxycholesterol C[5]-3β,7β-diol	$C_{27}H_{46}O_2$ 402.35	(5 ng/mL)[11]			
7-Hydroxycholesterol C[5]-3β,7-diol	$C_{27}H_{46}O_2$ 402.35			(234–745 ng/mg)[9]	
20S-Hydroxycholesterol C[5]-3β,20S-diol	$C_{27}H_{46}O_2$ 402.35				Rat (50 pg/mg rat)[12]
22R-Hydroxycholesterol C[5]-3β,22R-diol	$C_{27}H_{46}O_2$ 402.35			(45–90 pg/mg)[13]	
24S-Hydroxycholesterol C[5]-3β,24S-diol	$C_{27}H_{46}O_2$ 402.35	(80 ng/mL)[14] (70 ng/mL)[5] (60 ng/mL)[15,16]	(2.5 ng/mL)[5]	(5–15 ng/mg)[5]	Mouse (40–50 ng/mg)[4] Rat (15–25 ng/mg)[10]
25-Hydroxycholesterol C[5]-3β,25-diol	$C_{27}H_{46}O_2$ 402.35	(3 ng/mL)[11]			Rat Astrocytes[18]
27-Hydroxycholesterol C[5]-3β,27-diol	$C_{27}H_{46}O_2$ 402.35	(120 ng/mL)[15] (150 ng/mL)[14,16]	(1 ng/mL)[17]	(1–3 ng/mg)[4]	Mouse (5 ng/mg)[4]

Compound	Formula / Mass		
24-Oxohydroxycholesterol	$C_{27}H_{44}O_3$ 416.33	CYP46A1[19]	Rat ($\sim$125 pg/mg)[10]
C^5-3β,x-diol-24-one			
7α,25-Dihydroxycholest-4-en-3-one	$C_{27}H_{44}O_3$ 416.33		Rat Astrocytes[18]
C^4-7α,25-diol-3-one			
7α,27-Dihydroxycholest-4-en-3-one	$C_{27}H_{44}O_3$ 416.33		Rat Astrocytes[18]
C^4-7α,27-diol-3-one			
6,24-Dihydroxycholesterol	$C_{27}H_{46}O_3$ 418.34	CYP46A1[19]	Rat ($\sim$150 pg/mg)[10]
C^5-3β,6,24-triol			
7α,25-Dihydroxycholesterol	$C_{27}H_{46}O_3$ 418.34		Rat ($\sim$30 pg/mg)[10]
C^5-3β,7α,25-triol			Rat Astrocytes[18]
7α,27-Dihydroxycholesterol	$C_{27}H_{46}O_3$ 418.34		Rat ($\sim$30 pg/mg)[10]
C^5-3β,7α,27-triol			Rat Astrocytes[18]
24,25-Dihydroxycholesterol	$C_{27}H_{46}O_3$ 418.34	CYP46A1[19]	Rat ($\sim$125 pg/mg)[10]
C^5-3β,24,25-triol			Rat Astrocytes[18]
24,27-Dihydroxycholesterol	$C_{27}H_{46}O_3$ 418.34	CYP46A1[19]	Rat ($\sim$30 pg/mg)[10]
C^5-3β,24,27-triol			
25,27-Dihydroxycholesterol	$C_{27}H_{46}O_3$ 418.34		Rat ($\sim$75 pg/mg)[10]
C^5-3β,25,27-triol			Rat Astrocytes[18]
7α,25,27-Trihydroxycholesterol	$C_{27}H_{46}O_4$ 434.34		Rat Astrocytes[18]
C^5-3β,7α,25,27-tetraol			
3β-Hydroxycholest-5-en-27-oic acid	$C_{27}H_{44}O_3$ 416.33	(100 ng/mL)	Rat Astrocytes[18]
CA^5-3β-ol			
7α-Hydroxy-3-oxocholest-4-en-27-oic acid	$C_{27}H_{42}O_4$		Rat Astrocytes[18]
CA^4-7α-ol-3-one	430.31		
3β,7α-Dihydroxycholest-5-en-27-oic acid	$C_{27}H_{44}O_4$ 432.32		Rat Astrocytes[18]
CA^5-3β,7α-diol			
5,6-seco-sterol	$C_{27}H_{46}O_3$ 418.34	(146 pg/mg)[20]	($\sim$100 pg/mg)[10]
Aldol			($\sim$300 pg/mg)[10]

(*Continued*)

Table 3.2 *(Continued)*

Sterol[1]	Chemical formula Mass	Human plasma	Human CSF	Human brain	Rodent brain
Chenodeoxycholic acid[21] BA-3α,7α-diol	$C_{24}H_{40}O_4$ 392.29				(600 pg/mg)[21]
Deoxycholic acid[21] BA-3α,12α-diol	$C_{24}H_{40}O_4$ 392.29				(25 pg/mg)[21]
Cholic acid[21] BA-3α,7α,12α-triol	$C_{24}H_{40}O_5$ 408.29				(50 pg/mg)[21]

[1] See Schemes 3.1–3.4 for structures.
[2] Cholesterol and desmosterol are precursors of oxysterols.
[3] Other precursors of cholesterol, besides desmosterol, have been identified in rat brain.[87,88] Desmosterol concentration from ref. 87.
[4] From ref. 48.
[5] From ref. 52.
[6] Sulphated sterol, from ref. 18.
[7] Data from ref. 70.
[8] Data from ref. 55.
[9] Data for Cerebellar cortex, from ref. 89.
[10] Data from ref. 79.
[11] Data from ref. 90.
[12] Data from ref. 91.
[13] Data from ref. 92.
[14] Data from ref. 75.
[15] Data from ref. 74.
[16] Data from ref. 45.
[17] Data from ref. 54.
[18] Astrocytes show 25-hydroxylase activity when incubated with 27- and 24-hydroxycholesterol, and 7α-hydroxylase activity when incubated with 25- and 27-hydroxycholesterol,[30] while 25-hydroxycholesterol, 7α,25-dihydroxycholesterol, and 7α,25-dihydroxycholest-4-en-3-one are formed from endogenous precursors.
[19] Recombinant CYP46A1 shows 25- and 27-hydroxylase activity when incubated with 24S-hydroxycholesterol. HEK293 cells transfected with CYP46A1 can also hydroxylate the steroid nucleus.[32]
[20] Concentration of combined 5,6-*seco*-sterol and its aldol.[24]
[21] Bile acid data from ref. 31.

autoxidation, or by reaction with ozone (Schemes 3.3 and 3.4). Interestingly, Wentworth and colleagues recently demonstrated that ozone is formed by antibody-catalysed oxidation of water[22] and can react with cholesterol to form oxysterols.[23–26] Additionally, oxysterols are products of environmental exposure to ozone.[27] The sites in cholesterol most susceptible to autoxidation are the allylic C-7 carbon and the tertiary C-25 carbon, leading to the formation of a mixture of 7α-hydroxycholesterol (C^5-3β,7α-diol), 7β-hydroxycholesterol (C^5-3β,7β-diol) and 7-oxocholesterol (C^5-3β-ol-7-one) or 25-hydroxycholesterol (C^5-3β,25-diol) (Scheme 3.3). In humans, oxysterols are formed enzymatically from cholesterol by the action of cytochrome P450 (CYP) enzymes or by a cholesterol 25-hydroxylase (Schemes 3.3 and 3.4). CYP7A1 is a hepatic microsomal enzyme responsible for 7α-hydroxylation of cholesterol; CYP27A1 is mitochondrial, is expressed in most tissues and hydroxylates C-27; CYP46A1 is microsomal and is only present in brain, it hydroxylates C-24, and to a lesser extent C-25 and C-27; CYP3A4 is a hepatic microsomal enzyme and catalyses 4β-hydroxylation of cholesterol. Cholesterol 25-hydroxylase is responsible for 25-hydroxylation of cholesterol but is not a P450 enzyme. 7α-hydroxylation of cholesterol and 27-hydroxyation of cholesterol represent the initial steps in the "neutral" and "acidic" pathways in bile acid biosynthesis, respectively (Scheme 3.4). While the "neutral" pathway operates in the liver, the first step of the acidic pathway may be extrahepatic. Oxysterols with a 3β,7α-dihydroxy-5-ene structure can be oxidised to the corresponding 7α-hydroxy-4-en-3-ones by 3β-hydroxy-Δ^5-C_{27} steroid dehydrogenase/isomerase present in both endoplasmic reticulum and mitochondria; and 27-hydroxylated oxysterols can be converted to 27-carboxylic acids by the action of CYP27A1 or by an alcohol dehydrogenase both in the liver and extrahepatically. Examples of such metabolites occurring in human circulation are 7α-hydroxycholest-4-en-3-one (C^4-7α-ol-3-one), 3β-hydroxycholest-5-en-27-oic acid (CA^5-3β-ol), 3β,7α-dihydroxycholest-5-en-27-oic acid (CA^5-3β,7α-diol), and 7α-hydroxy-3-oxocholest-4-en-27-oic acid (CA^4-7α-ol-3-one). It should be noted that the enzyme responsible for 7α-hydroxylation of 3β-hydroxycholest-5-en-27-oic acid, and 27-hydroxycholesterol (C^5-3β,27-diol), CYP7B1, is distinct from that responsible for 7α-hydroxylation of cholesterol, *i.e.* CYP7A1. Because of their structures and since some of them have regulatory properties similar to those of the conventional oxysterols, it is logical to extend the term oxysterol to include cholestenoic acids (*i.e.* C_{27} acids) and bile alcohols (*i.e.* C_{27} alcohols). For example, oxysterols with regulatory properties include 7α-hydroxycholest-4-en-3-one, 7α,27-dihydroxycholest-4-en-3-one (C^4-7α,27-diol-3-one) and 27-hydroxycholesterol which have been found to inhibit HMG-CoA reductase and have been ascribed a regulatory role in cholesterol homeostasis.[28]

3.2.4 Neurosterols

The increased hydrophilicity of oxysterols compared to cholesterol facilitates their ability to cross membranes and move between cells and their intracellular compartments. This is particularly important in brain, as cholesterol itself is unable to cross the blood–brain barrier. Brain cells synthesise cholesterol *de novo*, and neurons metabolise cholesterol to 24S-hydroxycholesterol, which in tern can be metabolised further by both neurons and astrocytes to additional oxysterols. In analogy to the term neurosteroid, we use the term neurosterol to refer to sterols and oxysterols synthesised by or found in the nervous system. Zhang *et al.*[29,30] demonstrated that rat brain microsomes and cultures of cells of the nervous system, 7α-hydroxylate 25- and 27-hydroxycholesterol with their subsequent conversion to 3-oxo-Δ^4 steroids (Scheme 3.3b). A minor fraction of 27-hydroxycholesterol and its 7α-hydroxylated metabolites were further converted into 3β-hydroxycholest-5-en-27-oic acid, 3β,7α-dihydroxycholest-5-en-27-oic acid and 7α-hydroxy-3-oxocholest-4-en-27-oic acid (Scheme 3.3b, cf. Scheme 3.4). In contrast 24S-hydroxycholesterol (C^5-3β,24S-diol) was not found to be 7α-hydroxylated by brain microsomes, or cultures of Schwann cells, astrocytes or neurons, and it has subsequently been confirmed that CYP7B1 7-hydroxylates 24-hydroxycholesterol to only a minor degree, if at all. In astrocytes, 27-hydroxycholesterol and 24-hydroxycholesterol become 25-hydroxylated, and endogenous 7α,25-dihydroxycholest-4-en-3-one (C^4-7α,25-diol-3-one), 25-hydroxycholesterol, 7α,25-dihydroxycholesterol (C^5-3β,7α,25-triol) were reported (Scheme 3.3b, Table 3.2).[30] The possibility exists that the 24,25-diol is cleaved by oxidation into a C_{24} acid. In brain this could lead to the formation of C_{24} bile acids. Significantly, in 2004 Goto and colleagues[31] showed that cholic acid, chenodeoxycholic acid (BA-3α,7α-diol) and deoxycholic acid are present in rat brain. Chenodeoxycholic acid was found at high levels (600 ng/g wet weight), corresponding to about 30 times its plasma concentration. Recently, Mast *et al.*[32] have shown that CYP46A1, the enzyme responsible for 24S-hydroxylation of cholesterol in brain,[33] is also responsible for the metabolism of 24S-hydroxycholesterol to 24,25- and also to 24,27-dihydroxycholesterol (C^5-3β,24S,25-triol, C^5-3β,24S,27-triol), in both cultures of HEK293 cells transfected with human CYP46A1 cDNA, and an *in vitro* reconstituted system with recombinant CYP46A1 enzyme.

3.2.5 Biological Activity of Oxysterols

As discussed above, oxysterols have been found to be suppressors of cholesterol synthesis;[28] this was first demonstrated by Kandutsch and Chen in 1973.[34] This has lead to the formulation of the "oxysterol hypothesis", which suggests that many of the effects of cholesterol are due to oxysterols rather than cholesterol itself. *In vitro* oxysterols have a diverse effect on lipid metabolism. Oxysterols can behave as ligands for the LXRs, which heterodimerises with the retinoid X receptor (RXR) and activate transcription of many genes involved in lipid metabolism. For example, the gene coding for the rate-limiting enzyme in the degradation of cholesterol into bile acids, CYP7A1 (Scheme 3.4), is transcriptionally regulated by a mechanism that has been suggested to involve oxysterols and LXR, at least in rat.[35,36] Furthermore, oxysterols are involved in the post-translational control of sterol regulatory element binding proteins

(SREBPs)[37] which are themselves transcriptional factors for genes involved in cholesterol and fatty acid synthesis.[38,39]

As mentioned above, novel oxysterols have been found in human tissue formed from cholesterol by reaction with ozone. Evidence has been presented for their involvement in the mediation of inflammatory processes associated with the development of atherosclerotic lesions,[23] and in the initiation of protein misfolding as occurs in several neurological diseases.[24] Of these products, 3β-hydroxy-5-oxo-5,6-*seco*cholestan-6-al (5,6-*seco*-sterol), a ketoaldehyde, and its aldol condensation product, 3,5-dihydroxy-B-norcholestane-6-carboxyaldehyde, both contain a reactive aldehyde group (Scheme 3.3). Incubation of 5,6-*seco*-sterol, or its aldol, with amyloid β peptide (Aβ) leads to amyloidogenesis, possibly by condensation with amino groups in the peptide.[24] The levels of the 5,6-*seco*-sterol and its aldol were found to be surprisingly high in brain of subjects with Alzheimer's disease (AD) and also healthy controls {combined 5,6-*seco*-sterol and its aldol in AD brain 0.44 pmol/mg (184 ng/g) and 0.35 pmol/mg (146 ng/g) in control brain, respectively}. 5,6-*seco*-sterol has also been linked to Lewy body dementia, where combined levels of 5,6-*seco*-sterol and its aldol are elevated in the brain cortices of individuals with Lewy body dementia relative to those of age matched controls.[25,26] Additionally, these metabolites were shown to accelerate α-synuclein aggregation *in vitro*. Intra-neuronal deposition of α-synuclein as amyloid fibrils or Lewy bodies is the hallmark of this disease. In contrast to 5,6-*seco*-sterol and its aldol, 24S-hydroxycholesterol is reported to be protective against the formation of Aβ peptide,[40] the amyloidogenic peptide found in amyloid plaques in AD brain.

Biologically active oxysterols are also formed during ozonolysis of cholesterol in lung surfactant. Pulfer and Murphy[27] identified the major cholesterol-derived ozonolysis product in lung surfactant as 5β,6β-epoxycholesterol (C-3β-ol-5β,6β-epoxide) (Scheme 3.3). In studies of the metabolism of this oxysterol in lung epithelial cells they found small amounts of the expected metabolite cholestane-3β,5α,6β-triol (C-3β,5α,6β-triol), and more abundant levels of the unexpected metabolite 3β,5α-dihydroxycholestan-6-one (C-3β,5α-diol-6-one) (Scheme 3.3). Significantly, both 5β,6β-epoxycholesterol and 3β,5α-dihydroxycholestan-6-one were shown to be cytotoxic to human bronchial epithelial cells. Interestingly, 3β,5α-dihydroxycholestan-6-one, 5,6-*seco*-sterol and its aldol are isomeric and present a challenge for their correct identification. The formation of unexpected oxysterols as demonstrated by the reaction of ozone with cholesterol, once again highlights the need for unbiased methods for identifying and profiling steroids in biological matrices.

3.3 Analysis by Mass Spectrometry

Sterols and steroids can be analysed radioimmunoassays (RIA).[12,41,42] However, despite providing highly sensitive measurements, RIA suffers from lack of specificity and the necessity to pre-define the sterols/steroids to be investigated prior to analysis. In contrast, mass spectrometry when combined with gas or liquid chromatography (GC or LC) is exquisitely specific and allows the profile analysis of an undefined mixture of sterols or steroids with high sensitivity (sub pg) (see Chapter 1).[43] GC-MS has proved to be the "gold" standard for sterol/ steroid profile analysis, but as the twenty-first century progresses the advantages provided by LC-MS are now being exploited.

3.3.1 GC-MS Analysis

GC-MS analysis of steroids and sterols requires removal of any conjugating groups (*e.g.* sulfate esters) and derivatisation of the remaining polar functional groups to provide thermal stability and volatility to the analyte. To retain information with respect to conjugating groups, separation procedures can be incorporated prior to de-conjugation so as to separate neutral steroids from steroid sulfates, for example.[43]

3.3.1.1 Neurosterols

Neurosterols are usually analysed following derivatisation of alcohol groups to trimethylsilyl (TMS) ethers, oxo groups to methyl oximes, and acid groups to methyl esters (see Scheme 3.5, R1–R3 for derivatisation reactions) (see also Chapter 1).[30,32,44] The combination of GC with electron ionisation (EI)-MS provides unsurpassed chromatographic resolution, and first dimension mass spectra which contain a wealth of structural information. However, a significant disadvantage of GC-EI-MS is that molecular weight information is often not directly available, as molecular ions are not usually observed. However, the neutral loss characteristic of a derivatising group may allow the molecular weight of a compound to be calculated, and the availability of extensive EI-MS libraries and established retention indices aid in compound identification. GC-EI-MS is well suited to quantitative studies by stable isotope dilution methods. Stable isotope labelled versions of the analyte of interest should be added as early as possible in the analytical work-flow. Stable isotope labelled compounds are physically and chemically almost identical to their unlabelled equivalents, and in mass spectra give identical ion currents but at different m/z values (a reflection of the different isotopic content). Stable isotope labelled compounds have been extensively used for neurosterol quantification, particularly using the selected ion monitoring (SIM) mode.[45] In the SIM mode the ion current at selected m/z values only is monitored; this maximises sensitivity as it avoids scanning over the entire mass range. Björkhem, Diczfalusy and Lütjohann have extensively used this methodology for the quantification of neurosterols in brain, cerebrospinal fluid (CSF) and plasma (see Table 3.2).[46–55]

3.3.1.2 Neurosteroids

Neurosteroids have also been extensively analysed by GC-EI-MS and the inherent advantages of the methodology have been exploited in the identification

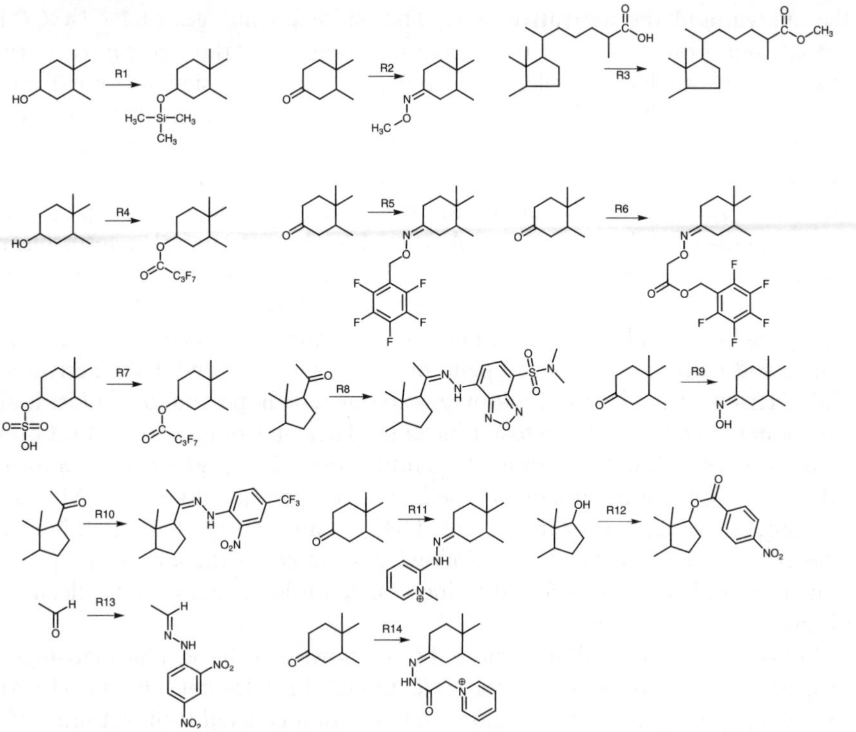

Scheme 3.5 Some useful derivatisation reactions for the mass spectrometric analysis of sterols and steroids.

of DHEA and pregnenolone in the sulfate ester fraction, and progesterone (P^4-3,20-dione), 5α-pregnane-3,20-dione (5α-P-3,20-dione) and allopregnanolone in the neutral fraction from rat brain.[12,15,16] As neurosteroids in brain are present at low levels (<20 ng/g in brain),[10,11] efforts have been made to develop more sensitive methods than GC-EI-MS (Table 3.1). An alternative method is negative chemical ionisation (NCI)-GC-MS (see also Chapter 1).[56,57] NCI methods are based on specifically derivatising the analyte of interest with a group of high electron affinity. The derivatised analyte is then preferentially ionised by electron capture so as to generate a negative ion in the ion source. The method often utilises fluorinated agents in the preparation of volatile derivatives with high electron affinities. For example, trifluoroacetic, pentafluoropropionic or heptafluorobutyric (HFB) anhydrides can be used to prepare acyl derivatives of hydroxyl groups, while carbonyl groups can be converted to oximes which can then be converted to pentafluorobenzyl oximes[57] or pentafluorobenzylcarboxymethoximes, for example (Scheme 3.5, R4–R6).[56] Ionisation proceeds with the capture of a secondary low-energy electron by the high electron affinity fluorinated group. Ionisation may lead to the formation of stable [M]$^-$ ions, or it may be dissociative[56,57] depending on the analyte and the derivative used. The major advantages of NCI-GC-MS is that ionisation is specific to compounds containing the electron capturing tag, and it provides excellent sensitivity in terms of signal-to-noise ratio when either a stable $[M]^-$ ion or a negatively charged fragment ion is monitored. Using carboxymethoximepentafluorobenzyl and TMS derivatives and monitoring $[M - 181]^-$ (loss of $CH_2C_6F_5$) fragment ions, Kim *et al.* achieved instrumental detection limits for neurosteroids in the pg range, and quantified androsterone (5α-A-3α-ol-17-one, ~ 50 pg/mL), testosterone (Δ^4-17β-ol-3-one, ~ 200 pg/mL), allopregnanolone (~ 50 pg/mL) and pregnenolone (~ 40 pg/mL) in human CSF using $[^2H_4]$dihydrotestosterone (5α-A-17β-ol-3-one) and $[^2H_4]$allopregnanolone as internal standards (Table 3.1).[56] Vallée and colleagues preferred pentafluorobenzyloxime and TMS derivatives and determined the levels of allopregnanolone, epiallopregnanolone (5α-P-3β-ol-20-one), pregnenolone, testosterone and DHEA in frontal cortex of rat brain to be near or below their limit of quantification (2.5 ng/g) when monitoring $[M - 178]^-$ ions for allopregnanolone and epiallopregnanolone, and $[M - 20]^-$ for pregnenolone, testosterone and DHEA, and using deuterium-labelled internal standards (Table 3.1).[57] Following swim stress the level of pregnenolone in frontal cortex was found to increase ten-fold, a change not reflected in plasma levels.

Although steroid sulfates have been shown to have pharmacological properties,[10,11] and were indirectly identified in RIA and by GC-EI-MS following hydrolysis of the steroid sulfate fraction isolated from rat brain,[15,16] intact neurosteroid sulfates have been difficult to observe in rat brain using MS.[17,18] Data from Liere *et al.* obtained using GC-MS indicates that the earlier identification of neurosteroids in the sulfate fraction from rat brain, was due to contamination of the fraction, and in fact endogenous levels are below the detection limit of 0.1 ng/g.[17] This low detection limit (2 pg on column) was achieved by derivatising steroid sulfates with HFB anhydride which results in both solvolysis of the sulfate ester group and formation of a HFB acid ester in its place (Scheme 3.5, R7). GC-MS detection was based on using SIM of $[M - 229]^+$ and $[M - 214]^+$ for pregnenolone, DHEA, and 5α-androstane-3α,17β-diol (5α-A-3α,17β-diol); and $[M]^+$ and $[M - 18]^+$ ions for allopregnanolone and epiallopregnanolone, HFB esters.

3.3.2 LC-MS Analysis

While GC-MS provides unsurpassed chromatographic resolution, high sensitivity, and informative first dimension mass spectra, the technique is unable to directly analyse polar and thermally labile steroids. This failing was evident in the GC-MS studies of sulfated neurosteroids discussed above. LC-MS methods, although giving inferior chromatographic performance, offer the advantage of being compatible with atmospheric pressure ionisation (API), which allows the analysis of intact conjugates.

3.3.2.1   Neurosteroids

Neurosteroids have been investigated by LC-MS and LC-tandem mass spectrometry (two-dimensional mass spectrometry, MS/MS) by a number of groups.[18,58–61] Liu and colleagues developed a nano-LC-electrospray (ES)–MS method for steroid sulfate analysis.[62] They validated the method on plasma and extended it to the identification of neurosteroid sulfates in rat brain.[18,62] The neurosteroid fraction from rat brain was separated on-line using a 350 mm × 100 μm (i.d.), 3 μm particle C_{18} nano column and analysed by negative ion ES-MS and ES-MS/MS. Steroid sulfates give abundant $[M - H]^-$ ions upon negative ion ES ionisation, and fragment in MS/MS experiments to give signals at m/z 97 corresponding to HSO_4^-.[63] This allowed the development of a precursor ion scan method in which MS_1 of the MS/MS instrument was scanned, while MS_2 was set to transmit fragment ions of m/z 97. In this way any compound eluting from the column and giving a precursor ion which would subsequently dissociate to give a fragment ion of m/z 97 is detected. This method provides both sensitivity and specificity, and when used in the multiple reaction monitoring (MRM) mode, where both the precursor and product ion masses are predefined in the MS/MS experiment, the on-column detection limit was as low as 0.1 pg. When neurosteroid sulfates from the sulfate fraction from rat brain were analysed only cholesterol sulfate (1.2 μg/g) was detected, no evidence was obtained for any endogenous neurosteroid sulfates including pregnenolone sulfate or DHEA sulfate above the detection limit of 0.3 ng/g, although internal standards added to brain homogenate were detected (Table 3.1). This result agreed with LC-ES-MS data presented by Mitamura *et al.*,[59] and enzyme-linked immunosorbent assay (ELISA) data of Higashi *et al.*,[64] who found levels of pregnenolone sulfate (500 pg/g by LC-MS, 50–500 pg/g by ELISA) in rat brain to be much lower than determined by GC-EI-MS (~ 10 ng/g).[16] Mitamura *et al.* used a reversed-phase-LC negative ion ES method which involved derivatisation of the oxo group of pregnenolone sulfate with 4-(N,N-dimethylaminosulfonyl)-7-hydrazino-2,1,3-benzoxadiazole and monitoring of the unfragmented $[M - H]^-$ ion (Scheme 3.5, R8).[59] The derivative enhanced sensitivity by an order of magnitude giving an on-column detection limit of ~ 100 pg.[59]

Neutral neurosteroids have also been extensively analysed by LC-MS and LC-MS/MS using derivatisation to enhance sensitivity. Liu *et al.* derivatised the oxo groups of neurosteroids with hydroxylamine so as to give steroid oximes (Scheme 3.5, R9).[18] Steroid oximes are more readily protonated in the electrospray process than underivatised analogues and give an enhancement in sensitivity of at least 20-fold.[65] Detection limits for testosterone, DHEA, pregnenolone and progesterone oximes in LC-ES-MRM experiments ranged from 0.1 pg to 0.5 pg.[18] Analysis of the neutral steroid fraction from rat brain gave levels of progesterone in the range 1–20 ng/g, pregnenolone 0.6–3.9 ng/g, epipregnanolone (5β-P-3β-ol-20-one) or epiallopregnanolone 0.05–2.5 ng/g, pregnanolone (5β-P-3α-ol-20-one) ~ 0.1 ng/g, allopregnanolone 0.5–11 ng/g, DHEA 0.05–0.11 ng/g and testosterone 0.04–0.5 ng/g. These levels are comparable to those reported by both Valée *et al.*[57] and Uzunov *et al.* determined using NCI-GC-MS.[66] Higashi and colleagues have also used an LC-MS approach incorporating derivatisation to enhance ionisation in the analysis of neutral neurosteroids in rat brain.[60] They found that by derivatising the oxo group of neurosteroids with 2-nitro-4-trifluoromethylphenylhydrazine to the corresponding hydrazone (Scheme 3.5, R10) and performing LC-electron capture atmospheric pressure chemical ionisation (ECAPCI)-MS they were able to enhance sensitivity by 20-fold, leading to an on-column detection limit of 1–6 pg. Although the derivatised steroids gave intense fragment-ions in MS/MS spectra, which could potentially be used for MRM analysis on MS/MS instruments, Higashi *et al.* performed their study on an ion trap instrument where the potential gain in sensitivity of the MRM scan is lost, although added specificity is still maintained. Higashi *et al.* sacrificed added specificity for sensitivity and performed their analysis on brain by LC-MS and monitoring $[M]^-$ ions. For control rats they found the pregnenolone level to be 7 ng/g and the progesterone level to be below 0.5 ng/g; however, for fixation stressed rats these levels increased to 60 ng/g and 13 ng/g, respectively.[60] Higashi and colleagues have also experimented with other derivatisation methods,[67–69] two of which have been exploited for the LC-MS analysis of testosterone and 5α-androstane-3α,17β-diol in rat brain.[61] Derivatisation of testosterone with 2-hydrazino-1-methylpyridine (HMP) and of 5α-androstane-3α,17β-diol with p-nitrobenzoyl chloride (Scheme 3.5, R11, R12) improved the detection limits by LC-MS by factors of 70 and 400, respectively. Derivatisation of the oxo group with HMP tags a positively charged group to the steroid giving an $[M]^+$ ion and enhancing analysis by LC-ES-MS. Using SIM of the $[M]^+$ ion the limit of quantification (LOQ) for testosterone was established at 60 pg/g of brain tissue. The brain testosterone level was determined to be ~ 1 ng/g which did not change appreciably as a result of immobilisation stress or ethanol administration. 5α-Androstane-3α,17β-diol only has alcohol groups available for derivatisation, and these can be reacted with p-nitrobenzoyl chloride to give a derivative with electron-capturing properties and suitable for ECAPCI. By monitoring the $[M]^-$ ion of the *bis* derivative in a LC-ECAPCI-MS experiment a LOQ of 0.2 ng/g was achieved, and the level of 5α-androstane-3α,17β-diol in rat brain determined to be 0.2 ng/g (Table 3.1).

3.3.2.2   Neurosterols

Neurosterols are also suitable for analysis by LC-MS with, or without, derivatisation. The neurosterol 24S-hydroxycholesterol is formed exclusively in brain, and its levels in plasma and CSF have been suggested as a marker for neurodegenerative disease.[52,54,71–73] With this in mind there has been considerable interest in the development of LC-MS assays for 24S-hydroxycholesterol and other oxysterols. Using atmospheric pressure chemical ionisation (APCI) as the method of ionisation, both 24S- and 27-hydroxycholesterol fail to give $[M+H]^+$ ions, but rather fragment in the ion source to give $[M+H-H_2O]^+$ and $[M+H-2H_2O]^+$ ions. The isomeric nature of these, and many other oxysterols, necessitates LC separation and identification is then based on retention time.[74] Using SIM of $[M + H–H_2O]^+$ and $[M + H–2H_2O]^+$ ions Burkard *et al.* achieved a quantification limit for plasma samples of 0.5 mL of 40 µg/L (8 ng on-column) and 25 µg/l (5 ng on-column) for 24S- and 27-hydroxycholesterol, respectively.[74] The normal levels of these oxysterols from healthy volunteers were determined to be ~ 60 ng/mL (range 39–91 ng/mL) and 120 ng/mL (range 67–199 ng/mL), respectively (Table 3.2). These values are of the same order of magnitude as found by other workers for these oxysterols in plasma using GC-EI-MS with SIM (24S-hydroxycholesterol ~ 60 ng/mL,[45,72] ~ 80 ng/mL,[75] 27-hydroxycholesterol ~ 150 ng/mL).[45,75]

To enhance the ionisation properties of neurosterols a number of groups have exploited derivatisation chemistry prior to LC-MS analysis.[24,25,76–79] One class of important neurosterols analysed in this manner are *seco*-sterols and aldol-sterols (Scheme 3.3), which are formed from the sterol structure by an initial ring opening. Recently, it has been suggested that antibodies catalyse ozone production during inflammation,[22] and that ozonolysis products of cholesterol including *seco*-sterols, are generated during the inflammatory component of atherosclerosis.[23] *Seco*-sterols have been detected in human brain, and have been shown to covalently modify Aβ peptide and accelerate amyloidogenesis *in vitro*.[24] 3β-Hydroxy-5-oxo-5,6-*seco*cholestan-6-al (5,6-*seco*-sterol) is a major product of cholesterol ozonolysis and can be converted to 3,5-dihydroxy-B-norcholestane-6-carboxyaldehyde in an aldolisation reaction catalysed by primary or secondary amines.[23] Zhang *et al.* identified 5,6-*seco*-sterol and its aldol in brain extracts following derivatisation with 2,4-dinitrophenylhydrazine (Scheme 3.5, R13), a procedure which can also result in aldolisation of the *seco*-sterol.[24,80] Analysis of the resulting hydrazones was performed by LC-ES-MS in the negative ion mode with SIM for the $[M – H]^-$ ion at *m/z* 597. Identification and quantification was made by comparison of mass, retention time and peak intensity with authentic standards. Zhang *et al.* found that the combined level of 5,6-*seco*-sterol and its aldol in brain from AD patients (0.44 pmol/mg, 184 ng/g) and age-matched controls (0.35 pmol/mg, 146 ng/g) were similar, but postulated that a transient increase in these sterols may be responsible for initiating Aβ amyloidogenesis.[24] In a follow-up study, applying similar methodology but using LC only for quantification, Bosco *et al.* found elevated levels of 5,6-*seco*-sterol and its aldol in the cortex of brain from patients with Lewy body dementia (0.213 µM, 89 ng/mL) relative to age-matched controls (0.093 µM, 39 ng/mL), and that the concentration of this metabolite was almost an order of magnitude higher in cells over expressing α-synuclein (0.57 µM, 238 ng/mL), than in cells not over expressing this protein (0.075 µM, 31 ng/mL).[25] At µM concentration 5,6-*seco*-sterol and its aldol accelerate α-synuclein aggregation *in vitro*. Bosco *et al.* hypothesised that oxidative stress produces cholesterol aldehydes that enable α-synuclein aggregation, leading to a pathogenic cycle.[25] It is noteworthy that Bosco *et al.* postulate the formation of 5,6-*seco*-sterol and its aldol in brain to be a result of reaction of cholesterol with reactive oxygen species, while earlier reports specified ozone as the oxidising agent in brain and atherosclerotic tissue.[23–25]

Additionally, the mass and elemental composition of 5,6-*seco*-sterol and its aldol (Scheme 3.3, 418.34470 Da, $C_{27}H_{46}O_3$) is equivalent to that of 3β,5α-dihydroxycholestan-6-one, a metabolite of 5β,6β-epoxycholesterol, a major product of the reaction between cholesterol and ozone,[26] and that the 6-oxo group can react with 2,4-dinitrophenylhydrazine to give a derivative with identical mass to the identified hydrazones (Scheme 3.5, R13). Further, as pointed out by Leland Smith in his humorous review "Oxygen, Oxysterols, Ouabain, and Ozone: A Cautionary Tale", 3β,5α-dihydroxycholestan-6-one is a common cholesterol autoxidation product.[81]

3.3.2.3   Girard P (GP) Hydrazone Derivatives for LC-MS

At the School of Pharmacy we have also developed derivatisation chemistry for the analysis of both neurosteroids and neurosterols.[76–79,82] For neurosteroids or neurosterols possessing a oxo group we derivatise with the Girard P (GP) reagent ([1-(2-hydrazino-2-oxoethyl)pyridinium chloride]) to give a GP-hydrazone (Scheme 3.5, R14). The GP-hydrazone group posses a charged quaternary nitrogen, and derivatised neurosteroids give very intense signals in both ES and matrix-assisted laser desorption/ionisation (MALDI) mass spectra. Neurosteroid-GP hydrazones can be detected at the sub pg level in mass spectra and give informative MS/MS spectra at the 50 pg level. MS/MS spectra of the $[M]^+$ ions from neurosteroid-GP hydrazones are dominated by $[M - 79]^+$ and $[M - 107]^+$ fragment ions, which correspond to the loss of pyridine and pyridine plus carbon monoxide from the derivatising group, respectively (Figure 3.1). For neurosteroids possessing a 3-oxo-Δ^4 (3-oxo-4-ene) group, characteristic fragments are observed at m/z 151, 163 and 177 corresponding to cleavage in the A and B rings (labelled *b_1-12, *b_3-28 and *b_2 respectively in Figure 3.1, upper panel). This allows the simple differentiation of *e.g.* testosterone from DHEA (Figure 3.1). Neurosteroids possessing a 20-oxo group also give characteristic fragment ions in the MS/MS spectra of their derivatives, an ion at m/z 125 being a marker of 20-oxoneurosteroids.

Many neurosterols do not posses an oxo group, *e.g.* 24S-hydroxycholesterol the most abundant oxysterol in brain,[83] and will not be derivatised by the GP reagent. However, neurosterols (and neurosteroids) with a 3β-hydroxy-Δ^5 group or 3β-hydroxy-5α-hydrogen stereochemistry when treated with cholesterol oxidase will be converted to 3-oxo-Δ^4 or 3-oxo sterols respectively, which are substrates for GP derivatisation.[84,85] Care must be taken in the selection of cholesterol oxidase and incubation time to avoid artefacts.[86] We have used cholesterol oxidase from both *Brevibacterium* and *Streptomyces sp.*, and our current preference is for the latter. Following treatment of neurosterols with cholesterol oxidase and GP hydrazine, we find an improvement in ES-MS sensitivity of three orders of magnitude for monohydroxycholesterols, *e.g.* 25-hydroxycholesterol.[76] Following oxidation of 3β-hydroxy-Δ^5 neurosterols and derivatisation of the resulting 3-oxo-Δ^4 group with GP-hydrazine the derivatives give the characteristic fragment ions corresponding to $[M - 79]^+$,

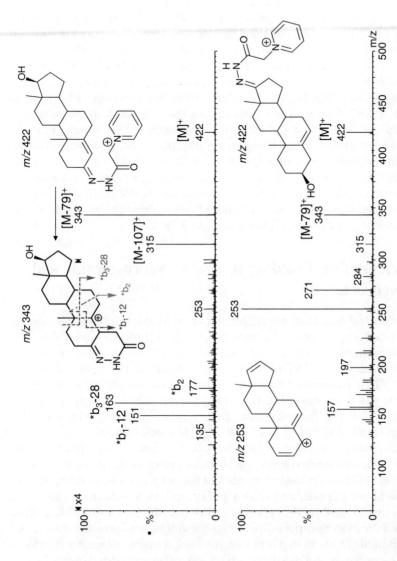

Figure 3.1 MS/MS spectra of testosterone-GP hydrazone *m/z* 422 (upper panel), and DHEA-GP hydrazone *m/z* 422 (lower panel). Spectra recorded on a Q-TOF (Waters) instrument. The *m/z* range 20–300 has been magnified by a factor of 4.

$[M - 107]^+$, and A-, and B-ring fragment ions at m/z 151 (*b_1–12), 163 (*b_3–28), 177 (*b_2) in the MS/MS spectra (Figure 3.2, upper panel). Incorporation of a further substituent in the AB-rings alters this pattern of fragment ions, *e.g.* the presence of a 7-hydroxy group changes the pattern of AB-ring fragment ions to m/z 151 (*b_1–12), 177 (*b_2) and 179 (*b_3–28) (Figure 3.2, lower panel), while only the ion at 177 is observed if a 6β-hydroxy group is present. The presence of substituents at other locations are also characterised by specific fragment ion patterns, *e.g.* in the MS/MS spectrum of oxidised/derivatised 24S-hydroxycholesterol a peak at m/z 353 is observed with elevated intensity. The spectra presented in Figure 3.2 were recorded on a Q-TOF type of mass spectrometer where MS_1 and MS_2 are arranged in series, and similar spectra are generated by other MS/MS spectrometers where the MS_1 and MS_2 are separated "in space". Currently we are using an ion trap mass spectrometer for neurosterol analysis, where the MS_1 and MS_2 stages are separated "in time" rather than "in space". When using an ion trap the capacity of the instrument to perform multiple stages of fragmentation "in time" can be exploited, and for GP derivatives maximum structural information is usually obtained by fragmenting the $[M - 79]^+$ ion in a MS^3 experiment *i.e.* $[M]^+ \rightarrow [M - 79]^+ \rightarrow$. Using this methodology we have confirmed 24S-hydroxycholesterol to be the most abundant oxysterol in rat brain (Figure 3.3).

3.4   Metabolite Profiling in Brain: Neurosteroids and Neurosterols

Neurosterols and neurosteroids are present in brain at low levels (ng/g to μg/g) against a high background of cholesterol (mg/g) (Table 3.1 and Table 3.2). Furthermore, neurosteroids/sterols, with the exception of the sulfate esters, are difficult to analyse by API techniques on account of their poor ionisation efficiency. These factors make it advisable to perform specific group separation procedures prior to mass spectrometric analysis, which is then preferably performed following derivatisation, to enhance sensitivity, and with LC separation to allow the analysis of individual isomeric compounds.

For neurosteroid profiling at the School of Pharmacy we adopt the sample preparation procedure developed by Sjövall's group at Karolinska Institutet (Figure 3.4).[18] Ethanol is used as the solvent for brain homogenisation, as it is a good solvent for steroids, and readily penetrates cell membranes. This initial solvent is then diluted to 70% ethanol and passed through a bed of C_{18} which will extract the most non-polar lipids. Neurosteroids, pass through the C_{18} bed in 70% ethanol. The next step is passage through a cation exchanger to remove cationic compounds, followed by separation of neurosteroids according to acidity on an anion exchanger. Weak acids (*e.g.* bile acids) and steroid sulfates are then analysed by LC-MS using negative ion ES. As steroid sulfates and bile acids are present in brain at low levels (Table 3.1 and Table 3.2), our preference is for low flow rate (< 1 μL/min) chromatography linked to negative ion nano-ES, so as to maximise analytical sensitivity (see Chapter 1). For steroid sulfate

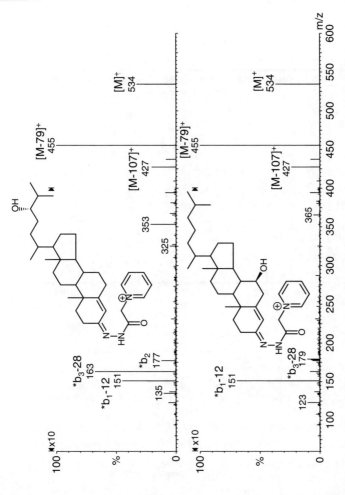

Figure 3.2 MS/MS spectra of oxidised and GP derivatised 24S-hydroxycholesterol m/z 534 (upper panel), and 7β-hydroxycholesterol m/z 534 (lower panel). Spectra were recorded on a Q-TOF (Waters) instrument. The m/z range 60–400 has been magnified by a factor of 10.

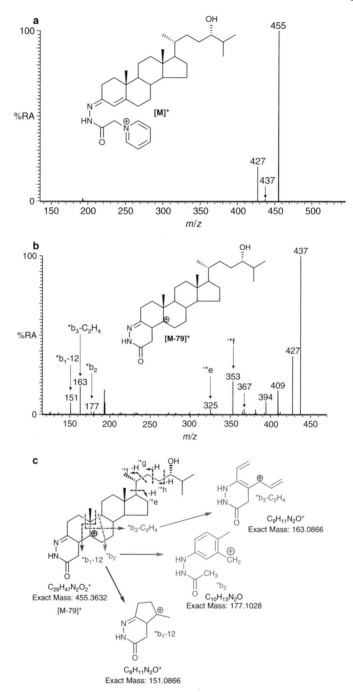

Figure 3.3   (a) MS^2 ([M]$^+$ →), and (b) MS^3 ([M]$^+$ → [M − 79]$^+$ →), spectra of oxidised and GP derivatised 24S-hydroxycholesterol *m/z* 534. Spectra were recorded on an LTQ–Orbitrap (Thermo Electron), with ion detection in the Orbitrap. (c) Fragmentation nomenclature.

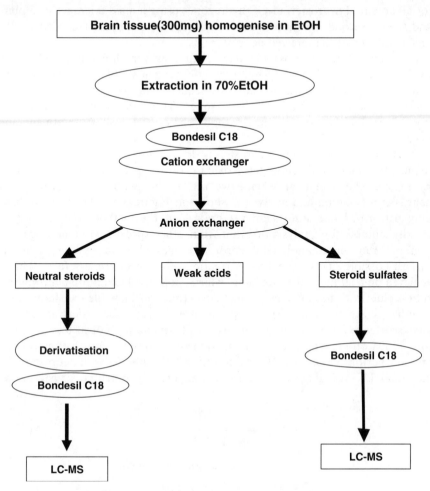

Figure 3.4   Extraction and isolation of neurosteroids from brain tissue.

analysis, advantage can be taken of precursor ion scanning on tandem quadrupole instruments as a major MS/MS fragment ion is observed at m/z 97. For neutral neurosteroids possessing an oxo group our preferred derivatisation reaction is with GP hydrazine to give GP-hydrazones (Scheme 3.5, R14). Again low flow rate LC-MS/MS can be exploited for maximum sensitivity. As steroid-GP hydrazones fragment in MS/MS to give abundant $[M-79]^{+}$ ions (Figure 3.1), tandem quadrupole instruments can exploit a neutral loss scan for 79 Da. This allows the identification of molecules derivatised with the GP group, for their subsequent product ion analysis and identification. When neurosteroid analysis is performed on instruments without a precursor ion or neutral loss scan facility, but offering exact mass capability, an initial LC-MS analysis is performed, followed by generation of reconstructed ion chromatograms (RICs) for all potential neurosteroid-GP hydrazones using a mass tolerance of 5 ppm, and then a second injection is made and product ion scans recorded on the precursor ions of the inferred neurosteroids.

Not all neurosteroids posses an oxo group; however, cholesterol oxidase will oxidise 3β-hydroxy-Δ^5 and 3β-hydroxy-5α-hydrogen steroids to their 3-oxo-Δ^4 and 3-oxo equivalents, which can be subsequently derivatised with GP-hydrazine and analysed as their GP-hydrazones. Thus, the neutral neurosteroid fraction is analysed twice: once without cholesterol oxidase treatment and once with.

As discussed above, most neurosterols do not posses an oxo group, hence, require treatment with cholesterol oxidase prior to derivatisation with GP-hydrazine. The analytical scheme we use at the School of Pharmacy for neurosterol isolation and analysis is shown in Figure 3.5.[79] To date, we have only performed our studies on rat brain, but the methodology should be equally suitable for analysis of brain from other species. After homogenisation of brain in ethanol, cholesterol is separated from oxysterols in the earliest possible step to avoid the formation of autoxidation artefacts. This is achieved on a straight-phase column, where cholesterol and more hydrophobic sterols elute with hexane/dichloromethane (Fraction 1), while oxysterols elute in ethylacetate (Fraction 2). Oxysterols are then oxidised, derivatised and analysed by LC-MS. Again advantage can be made of neutral loss scanning for –79 Da on tandem quadrupole instruments. However, most of our studies are now performed on Q-TOF or ion trap instruments. As with neurosteroids, an initial LC-MS analysis is performed to identify potential oxysterols by

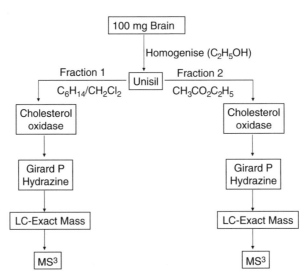

Figure 3.5   Workflow depicting the extraction, purification and analysis of oxysterols from rat brain.

generating RICs for the appropriate masses (5 ppm window for studies using instruments capable of exact mass measurement, *e.g.* LTQ-Orbitrap, Q-TOF). For example, Figure 3.6 shows RICs for oxidised/derivatised oxysterols based on the hydroxycholesterol (upper trace), dihydroxycholesterol (middle trace) and dihydroxycholestanol (lower trace) structures. Comparison of retention time with authentic standards revealed that the two peaks in the upper trace correspond to oxidised/derivatised 24S-hydroxycholesterol. Twin peaks arise on account of the formation of *syn* and *anti* derivatives. In the middle trace at least nine distinct peaks are observed corresponding to di-hydroxycholesterol isomers. By performing MS^3 ($[M]^+ \rightarrow [M-79]^+ \rightarrow$) scans, six of the isomers were identified, *i.e.* 24,25-, 24,27-, 25,27-, 6,24-, 7α,25- and 7α,27-dihydroxycholesterols (Figure 3.7, Table 3.2). This data agrees well with that of Zhang *et al.*[30] who studied the metabolism of 24-, 25- and 27-hydroxycholesterol in rat astrocytes, and that of Mast[32] who identified the products of CYP46A1 catalysed reactions (see Section 3.2.4 of this chapter and Table 3.2).

5,6-*Seco*-sterol and its aldol both posses oxo groups and will also be derivatised by the GP reagent. The RIC for their exact mass is shown in Figure 3.6 (lower trace). This chromatogram was recorded on the oxysterol fraction (Fraction 2) isolated from rat brain according to the procedure detailed in Figure 3.5. The components present in chromatographic peaks at 9.23 and 10.03 min, when subjected to MS^2 gave spectra identical to authentic samples of 3β-hydroxy-5-oxo-5,6-*seco*cholestan-6-al and 3,5-dihydroxy-B-norcholestane-6-carboxyaldehyde, respectively (Figure 3.8), and the retention times of 9.23 and 10.03 min also coincide with those of the respective authentic standards. This data confirms the presence of 5,6-*seco*-sterol and its aldol in rat brain.

Like 5,6-*seco*-sterol and its aldol, other oxysterols are present in brain which naturally posses an oxo group, *e.g.* 7α,25-dihydroxycholest-4-en-3-one, 7α,27-dihydroxycholest-4-en-3-one and 7α-hydroxy-3-oxocholest-4-en-27-oic acid (Scheme 3.3). Using the analytical scheme presented in Figure 3.5, these oxysterols containing a "natural" oxo group, will not be differentiated from their 3β-hydroxy-Δ^5 analogues. However, this drawback can be simply overcome by performing a parallel experiment in the absence of cholesterol oxidase.

3.5 Linking Proteomics with Lipidomics

In this chapter we have discussed at length the mass spectrometric identification of sterols and steroids in brain. While some of them can enter the brain across the blood–brain barrier, evidence suggests that many neurosteroids and neurosterols are synthesised in brain itself.[10,11,33,93] Hence, enzymes of the cholesterol and steroid biosynthetic pathways (Schemes 3.1–3.3) must be present in brain. Although many of the enzymes of these pathways have been identified at the mRNA level, or by activity,[94] surprisingly few have

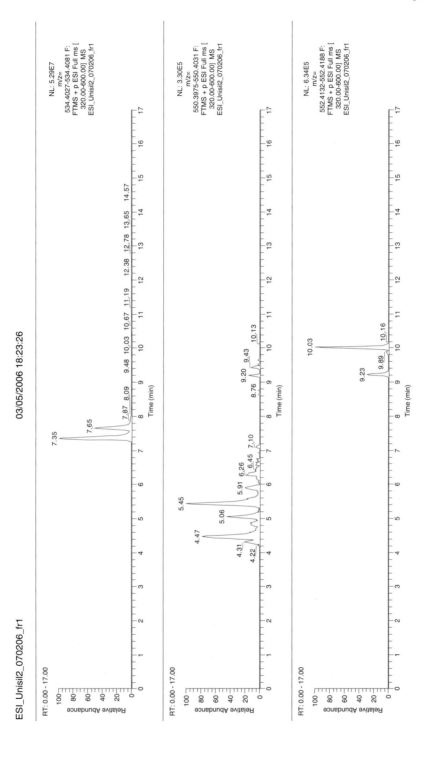

been identified at the protein level. With the advent of the proteomic revolution this deficit is being reversed as numerous studies generating protein lists are currently being generated.[95] Perhaps more interesting are metabolite-led proteomic studies, where the effect of a metabolite on a proteome is investigated. We are performing such studies at the School of Pharmacy, where we have incubated rat neuron cells with oxysterols, for example, and monitored the quantitative changes in protein expression in response. Using such studies, much information on the effect of neurosterols on biological pathways can be gleaned. Below we illustrate this, by presenting data on the effect of 24S-hydroxycholesterol on protein expression in rat cortical neurons.

24S-Hydroxycholesterol is formed in neurons by the oxidation of cholesterol in a reaction catalysed by CYP46A1 (Scheme 3.3).[33] We treated rat cortical neurons with 10 µM 24S-hydroxycholesterol and monitored the proteomic changes in comparison to cells treated with vehicle. 24S-Hydroxycholesterol was found to down-regulate the expression of HMG-CoA synthase, the enzyme responsible for the conversion of acetoacetyl-CoA to HMG-CoA in the cholesterol synthesis pathway (Scheme 3.1). In contrast, the expression of apolipoprotein E (apo-E) was up-regulated.

Enzymes of the cholesterol synthesis pathway are transcriptionally regulated by SREBPs, which also regulate the expression of enzymes involved in the synthesis of fatty acids and triglycerides.[38,39] However, the expression of enzymes of the fatty acid and triglyceride synthesis pathways were not found to be affected by 24S-hydroxycholesterol. 24S-Hydroxycholesterol is a ligand for the LXR,[96] the β form of which is expressed in brain.[97] ApoE is under transcriptional regulation of LXR,[98] as is SREBP-1c, but not SREBP-1a or -2.[99] Thus, the up-regulation of apoE in neurons observed when they are treated with 24S-hydroxycholesterol can be explained as an LXR-regulated event. What then accounts for down-regulation of the enzymes in the cholesterol synthesis pathway? The expression of these enzymes is regulated by SERBPs 1a and 2. SREBPs are synthesised as inactive precursors bound to

---

Figure 3.6 Upper trace, RIC for m/z 534.4054 (± 5 ppm); central trace, RIC for m/z 550.4003 (± 5 ppm); and lower trace RIC m/z 552.4160 (± 5 ppm), corresponding to oxidised/derivatised monohydroxycholesterols, dihydroxycholesterols and dihydroxycholestanols extracted from rat brain. Chromatographic separation was performed on a Finnigan Surveyor HPLC system utilising a Hypersil GOLD reversed-phase column (1.9 µm particles, 50 × 2.1 mm) from Thermo Electron. Mobile phase A consisted of 50% methanol containing 0.1% formic acid, and mobile phase B consisted of 95% methanol containing 0.1% formic acid. After 1 min at 20% B, the proportion of B was raised to 80% B over the next 7 min, and maintained at 80% B for a further 5 min, before returning to 20% B in 6 s and re-equilibration for a further 3 min 54 s, giving a total run time of 17 min. The flow rate was maintained at 200 µL min and eluent directed to the API source of a LTQ-Orbitrap mass spectrometer. The RICs were recorded by Fourier Transform in the Orbitrap analyser.

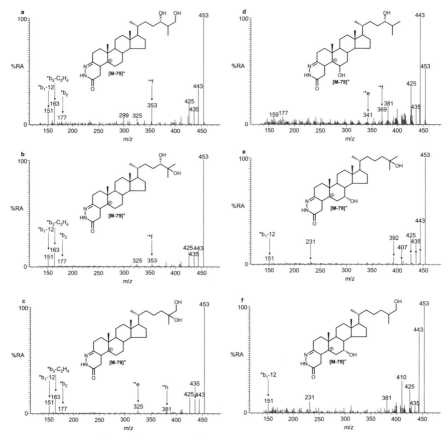

Figure 3.7 MS^3 ([M]$^+$ → [M − 79]$^+$ →) spectra of chromatographic peaks (m/z 550.4003) eluting at (a) 4.31 min corresponding to 24,27-dihydroxycholesterol; (b) 4.47 min corresponding to 24,25-dihydroxycholesterol; (c) 5.06 min corresponding to 25,27-dihydroxycholesterol; (d) 5.45 min corresponding to 6,24-dihydroxycholesterol; (e) 5.91 min corresponding to 7α,25-dihydroxycholesterol; and (f) 6.29 min corresponding to 7α,27-dihydroxycholesterol; in the chromatogram shown in Figure 3.6 (central trace.) MS^3 spectra recorded on the LTQ detector of an LTQ-Orbitrap.

the endoplasmic reticulum (ER) and must be escorted to the Golgi by SREBP cleavage-activating protein (SCAP) to be proteolytically cleaved to nuclear SREBP (nSREBP), which translocates to the nucleus, where it activates transcription by binding to nonpalindronic sterol response elements (SREs) in the promoter regions of target genes.[39] When cholesterol builds up in ER membranes, the sterol binds to SCAP and triggers a conformational change that causes SCAP to bind to INSIG (insulin-induced gene) proteins, so they can no longer escort SREBPs to the Golgi for processing to the active form. Adams *et al.* have shown that 25-hydroxycholesterol also inhibits cholesterol

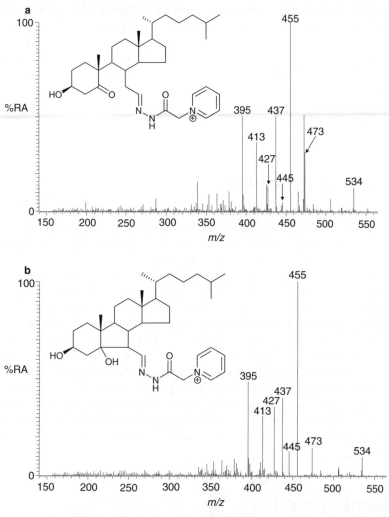

Figure 3.8 The MS^2 spectra of chromatographic peaks (*m/z* 552.4160) eluting at (a) 9.23 min and corresponding to 3β-hydroxy-5-oxo-5,6-*seco*cholestan-6-al; and (b) 10.03 min and corresponding to 3,5-dihydroxy-B-norcholestane-6-carboxyaldehyde, in the chromatogram shown in Figure 3.6 (lower trace). MS^2 spectra recorded on the LTQ detector of an LTQ-Orbitrap.

synthesis via SREBPs.[100] While cholesterol has been shown in cross-linking experiments to bind directly to SCAP and trigger conformational change, 25-hydroxycholesterol does not bind directly or cause a detectable conformational change to SCAP. Instead it is proposed that 25-hydroxycholesterol interacts indirectly with SCAP and INSIG through an oxysterol putative sensing protein. We suggest that 24S-hydroxycholesterol behaves in a similar manner to 25-hydroxycholesterol, interacting with an oxysterol-binding protein so as to bind SCAP to INSIG and prevent SREBP-1a and 2 from reaching the Golgi for processing to nSREBPs. Our data also suggests that 24S-hydroxycholesterol in neurons has a greater effect on SREBPs 1a and 2 than 1c. This can be reconciled with the fact that SREBP-1c regulates enzymes involved in fatty acid synthesis, and that one way of removing excess free sterol is by ester formation with fatty acids. A second pathway for removal of excess sterol is via transport involving apoE.

3.6　Future Prospective

In this chapter we have highlighted the presence of neurosteroids and neuro-sterols in brain. Many of these compounds are present at very low levels (ng/g to μg/g), and it is only in the last decade that suitable analytical techniques have become available for the reliable measurement of these compounds when present at low abundance. The challenge is now to identify and quantify neurosteroid/sterols in distinct brain regions and to investigate how their levels change with age or stress, for example, or in neurodegenerative disease. It is likely that novel neurosteroid/sterols are still to be identified, and the biological effects of these and many already known neurosteroid/sterols must be investigated. While many of the enzymes responsible for steroidogenesis have been identified in brain at the mRNA level, the abundance and location of the expressed proteins has yet to be explored. While brain proteome projects will go some way to identifying these enzymes on the protein level, the challenge of investigating the levels of these proteins in different brain regions and under different physiological states will entertain experimentalists for some time yet.

Acknowledgement

This work was supported by funding from UK Biotechnology and Biological Sciences Research Council (BBSRC grant no. BB/C515771/1 and BB/C511356/1), and The School of Pharmacy.

Appendix Sterol, steroid and bile acid molecular frame work. Cholesterol, 3β-hydroxycholest-5-en-27-oic acid and cholic acid are drawn with accompanying stereochemistry. Some confusion can arise concerning the nomenclature used to describe 27-hydroxycholesterol, which was previously denoted as 26-hydroxycholesterol. According to rules of priority of numbering the correct description is of 27-hydroxycholesterol is 25R,26-hydroxycholesterol; however, as the medical community uses the name 27-hydroxycholesterol, we will use this name in this chapter.

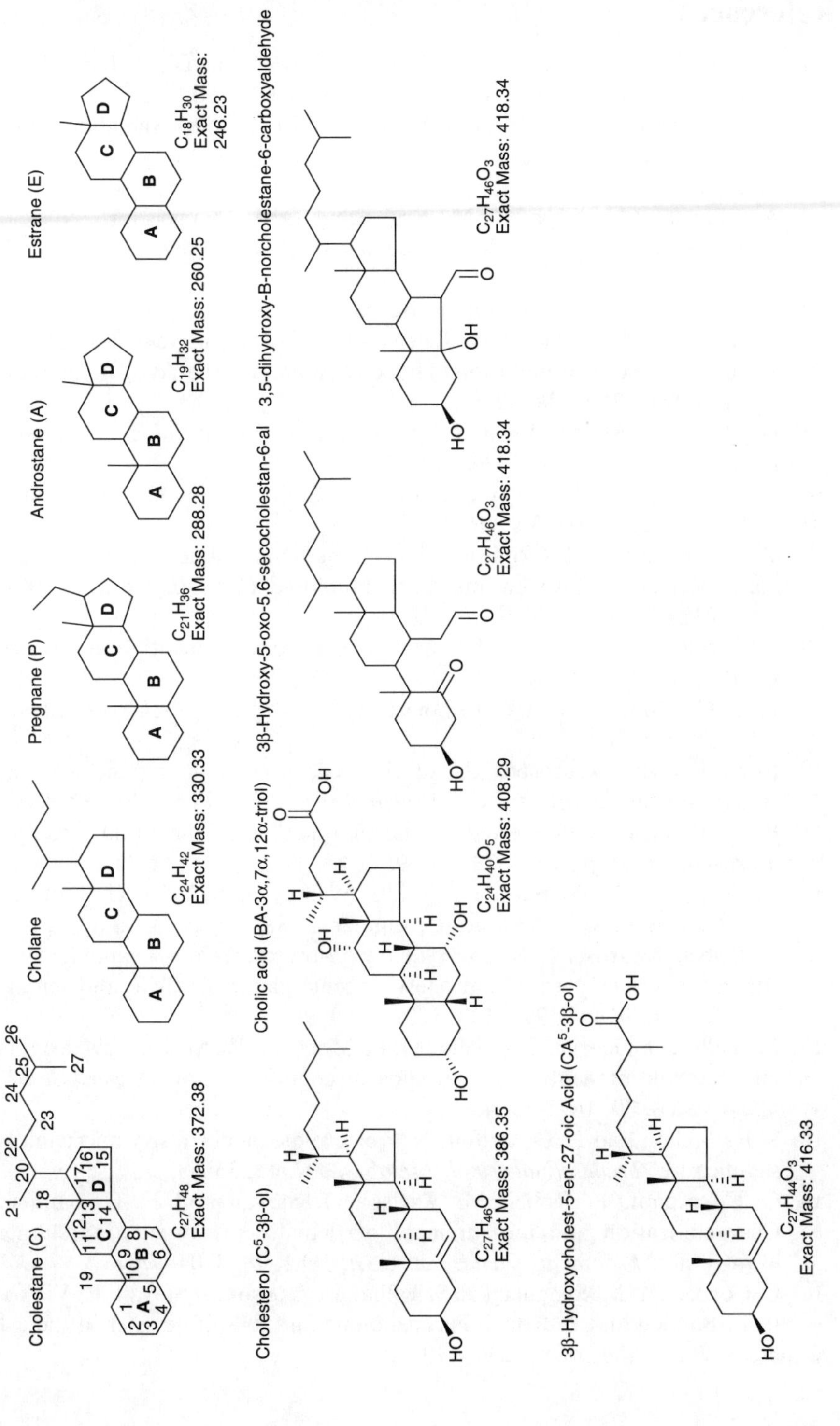

Estrane (E)

$C_{18}H_{30}$
Exact Mass: 246.23

Androstane (A)

$C_{19}H_{32}$ Exact Mass: 260.25

$C_{19}H_{32}$
Exact Mass: 288.28

Pregnane (P)

$C_{21}H_{36}$
Exact Mass: 330.33

Cholane

$C_{24}H_{42}$
Exact Mass: 372.38

Cholestane (C)

$C_{27}H_{48}$
Exact Mass: 372.38

3,5-dihydroxy-B-norcholestane-6-carboxyaldehyde

$C_{27}H_{46}O_3$
Exact Mass: 418.34

3β-Hydroxy-5-oxo-5,6-secocholestan-6-al

$C_{27}H_{46}O_3$
Exact Mass: 418.34

Cholic acid (BA-3α,7α,12α-triol)

$C_{24}H_{40}O_5$
Exact Mass: 408.29

Cholesterol (C⁵-3β-ol)

$C_{27}H_{46}O$
Exact Mass: 386.35

3β-Hydroxycholest-5-en-27-oic Acid (CA⁵-3β-ol)

$C_{27}H_{44}O_3$
Exact Mass: 416.33

References

1. *Steroid Analysis*, H. L. J. Makin, D. B. Gower and D. N. Kirk, Eds., Blackie Academic & Professional, Glasgow, 1995.
2. P. A. Edwards and J. Ericsson, Sterols and isoprenoids: signaling molecules derived from the cholesterol biosynthetic pathway, *Annu. Rev. Biochem.*, 1999, **68**, 157–185.
3. A. Chawla, J. J. Repa, R. M. Evans and D. J. Mangelsdorf, Nuclear receptors and lipid physiology: Opening the X-files, *Science*, 2001, **294**, 1866–1870.
4. D. S. Geller, A. Fahri, N. Pinkerton, M. Fradley, M. Moritz, A. Spitzer, G. Meinke, T. F. Tsai, P. B. Sigler and R. P. Lifton, Activating mineralocorticoid receptor mutation in hypertension exacerbated by pregnancy, *Science*, 2000, **289**, 119–123.
5. H. P. Rang, M. M. Dale, J. M. Ritter, In *Pharmacology*, Churchill Livingston, New York, 1996, 323.
6. R. I. Kelley and R. C. Hennekam, The Smith-Lemli-Opitz syndrome, *J. Med. Genet.*, 2000, **37**, 321–335.
7. Y. Wang and W. J. Griffiths, Modern methods of bile acid analysis by mass spectrometry: A view into the metabolome, *Curr. Anal. Chem.*, 2007, **3**, 103–126.
8. J. Sjövall, Fifty years with bile acids and steroids in health and disease, *Lipids*, 2004, **39**, 703–722.
9. P. T. Clayton, Disorders of cholesterol biosynthesis, *Arch. Dis. Child.*, 1998, **78**, 185–189.
10. E. E. Baulieu, Neurosteroids: of the nervous system, by the nervous system, for the nervous system, *Recent. Prog. Horm. Res.*, 1997, **52**, 1–32.
11. E. E. Baulieu, Neurosteroids: a novel function of the brain, *Psychoneuroendocrinology*, 1998, **23**, 963–987.
12. C. Corpéchot, J. Young, M. Calvel, C. Wehrey, J. N. Veltz, G. Trouyer, M. Mouren, V. V. K. Prasad, C. Banner, J. Sjövall, E. -E. Baulieu and P. Robel, Neurosteroids, 3α-Hydroxy-5α-pregnane-20-one and its precursors in brain, plasma, and steroidogenic glands of male and female rats, *Endocrinology*, 1993, **133**, 1003–1009.
13. M. Vallée, O. George, S. Vitiello, M. Le Moel and W. Mayo, New insights into the role of neuroactive steroids in cognitive aging, *Exper. Gerontology*, 2004, **39**, 1695–1704.
14. S. H. Mellon and L. D. Griffin, Neurosteroids, biochemistry and clinical significance, *Trends Endocrinol. Metab.*, 2002, **13**, 35–43.
15. C. Corpéchot, P. Robel, M. Axelson, J. Sjövall and E. E. Baulieu, Characterization and measurement of dehydroepiandrosterone sulfate in rat brain, *Proc. Natl. Acad. Sci. USA*, 1981, **78**, 4704–4707.
16. C. Corpéchot, M. Synguelakis, S. Talha, M. Axelson, J. Sjövall, R. Vihko, E. E. Baulieu and P. Robel, Pregnenolone and its sulfate ester in the rat brain, *Brain. Res.*, 1983, **270**, 119–125.

17. P. Liere, A. Pianos, B. Eychenne, A. Cambourg, S. Liu, W. Griffiths, M. Schumacher, J. Sjövall and E. E. Baulieu, Novel lipoidal derivatives of pregnenolone and dehydroepiandrosterone and absence of their sulfated counterparts in rodent brain, *J. Lipid. Res.*, 2004, **45**, 2287–2302.

18. S. Liu, J. Sjövall and W. J. Griffiths, Neurosteroids in rat brain: extraction, isolation, and analysis by nanoscale liquid chromatography-electrospray mass spectrometry, *Anal. Chem.*, 2003, **75**, 5835–5846.

19. M. D. Majewska, N. L. Harrison, R. D. Schwartz, J. L. Barker and S. M. Paul, Steroid hormone metabolites are barbiturate-like modulators of the GABA receptor, *Science*, 1986, **232**, 1004–1007.

20. M. R. Bowlby, Pregnenolone sulfate potentiation of *N*-methyl-D-aspartate receptor channels in hippocampal neurons, *Mol. Pharmacol.*, 1993, **43**, 813–819.

21. R. P. Irwin, S. Z. Lin, M. A. Rogawski, R. H. Purdy and S.M. Paul, Steroid potentiation and inhibition of *N*-methyl-D-aspartate receptor-mediated intracellular Ca^{2+} responses: structure-activity studies, *J. Pharmacol. Exp. Ther.*, **271**, 677–682.

22. P. Wentworth, J. E. McDunn, A. D. Wentworth, C. Takeuchi, J. Nieva, T. Jones, C. Bauttista, J. M. Ruedi, A. Gutierrez, K. A. Janda, B. M. Babior, A. Eschenmoser and R. A. Lerner, Evidence for antibody-catalyzed ozone formation in bacterial killing and inflammation, *Science*, 2002, **298**, 2195–2199.

23. P. Wentworth Jr., J. Nieva, C. Takeuchi, R. Galve, A. D. Wentworth, R. B. Dilley, G. A. DeLaria, A. Saven, M. M. Babior, K. D. Janda, A. Eschenmoser and R. A. Lerner, Evidence for ozone formation in human atherosclerotic arteries, *Science*, 2003, **302**, 1053–1056.

24. Q. Zhang, E. T. Powers, J. Nieva, M. E. Huff, M. A. Dendle, J. Bieschke, C. G. Glabe, A. Eschenmoser, P. Wentworth Jr., R. A. Lerner and J. W. Kelly, Metabolite-initiated protein misfolding may trigger Alzheimer's disease, *Proc. Natl. Acad. Sci. USA*, 2004, **101**, 4752–4757.

25. D. A. Bosco, D. M. Fowler, Q. Zhang, J. Nieva, E. T. Powers, P. Wentworth Jr., R. A. Lerner and J. W. Kelly, Elevated levels of oxidized cholesterol metabolites in Lewy body disease brains accelerate alpha-synuclein fibrilization, *Nat. Chem. Biol.*, 2006, **2**, 249–253.

26. J. Bieschke, Q. Zhang, D. A. Bosco, R. A. Lerner, E. T. Powers, P. Wentworth Jr. and J. W. Kelly, Small molecule oxidation products trigger disease-associated protein misfolding, *Acc. Chem. Res.*, 2006, **39**, 611–619.

27. M. K. Pulfer and R. C. Murphy, Formation of biologically active oxysterols during ozonolysis of cholesterol present in lung surfactant, *J. Biol. Chem.*, 2004, **279**, 26331–26338.

28. M. Axelson, O. Larsson, J. Zhang, J. Shoda and J. Sjövall, Structural specificity in the suppression of HMG-CoA reductase in human fibroblasts by intermediates in bile acid biosynthesis, *J. Lipid. Res.*, 1994, **36**, 290–298.

29. J. Zhang, Y. Akwa, E. E. Baulieu and J. Sjövall, 7 Alpha-hydroxylation of 27-hydroxycholesterol in rat brain microsomes, *C. R. Acad. Sci. Paris*, 1995, **318**, 345–349.

30. J. Zhang, Y. Akwa, M. El-Etr, E. E. Baulieu and J. Sjövall, Metabolism of 27-, 25- and 24-hydroxycholesterol in rat glial cells and neurons, *Biochem. J.*, 1997, **322**, 175–184.

31. N. Mano, T. Goto, M. Uchida, K. Nishimura, M. Ando, N. Kobayashi and J. Goto, Presence of protein-bound unconjugated bile acids in the cytoplasmic fraction of rat brain, *J. Lipid. Res.*, 2004, **45**, 295–300.

32. N. Mast, R. Norcross, U. Andersson, M. Shou, K. Nakayama, I. Bjorkhem and I. A. Pikuleva, Broad substrate specificity of human cytochrome P450 46A1 which initiates cholesterol degradation in the brain, *Biochemistry*, 2003, **42**, 14284–14292.

33. E. G. Lund, J. M. Guileyardo and D. W. Russell, cDNA cloning of cholesterol 24-hydroxylase, a mediator of cholesterol homeostasis in the brain, *Proc. Natl. Acad. Sci. USA*, 1999, **96**, 7238–7243.

34. A. A. Kandutsch and H. W. Chen, Inhibition of sterol synthesis in cultured mouse cells by 7α-hydroxycholesterol, 7β-hydroxycholesterol and 7-ketocholesterol, *J. Biol. Chem.*, 1973, **248**, 8408–8417.

35. D. W. Russell, Nuclear orphan receptors control cholesterol catabolism, *Cell*, 1999, **97**, 539–542.

36. J. J. Repa and D. J. Mangelsdorf, The role of orphan nuclear receptors in the regulation of cholesterol homeostasis, *Annu. Rev. Cell Dev. Biol.*, 2000, **16**, 459–481.

37. C. M. Adams, J. Reitz, J. K. De Brabander, J. D. Feramisco, L. Li, M. S. Brown and J. L. Goldstein, Cholesterol and 25-hydroxycholesterol inhibit activation of SREBPs by different mechanisms, both involving SCAP and Insigs, *J. Biol. Chem.*, 2004, **279**, 52772–52780.

38. J. L. Goldstein, R. A. DeBose-Boyd and M. S. Brown, Protein sensors for membrane sterols, *Cell*, 2006, **124**, 35–46.

39. J. D. Horton, J. L. Goldstein and M. S. Brown, SREBPs: activators of the complete program of cholesterol and fatty acid synthesis in the liver, *J. Clin. Invest.*, 2002, **109**, 1125–1131.

40. J. Brown 3rd, C. Theisler, S. Silberman, D. Magnuson, N. Gottardi-Littell, J. M. Lee, D. Yager, J. Crowley, K. Sambamurti, M. M. Rahman, A. B. Reiss, C. B. Eckman and B. Wolozin, Differential expression of cholesterol hydroxylases in Alzheimer's disease, *J. Biol. Chem.*, 2004, **279**, 34674–34681.

41. A. A. Alomary, R. L. Fitzgerald and R. H. Purdy, Neurosteroid analysis, *Int. Rev. Neurobiol.*, 2001, **46**, 97–115.

42. M. Vallée, W. Mayo, M. Darnaudery, C. Corpéchot, J. Young, M. Koehl, M. Le Moal, E. E. Baulieu, P. Robel and H. Simon, Neurosteroids: deficient cognitive performance in aged rats depends on low pregnenolone sulfate levels in the hippocampus, *Proc. Natl. Acad. Sci. USA*, 1997, **94**, 14865–14870.

43. W. J. Griffiths, C. Shackleton and J. Sjövall, Steroid analysis, In *Encyclopedia of Mass Spectrometry, Volume 3*, R. Caprioli, Ed., Elsevier, Oxford, 2005.
44. L. Puglielli, A. L. Friedlich, K. D. Setchell, S. Nagano, C. Opazo, R. A. Cherny, K. J. Barnham, J. D. Wade, S. Melov, D. M. Kovacs and A. I. Bush, Alzheimer disease beta-amyloid activity mimics cholesterol oxidase, *J. Clin. Invest.*, 2005, **115**, 2556–2563.
45. S. Dzeletovic, O. Breuer, E. Lund and U. Diczfalusy, Determination of cholesterol oxidation products in human plasma by isotope dilution-mass spectrometry, *Anal. Biochem.*, 1995, **225**, 73–80.
46. I. Björkhem, D. Lütjohann, O. Breuer, A. Sakinis and A. Wennmalm, Importance of a novel oxidative mechanism for elimination of brain cholesterol. Turnover of cholesterol and 24(S)-hydroxycholesterol in rat brain as measured with $^{18}O_2$ techniques in vivo and in vitro, *J. Biol. Chem.*, 1997, **272**, 30178–30184.
47. I. Björkhem, D. Lütjohann, U. Diczfalusy, L. Stahle, G. Ahlborg and J. Wahren, Cholesterol homeostasis in human brain: turnover of 24S-hydroxycholesterol and evidence for a cerebral origin of most of this oxysterol in the circulation, *J. Lipid. Res.*, 1998, **39**, 1594–1600.
48. M. Heverin, N. Bogdanovic, D. Lütjohann, T. Bayer, I. Pikuleva, L. Bretillon, U. Diczfalusy, B. Winblad and I. Björkhem, Changes in the levels of cerebral and extracerebral sterols in the brain of patients with Alzheimer's disease, *J. Lipid. Res.*, 2004, **45**, 186–193.
49. D. Lütjohann, O. Breuer, G. Ahlborg, I. Nennesmo, A. Siden, U. Diczfalusy and I. Björkhem, Cholesterol homeostasis in human brain: evidence for an age-dependent flux of 24S-hydroxycholesterol from the brain into the circulation, *Proc. Natl. Acad. Sci. USA*, 1996, **93**, 9799–9804.
50. D. Lütjohann, A. Brzezinka, E. Barth, D. Abramowski, M. Staufenbiel, K. von Bergmann, K. Beyreuther, G. Multhaup and T. A. Bayer, Profile of cholesterol-related sterols in aged amyloid precursor protein transgenic mouse brain, *J. Lipid. Res.*, 2002, **43**, 1078–1085.
51. D. Lütjohann, M. Stroick, T. Bertsch, S. Kuhl, B. Lindenthal, K. Thelen, U. Andersson, I. Björkhem, Kv. K. Bergmann and K. Fassbender, High doses of simvastatin, pravastatin, and cholesterol reduce brain cholesterol synthesis in guinea pigs, *Steroids*, 2004, **69**, 431–438.
52. D. Lütjohann, Cholesterol metabolism in the brain: importance of 24S-hydroxylation, *Acta Neurol. Scand. Suppl.*, 2006, **185**, 33–42.
53. V. Leoni, T. Masterman, P. Patel, S. Meaney, U. Diczfalusy and I. Björkhem, Side chain oxidized oxysterols in cerebrospinal fluid and the integrity of blood-brain and blood-cerebrospinal fluid barriers, *J. Lipid. Res.*, 2003, **44**, 793–799.
54. V. Leoni, M. Shafaati, A. Salomon, M. Kivipelto, I. Björkhem and L. O. Wahlund, Are the CSF levels of 24S-hydroxycholesterol a sensitive biomarker for mild cognitive impairment? *Neurosci. Lett.*, 2006, **397**, 83–87.

55. V. Leoni, D. Lutjohann and T. Masterman, Levels of 7-oxocholesterol in cerebrospinal fluid are more than one thousand times lower than reported in multiple sclerosis, *J. Lipid Res.*, 2005, **46**, 191–195.

56. Y. S. Kim, H. Zhang and H. Y. Kim, Profiling neurosteroids in cerebrospinal fluids and plasma by gas chromatography/electron capture negative chemical ionization mass spectrometry, *Anal. Biochem.*, 2000, **277**, 187–195.

57. M. Vallée, J. D. Rivera, G. F. Koob, R. H. Purdy and R. L. Fitzgerald, Quantification of neurosteroids in rat plasma and brain following swim stress and allopregnanolone administration using negative chemical ionization gas chromatography/mass spectrometry, *Anal. Biochem.*, 2000, **287**, 153–166.

58. K. Shimada and Y. Mukai, Studies on neurosteroids. VII. Determination of pregnenolone and its 3-stearate in rat brains using high-performance liquid chromatography-atmospheric pressure chemical ionization mass spectrometry, *J. Chromatogr. B Biomed. Sci. Appl.*, 1998, **714**, 153–160.

59. K. Mitamura, M. Yatera and K. Shimada, Quantitative determination of pregnenolone-3-sulfate in rat brains using liquid chromatography/electrospray ionization-mass spectrometry, *Anal. Sci.*, 1999, **15**, 951–955.

60. T. Higashi, N. Takido and K. Shimada, Studies on neurosteroids XVII. Analysis of stress-induced changes in neurosteroid levels in rat brains using liquid chromatography-electron capture atmospheric pressure chemical ionization-mass spectrometry, *Steroids*, 2005, **70**, 1–11.

61. T. Higashi, Y. Ninomiya, N. Iwaki, A. Yamauchi, N. Takayama and K. Shimada, Studies on neurosteroids XVIII LC-MS analysis of changes in rat brain and serum testosterone levels induced by immobilization stress and ethanol administration, *Steroids*, 2006, **71**, 609–617.

62. S. Liu, W. J. Griffiths and J. Sjövall, Capillary liquid chromatography/electrospray mass spectrometry for analysis of steroid sulfates in biological samples, *Anal. Chem.*, 2003, **75**, 791–797.

63. W. J. Griffiths, Tandem mass spectrometry in the study of fatty acids, bile acids, and steroids, *Mass Spectrom. Rev.*, 2003, **22**, 81–152.

64. T. Higashi, Y. Daifu, T. Ikeshima, T. Yagi and K. Shimada, Studies on neurosteroids XV. Development of enzyme-linked immunosorbent assay for examining whether pregnenolone sulfate is a veritable neurosteroid, *J. Pharm. Biomed. Anal.*, 2003, **30**, 1907–1917.

65. S. Liu, J. Sjövall and W. J. Griffiths, Analysis of oxosteroids by nanoelectrospray mass spectrometry of their oximes, *Rapid Commun. Mass Spectrom.*, 2000, **14**, 390–400.

66. D. P. Uzunov, T. B. Cooper, E. Costa and A. Guidotti, Fluoxetine-elicited changes in brain neurosteroid content measured by negative ion mass fragmentography, *Proc. Natl. Acad. Sci. USA*, 1996, **93**, 12599–12604.

67. T. Higashi and K. Shimada, Derivatization of neutral steroids to enhance their detection characteristics in liquid chromatography–mass spectrometry, *Anal. Bioanal. Chem.*, 2004, **378**, 875–882.

68. T. Higashi, A. Yamauchi and K. Shimada, 2-hydrazino-1-methylpyridine: a highly sensitive derivatization reagent for oxosteroids in liquid chromatography-electrospray ionization-mass spectrometry, *J. Chromatogr. B Analyt. Technol. Biomed. Life Sci.*, 2005, **825**, 214–222.

69. T. Higashi, N. Takido and K. Shimada, Detection and characterization of 20-oxosteroids in rat brains using LC-electron capture APCI-MS after derivatization with 2-nitro-4-trifluoromethylphenylhydrazine, *Analyst*, 2003, **128**, 130–133.

70. C. Shackleton, Genetic disorders of steroid metabolism diagnosed by mass spectrometry, In *Laboratory Guide to the Methods in Biochemical Genetics*, 2007.

71. L. Bretillon, A. Siden, L. O. Wahlund, D. Lütjohann, L. Minthon, M. Crisby, J. Hillert, C. G. Groth, U. Diczfalusy and I. Björkhem, Plasma levels of 24S-hydroxycholesterol in patients with neurological diseases, *Neurosci. Lett.*, 2000, **293**, 87–90.

72. D. Lütjohann, A. Papassotiropoulos, I. Björkhem, S. Locatelli, M. Bagli, R. D. Oehring, U. Schlegel, F. Jessen, M. L. Rao, K. von Bergmann and R. Heun, Plasma 24S-hydroxycholesterol (cerebrosterol) is increased in Alzheimer and vascular demented patients, *J. Lipid. Res.*, 2000, **41**, 195–198.

73. P. Schönknecht, D. Lütjohann, J. Pantel, H. Bardenheuer, T. Hartmann, K. von Bergmann, K. Beyreuther and J. Schroder, Cerebrospinal fluid 24S-hydroxycholesterol is increased in patients with Alzheimer's disease compared to healthy controls, *Neurosci. Lett.*, 2002, **324**, 83–85.

74. I. Burkard, K. M. Rentsch and A. von Eckardstein, Determination of 24S- and 27-hydroxycholesterol in plasma by high-performance liquid chromatography-mass spectrometry, *J. Lipid. Res.*, 2004, **45**, 776–781.

75. A. Babiker and U. Diczfalusy, Transport of side-chain oxidized oxysterols in the human circulation, *Biochim. Biophys. Acta*, 1998, **1392**, 333–339.

76. W. J. Griffiths, Y. Wang, G. Alvelius, S. Liu, K. Bodin and J. Sjövall, Analysis of oxysterols by electrospray tandem mass spectrometry, *J. Am. Soc. Mass Spectrom.*, 2006, **17**, 341–362.

77. Y. Wang, M. Hornshaw, G. Alvelius, K. Bodin, S. Liu, J. Sjövall and W. J. Griffiths, Matrix-assisted laser desorption/ionization high-energy collision-induced dissociation of steroids: analysis of oxysterols in rat brain, *Anal. Chem.*, 2006, **78**, 164–173.

78. Y. Wang, M. Karu and W. J. Griffiths, Analysis of neurosterols and neurosteroids by mass spectrometry, *Biochimie.*, 2007, **89**, 182–191.

79. K. Karu, M. Hornshaw, G. Woffendin, K. Bodin, M. Hamberg, G. Alvelius, J. Sjövall, J. Turton, Y. Wang and W. J. Griffiths, Liquid Chromatography Combined with Mass Spectrometry Utilising High-Resolution, Exact Mass, and Multi-Stage Fragmentation for the Identification of Oxysterols in Rat Brain, *J. Lipid Res.*, 2007, **48**, 976–987.

80. K. Wang, E. Bermudez and W. A. Pryor, The ozonation of cholesterol: separation and identification of 2,4-dinitrophenylhydrazine derivatization products of 3beta-hydroxy-5-oxo-5,6-secocholestan-6-al, *Steroids*, 1993, **58**, 225–229.

81. L. L. Smith, Oxygen, oxysterols, ouabain, and ozone: a cautionary tale, *Free Radic. Biol. Med.*, 2004, **37**, 318–324.

82. W. J. Griffiths, S. Liu, G. Alvelius and J. Sjövallm, Derivatisation for the characterisation of neutral oxosteroids by electrospray and matrix-assisted laser desorption/ionisation tandem mass spectrometry: the Girard P derivative, *Rapid Commun. Mass Spectrom.*, 2003, **17**, 924–935.

83. A. K. Dhar, J. I. Teng and L. L. Smith, Biosynthesis of cholest-5-ene-3beta, 24-diol (cerebrosterol) by bovine cerebral cortical microsomes, *J. Neurochem.*, 1993, **21**, 51–60.

84. C. J. Brooks, W. J. Cole, T. D. Lawrie, J. MacLachlan, J. H. Borthwick and G. M. Barrett, Selective reactions in the analytical characterisation of steroids by gas chromatography-mass spectrometry, *J. Steroid. Biochem.*, 1983, **19**, 189–201.

85. J. MacLachlan, A. T. Wotherspoon, R. O. Ansell and C. J. Brooks, Cholesterol oxidase: sources, physical properties and analytical applications, *J. Steroid. Biochem. Mol. Biol.*, 2000, **72**, 169–195.

86. J. I. Teng and L. L. Smith, Sterol peroxidation by Pseudomonas fluorescens cholesterol oxidase, *Steroids*, 1996, **61**, 627–633.

87. J. Dorszewska and Z. Adamczewska-Goncerzewicz, Patterns of free and esterified sterol fractions of the cerebral white matter in severe and moderate experimental hypoxia, *Med. Sci. Monit.*, 2000, **6**, 227–231.

88. M. Wender, Z. Adamczewska-Goncerzewicz and J. Doroszewska, Transformation of sterols pattern in course of late development of rat brain, *Folia Neuropathol.*, 1995, **33**, 31–34.

89. H. Miyajima, J. Adachi, S. Kohno, Y. Takahashi, Y. Ueno and T. Naito, Increased oxysterols associated with iron accumulation in the brains and visceral organs of acaeruloplasminaemia patients, *QJM*, 2001, **94**, 417–422.

90. S. Meaney, *Studies on oxysterols: Origins, properties and roles;* Karolinska Institute: Stockholm, 2003.

91. Y. Y. Lin, M. Welch and S. Lieberman, The detection of 20S-hydroxycholesterol in extracts of rat brains and human placenta by a gas chromatograph/mass spectrometry technique, *J. Steroid. Biochem. Mol. Biol.*, 2003, **85**, 57–61.

92. Z. X. Yao, R. C. Brown, G. Teper, J. Greeson and V. Papadopoulos, 22R-Hydroxycholesterol protects neuronal cells from beta-amyloid-induced cytotoxicity by binding to beta-amyloid peptide, *J. Neurochem.*, 2002, **83**, 1110–1119.

93. J. M. Dietschy and S. D. Turley, Cholesterol metabolism in the central nervous system during early development and in the mature animal, *J. Lipid Res.*, 2004, **45**, 1375–1397.

94. B. Stoffel-Wagner, Neurosteroid metabolism in the human brain, *Eur. J. Endocrinol.*, 2001, **145**, 669–679.

95. M. Hamacher, R. Apweiler, G. Arnold, A. Becker, M. Bluggel, O. Carrette, C. Colvis, M. J. Dunn, T. Frohlich, M. Fountoulakis, A. van Hall, F. Herberg, J. Ji, H. Kretzschmar, P. Lewczuk, G. Lubec, K. Marcus, L. Martens, N. Palacios Bustamante, Y. M. Park, S. R. Pennington, J. Robben, K. Stuhler, K. A. Reidegeld, P. Riederer, J. Rossier, J. C. Sanchez, M. Schrader, C. Stephan, D. Tagle, H. Thiele, J. Wang, J. Wiltfang, J. S. Yoo, C. Zhang, J. Klose and H. E. Meyer, HUPO Brain Proteome Project: Summary of the pilot phase and introduction of a comprehensive data reprocessing strategy, *Proteomics*, 2006, **6**, 5674.

96. B. A. Janowski, P. J. Willy, T. R. Devi, J. R. Falck and D. J. Mangelsdorf, An oxysterol signalling pathway mediated by the nuclear receptor LXR alpha, *Nature*, 1996, **383**, 728–731.

97. M. Teboul, E. Enmark, Q. Li, A. C. Wikstrom, M. Pelto-Huikko and J. A. Gustafsson, OR-1, a member of the nuclear receptor superfamily that interacts with the 9-cis-retinoic acid receptor, *Proc. Natl. Acad. Sci. USA*, 1995, **92**, 2096–2100.

98. K. Abildayeva, P. J. Jansen, V. Hirsch-Reinshagen, V. W. Bloks, A. H. Bakker, F. C. Ramaekers, J. de Vente, A. K. Groen, C. L. Wellington, F. Kuipers and M. Mulder, 24(S)-hydroxycholesterol participates in a liver X receptor-controlled pathway in astrocytes that regulates apolipoprotein E-mediated cholesterol efflux, *J. Biol. Chem.*, 2006, **281**, 12799–12808.

99. J. J. Repa, G. Liang, J. Ou, Y. Bashmakov, J. M. Lobaccaro, I. Shimomura, B. Shan, M. S. Brown, J. L. Goldstein and D. J. Mangelsdorf, Regulation of mouse sterol regulatory element-binding protein-1c gene (SREBP-1c) by oxysterol receptors, LXRalpha and LXRbeta, *Genes Dev.*, 2000, **14**, 2819–2830.

100. C. M. Adams, J. Reitz, J. K. De Brabander, J. D. Feramisco, L. Li, M. S. Brown and J. L. Goldstein, Cholesterol and 25-hydroxycholesterol inhibit activation of SREBPs by different mechanisms, both involving SCAP and Insigs, *J. Biol. Chem.*, 2004, **279**, 52772–52780.

CHAPTER 4
Phospholipid Profiling

ANTHONY D. POSTLE

Division of Infection, Inflammation and Repair, School of Medicine, Mailpoint 803, Room LF66, Level F, South Block, Southampton General Hospital, Tremona Road, Southampton SO16 6YD, UK

4.1 General Overview of Lipidomic Characterisation of Membrane Function

The lipid components of cell membranes are critically important for a wide range of cell functions. These functions range from maintaining the structural integrities of the cell and intracellular organelles, through providing an optimal physicochemical environment for the actions of a host of integral and peripheral membrane proteins, to being intimately involved in many of the critical signalling processes that regulate the cell's response to its environment. It is important to recognise that, with a few notable exceptions, it is the pattern of membrane lipid composition that determines these functions rather than any individual component. Moreover, acquiring a comprehensive understanding of how cells regulate their membrane compositions and functions requires detailed analysis in terms of intact individual molecular species, the biologically relevant molecules, rather than simple analysis of fatty acid compositions. The advent of modern mass spectrometry applications has revolutionised our ability to characterise and quantify such individual lipid species and it is now possible to provide comprehensive phospholipid profiles of very small numbers of cells.

Phospholipids are the major class of cell membrane lipid, together with sterols and glycolipids. They are amphipathic molecules with hydrophobic and hydrophilic regions and it is this amphipathic nature, which enables phospholipid molecules to assemble into bilayer and hexagonal membrane structures, which are critically important for the functional viability of all eukaryote cells. Cellular membranes separate the intracellular milieu from the extracellular environment and facilitate the formation of specialised intracellular organelles.

For many years, phospholipids were considered to be important but relatively inert structural components of the cell. More recently, the central role of membrane phospholipid composition and turnover in the regulation of a wide range of cellular functions has become widely recognised. For instance, all membrane receptor events take place within a phospholipid-rich environment, and it is therefore not surprising that cells have adopted hydrolysis of membrane phospholipids as a major signalling mechanism.

4.2 Phospholipid Structures

There are two major classes of phospholipid, depending on whether they contain a glycerol or sphingosyl backbone. Glycerophospholipids are based on phosphatidic acid (3-*sn*-phosphatidic acid), with the identity of the esterified group X defining the class of phospholipid (Figure 4.1). The most common headgroups are the nitrogenous bases choline and ethanolamine, the amino acid serine and the polyalcohols *myo*-inositol and glycerol. Within each of these phospholipid classes there is a distribution of individual molecular species, defined by the combination of esterified fatty acids attached to the glycerol. Generally, membrane glycerophospholipids tend to have palmitoyl (16:0) or stearoyl (18:0) at their *sn*-1 position and unsaturated fatty acids esterified at *sn*-2. This relatively restricted distribution reduces the overall complexity of phospholipid composition to some extent but, given that there are more than 10 different phospholipid classes each of which can contain combinations of 20 or more fatty acids, there are still potentially up to 1000 different phospholipids that can be identified in any given cell type. Sphingophospholipids contain sphingosine *(trans*-D-erythro-1,3-dihyroxy-2-amino-4-octadecene). Sphingomyelin is the most abundant sphingophospholipid class, and is the phosphorylcholine ester of *N*-acylsphingosine, otherwise called ceramide. Sphingophospholipids are important components of all cell membranes and are structurally and metabolically closely related to glycosphingolipids such glycosylceramides, gangliosides and cerebrosides. They contain principally saturated and monounsaturated fatty acids, with a simpler overall composition than glycerophospholipids.

Phospholipid molecular species compositions are maintained within relatively narrow ranges *in vivo* and are cell type specific. For instance, membrane lipid compositions of hepatocytes, lung epithelial cell and neurons are distinct and different from each other and cells expend a considerable amount of energy preserving their characteristic membrane compositions. While alterations to membrane lipid compositions have been linked to a wide range of diseases, including atherosclerosis, the metabolic syndrome, allergy, respiratory failure, neurological conditions and various cancers, the extent of such changes *in vivo* in response to diet or disease or during development are usually relatively subtle.

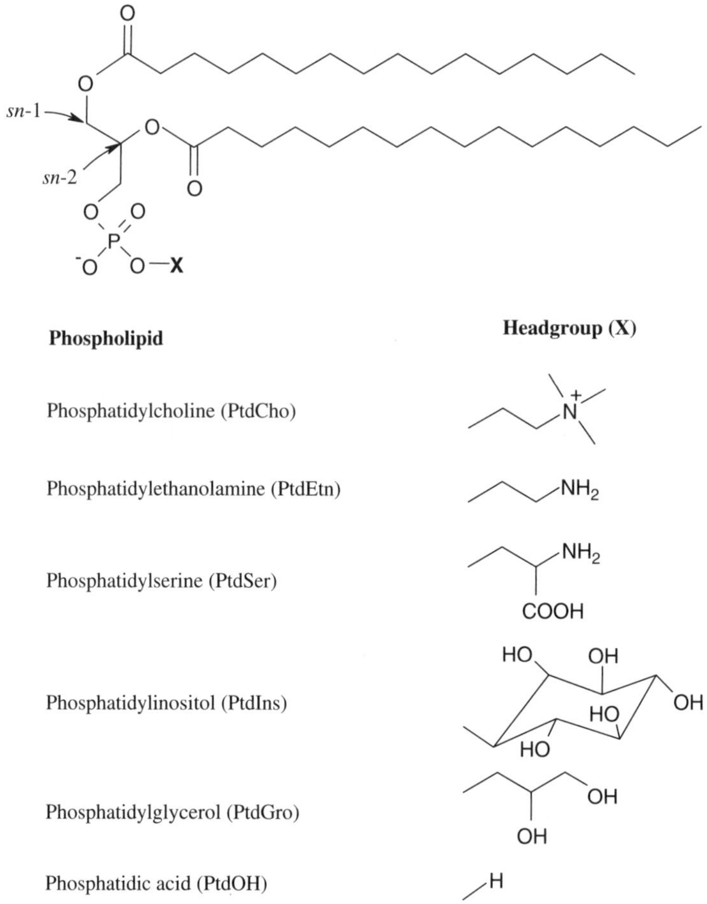

Figure 4.1 The class of phospholipid is defined by the nature of the nitrogenous base or polyol esterified to the phosphate group (X). The species distribution within any phospholipid class is determined by the fatty acyl substitutes at the *sn*-2 positions of the glycerol backbone. The dipalmitoyl shown here would be designated PC16:0/16:0 if X was choline. If arachidonic acid was esterified at *sn*-2, the molecule would be designated PC16:0/20:4. In the diacyl species shown above, fatty acids are attached by ester linkages. For *sn*-1-alkyl-2-acyl species, the *sn*-1 fatty acid is attached by an ether bond. For *sn*-1-alkenyl-2-acyl species, the *sn*-1 fatty acid is attached by a vinyl ether linkage.

4.3 Electrospray Ionisation Mass Spectrometry of Phospholipids

Electrospray ionisation mass spectrometry (ESI-MS) has over recent years become established as the method for determining patterns of individual molecular species of cell membrane lipids.[1,2] Compared with alternative techniques such as gas chromatography–mass spectrometry (GC-MS) and high-performance liquid chromatography (HPLC), ESI-MS offers an unrivalled combination of sensitivity, resolution and sample throughput. In addition to quantifying membrane lipid compositions, ESI-MS in combination with incorporations of lipid precursors labelled with stable isotopes permit detailed analysis of membrane lipid synthesis and dynamics both in cultured cells[3,4] and *in vivo*.[5] Finally, it is the most appropriate experimental approach to determine endogenous intracellular signalling lipids such as polyphosphoinositides, diacylglycerols, lysophospholipids, ceramides, sphingosine-1-phosphate and eicosanoids.

In principle, ESI-MS just adds or subtracts a proton from intact molecules and then resolves the resulting ions on the basis of their mass:charge (m/z) ratio. The majority of lipids are singly charged, so m/z is generally equivalent to molecular mass. Single quadrupole analysis, which just scans for m/z, will detect some lipids that prefer to be positively charged such as PtdCho, sphingomyelin, triacylglycerides (TAG) and diacylglycerides (DAG) and others that are preferentially negatively ionised such as PtdIns, PtdSer and other acidic phospholipids. Greater structural information can be provided by tandem mass spectrometry (MS/MS) using a triple quadrupole mass spectrometer. This instrument design has two identical mass analysers (MS1 and MS2) separated within the instrument by a collision cell. Ions selected at MS1 are partially broken down in the collision cell using argon gas and then the generated fragments resolved by MS2. A PtdCho molecule with mass 786.7 could be either stearoyl-linoleoyl (18:0/18:2) or dioleoyl (18:1/18:1), which have the same numbers of carbon atoms and double bonds. MS/MS fragmentation of this single ion under appropriate negative ionisation conditions will generate fatty acid fragment ions, which will give an unambiguous structural assignment (Figure 4.2).

MS/MS can also provide selective scans for whole classes of lipid. For instance, all PtdCho molecules in positive ionisation fragment to give a phosphorylcholine product ion of m/z 184. A precursor ion scan of m/z 184 which looks for ions at MS1 that fragment to give ions with m/z 184 at MS2 will generate a spectrum of PtdCho, irrespective of any other components in the sample. Similar diagnostic scans are established for all the other major classes of phospholipid and neutral lipid. Finally, these diagnostic scans enable rates of synthesis of individual molecular species of different lipids to be quantified using precursors labelled with stable isotopes. For instance, when cells are incubated with choline containing 9 deuterium atoms (*methyl*D$_9$-choline) a proportion of PtdCho will by synthesised 9 mass units higher than the endogenous material. Consequently, just as a precursor scan of m/z 184 generates a spectrum of endogenous PtdCho, newly synthesised PtdCho will be detected by a precursor scan of m/z 193 (Figure 4.3). Analogous methods are established for quantifying synthetic rates of all the major phospholipid classes. Apart from their sensitivity and detail of information provided, one major advantage of these methods is they can be readily used *in vivo* as no radioactivity is involved.

One major drawback of this experimental approach for MS analysis of lipid composition and synthesis is that precursor or neutral loss scans can only

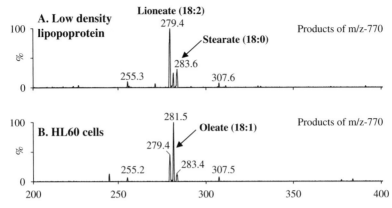

Figure 4.2 MS/MS fragmentation of PtdCho purified by solid-phase extraction from chloroform:methanol extracts of (a) human low density lipoprotein (LDL) or (b) HL60 promyelocytic cancer cells. The PtdCho species with m/z +786.7 ionised under negative ionisation by loss of CH_3 to give demethylated precursor ion of m/z –770.7. This ion from samples contained a mixture of molecular species which for LDL was predominately PC18:0/18:2 and for HL60 cells was PC18:1/18:1.

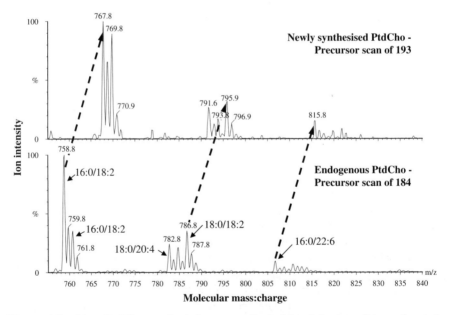

Figure 4.3 Liver PtdCho synthesis by mouse liver. After injection of 1 mg (*methyl*-D9)-choline Cl, total lipids were extracted using chloroform/methanol. Analysis by direct injection ESI-MS/MS selectively detected endogenous PtdCho molecular species by precursor scan of m/z +184, while the pattern of newly synthesised species was shown by precursor scan of m/z +193.

provide information in terms of total numbers of fatty acyl carbons and unsaturated double bonds in a given phospholipid molecule. Confirmation of molecular species identities by product ion scanning using a triple quadrupole instrument then requires separate scans to characterise each parent ion. However, the use of hybrid quadrupole–time of flight (QToF) mass spectrometers enables this limitation to be circumvented by multiple precursor scanning.[6] In this mode, individual masses are selected sequentially in the quadrupole, and fragments generated in the collision cell are then resolved by ToF analysis. This analytical approach in effect generates a two-dimensional map, providing information about the composition and positional specificity of all the fatty acyl moieties of all the phospholipids in a lipid extract of cells. Limitations of the approach are the long scan times required, from 20 to 50 minutes depending on the mass ranges selected, and the essentially qualitative nature of the compositional data generated. Nevertheless, a combination of triple quadrupole and QToF MS/MS analyses can provide precise detail of phospholipid composition and concentration.

4.4 Phospholipid Metabolism

Membrane homeostasis acts to maintain the physicochemical properties of the membrane within the range compatible with cell viability. Providing membrane fluidising components, such as polyunsaturated fatty acids, to a cell in culture is always accompanied by compensating changes to concentrations of more rigid components such as disaturated phospholipid and cholesterol. There is some evidence for such mechanisms *in vivo*, but the potential for more extensive alterations are limited due to the regulation of cellular lipid nutrition by hepatic metabolism. Hydrophobic lipid nutrients are supplied to extra-hepatic tissues as components of lipoproteins or bound to protein carriers, principally albumin, and the specificity of this process is determined by the properties of lipid synthesis and section by the hepatocyte. Consequently, for any individual cell type, membrane lipid compositions will be the result of interactions between cellular nutrition, cell architecture and gene expression, and conversely membrane lipid composition is a potential major modifier of gene expression. Within any body compartment with a common nutritional supply, such as the circulation, phenotypic differences determine, for instance, that endothelial cells, lymphocytes, neutrophils and erythrocytes maintain membrane lipid compositions that are unique to that cell type. Additionally, lipid molecular compositions of different phospholipid classes and different sub-cellular membranes are characteristic of, and tightly regulated by, each individual cell type. For instance, PtdCho and PtdIns molecular species compositions are always very different; similarly, lipid compositions of the plasma membrane, endoplasmic reticulum and nuclear envelope are very distinct. While most of the enzymes and transport proteins of cellular lipid metabolism have been identified, purified and cloned, virtually nothing is understood about how their activities interact to coordinate the regulation of these cell-specific membrane lipid compositions. With a few notable exceptions, the molecular bases whereby cells sense their membrane compositions are not clear, but are critically important to ensure that precisely the correct amounts of each individual lipid are synthesised to form new membrane when a cell divides and to maintain the appropriate balance of different lipid contents within the membrane.

All these factors impact on the experimental design of phospholipid profile studies of the roles of membrane lipids in health and disease. A number of factors must be considered.

- Restricted analysis of one individual membrane lipid species in isolation, however theoretically important, is not only inappropriate but is also statistically invalid. As contributions of all individual components to overall membrane structure and function are inter-dependent, comprehensive, integrated data-driven strategies are the only practical approaches in this field of research.
- The extent of disease or diet-related alterations to membrane lipid compositions and function will be relatively modest *in vivo*, with the clear implication that information must be provided in terms of pattern recognition and network analysis.
- For any model cell culture system to be valid, both basal membrane lipid composition and the extent of any altered composition *in vitro* must be maintained within physiologically relevant boundaries. For instance, mammalian cells *in vivo* maintain PtdIns compositions that are uniquely enriched in the single stearoyl arachidonoyl (PI18:0/20:4) molecular species. PI18:0/20:4 is in turn the precursor of PI-4,5,-bisphosphate ($PIP_2$18:0/20:4), the substrate for the phospholipase C (PLC)-mediated generation of the specific DAG (DAG18:0/20:4) pool responsible for activation of protein kinase C (PKC) isoforms. Cells lose PtdIns and PIP_2 molecular specificities in culture; the composition of the DAG signalling pool can then be readily modified by for example providing exogenous docosahexaenoate (DHA), but such changes will not necessarily be physiologically or pathologically relevant.
- Detailed sub-cellular phospholipid analysis and the use of specific pathway inhibitors will be essential to distinguish, for instance, signalling DAG species from the bulk of cell DAG associated with phospholipid synthesis.
- Modulation of individual cell membrane lipids must be considered in the context of whole body lipid metabolism. This must include analyses of composition, synthesis and turnover of the lipid components of the different families of plasma lipoproteins.

4.5   Phospholipids of Lung Surfactant

ESI-MS phospholipid profiling has been applied to a wide range of cell types, experimental conditions and different organisms. This chapter will concentrate instead on the application of ESI-MS methodologies to the characterisation of lung surfactant phospholipids in health, disease and development, across animal species, partly because this is the best described model and also because ESI-MS has fundamentally altered concepts of the mechanisms underlying surfactant function.

4.5.1 Lung Structure and Surfactant Function

The mammalian lung is constructed of a series of bifurcating airways starting with the trachea (windpipe) and ending some 16 divisions later in the terminal airways. These terminal airways then lead into the gas exchange sacs (alveoli) which consist of a thin epithelial layer separating the air spaces from the blood capillary bed. There are some 4.8×10^8 alveoli in the adult human lung with a typical diameter on 200 µm (Figure 4.4).[7] Their consequent large surface area:volume ratio facilitates the high demand for oxygen uptake by the lungs, but at a cost of dealing with surface tension forces. The Laplace equation ($P =$ surface tension/radius) shows that collapsing pressure is inversely proportional to radius at constant surface tension. If the lungs were lined simply with saline at a surface tension of 72 mN/m, it would be impossible to inflate the alveoli and they would all collapse into larger vacuole structures. This is precisely what happens when very preterm infants (<1500 g birth weight, born ≥ 10 weeks premature) are born with immature lungs and surfactant

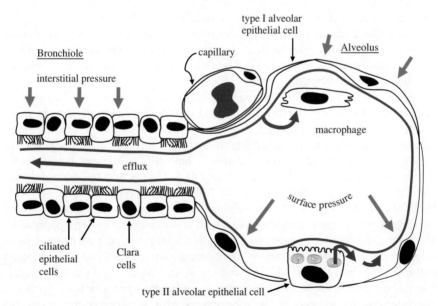

Figure 4.4 Schematic representation of the architecture of the peripheral lung. Pulmonary surfactant is synthesised and secreted by the type II cell of the alveolus, the gas exchange region of the lungs. The high surface pressure generated by surfactant at the air:liquid interface prevents alveolar collapse and infiltration of fluid into the air spaces.

deficiency, leading to respiratory failure requiring mechanical ventilation at high inspired oxygen concentrations and often associated with severe morbidity and mortality.

Lung surfactant is the phospholipid-rich material secreted by the type II epithelial cell of the lung alveolus that opposes surface tension forces within the lungs, prevents alveolar and terminal airway collapse when breathing out and reduces the work of breathing. PtdCho is the major component of lung surfactant, followed by phosphatidylglycerol (PtdGro), PtdIns, cholesterol and a number of surfactant-specific proteins (surfactant proteins A, B, C and D). The major PtdCho component of human lung surfactant is dipalmitoyl PtdCho (PC16:0/16:0), comprising between 45 and 55% to total PtdCho.[8,9] The importance of PC16:0/16:0 for function of lung surfactant was recognised almost 60 years ago[10] and has become arguably the best defined physiological role for any individual molecular species of phospholipid.

As a consequence of this early recognition of the important functional role of PC16:0/16:0, there has been relatively little research of interest in the other individual phospholipid molecular species of surfactant. When PC16:0/16:0 was spread on a Langmuir trough, it was capable of generating very high surface pressures on area compression and mimicked many of the surface properties of surfactant. The development of simple robust techniques to quantify disaturated PtdCho (DSPC) as the residue after the oxidative destruction of unsaturated species[11] led to the general identification of DSPC as PC16:0/16:0 and consequently little attention was paid in clinical studies to the precise identities of the other phospholipid species. This concentration on the central importance of PC16:0/16:0 in surfactant function has been reinforced by more recent observations that an effective exogenous surfactant can be constructed from this single PtdCho species together with acidic phospholipids and hydrophilic surfactant peptides.[12]

The advent of ESI-MS lipidomic methodologies in recent years has enabled surfactant phospholipid compositions and metabolism to be probed in much greater detail, and has completely changed our understanding of how phospholipid-based surface films withstand the high surface pressures generated within the lung alveoli. Traditional theory held that, as the gel:liquid crystal transition temperature of PC16:0/16:0 was 41.5 °C, it would be in a rigid, solid state at body temperature (37.5 °C).[13,14] This rigid molecular layer was then thought to withstand lateral compression and in effect to splint open the alveoli on expiration. A number of anomalies with this theory have been apparent for some time. In many circumstances, alveolar collapse is part of normal physiology, perhaps most apparent in diving mammals which have to completely collapse their entire lungs when swimming at any depth. Additionally, surface films of monounsaturated PtdCho species can under appropriate compression conditions generate meta-stable monolayers with surface tension properties comparable to native surfactant.[15] Finally, further enrichment of commercially available artificial surfactant preparations with more PC16:0/16:0 resulted in inferior surface tension lowering properties at dynamic compression rates comparable to breathing rates in people.[16]

4.5.2 ESI/MS Analysis of Lung Surfactant Phospholipid

Typical ESI scans of a total lipid extract of surfactant purified from human amniotic fluid are shown in Figure 4.5, illustrating the preference for detecting PtdCho under positive and PtdGro and PtdIns under negative ionisation. The

A. Positive ionisation

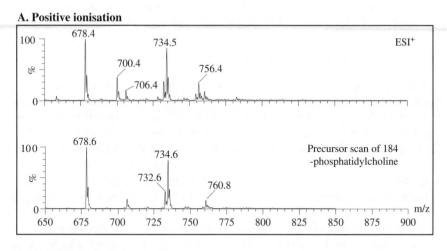

B. Negative ionisation

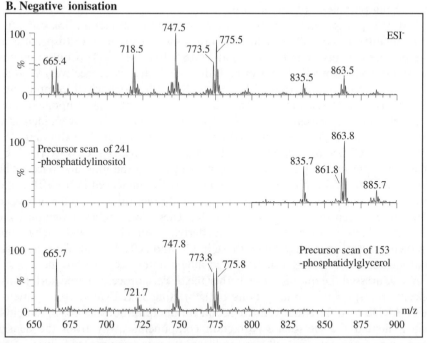

Figure 4.5 ESI-MS of human lung surfactant phospholipids. (a) Positive ionisation preferentially detects PtdCho species, while (b) the acidic phospholipids PtdIns and PtdGro are readily quantified under negative ionisation conditions.

ESI$^+$ spectrum contains not only the molecular ion peak (*e.g.* PC16:0/16:0 at m/z 734) but also a variable formation of sodium adducts at $[M+22]^+$ (*e.g.* m/z 756). The diagnostic precursor scan of m/z 184 only detects the molecular ion and is more straightforward to interpret. There is little comparable adduct formation of the acidic phospholipids in the ESI$^-$ spectrum and this can be used for quantification, with confirmation of identities by precursor scans of m/z 153 (glycerophosphate) and m/z 241 (inositol phosphate – H_2O).

4.5.2.1 *Surfactant Phospholipid and Lung Disease*

ESI-MS analysis of phospholipids has been applied to a variety of human lung diseases, including asthma, cystic fibrosis (CF), viral infections and acute respiratory distress syndrome (ARDS). As with many biological systems, lung surfactant is not a homogenous substance. Previous analyses of phospholipid fatty acid compositions of bronchoalveolar lavage fluid (BALF) obtained from cystic fibrosis patients by bronchoscopy for instance have shown decreased palmitate and increased arachidonate, but the relative contributions of surfactant and other sources of phospholipid were not determined. The additional sensitivity provided by ESI-MS has allowed these processes to be examined in more detail and one common theme to emerge from these studies is that the altered patterns of the different phospholipid classes can be related to the underlying pathology of the disease.

ESI/MS phospholipid analyses of human lung surfactant, plasma and inflammatory leucocytes (neutrophils) are shown in Figure 4.6. Inspection of the precursor scans of 184 shows that PC16:0/16:0 (m/z 734) is only a minor component of plasma and neutrophil PtdCho, which have respectively PC16:0/18:2 (m/z 758) and PC16:0/18:1 (m/z 760) as their principal components. Asthma is characterised by a substantial exudation of plasma components into the airways and it is clear from Figure 4.7a that this is reflected by an increased component of plasma phospholipid (PC16:0/18:2) recovered in the lungs.[8,17] CF[18] and ARDS by contrast are inflammatory diseases characterised by massive recruitment and activation of neutrophils in the lung air spaces. The corresponding BALF phospholipid changes involve increased PC16:0/18:1 and other PtdCho species characteristic of cell membranes rather than the increased PC16:0/18:2 seen in asthma (Figure 4.7b). These two examples demonstrate that the changes observed in asthma derive from altered endothelial and epithelial cell permeability, while those in CF and ARDS result from necrosis and membrane shedding from activated inflammatory cells within the lungs.

We addressed the question of whether the disease process altered surfactant-specific phospholipid compositions by subjecting BALF fluid from patients with ARDS to differential centrifugation on discontinuous sucrose gradients, which floated the less dense surfactant components away from the denser membrane fragments. ARDS is a condition of severe respiratory failure, often due to endotoxic shock, pneumonia or near drowning, and the phospholipid profile of BALF from such patients was dramatically altered compared with healthy controls (Figure 4.7b). The PtdCho composition of membrane

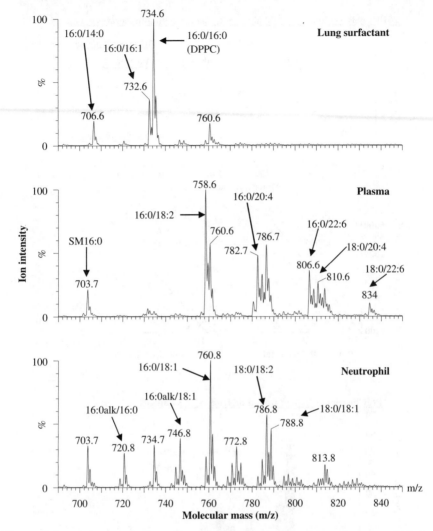

Figure 4.6 Differences in ESI-MS spectra of PtdCho demonstrate how accumulation of blood plasma or membranes derived from neutrophil white blood cells can be distinguished from lung surfactant.

fragments recovered from the bottom fraction of the sucrose gradient resembled that of total BALF, while the surfactant fraction was substantially enriched in PC16:0/16:0 and other species characteristic of surfactant. This result showed clearly both that there was negligible PtdCho exchange between surfactant and membrane material and that most probably even this severe respiratory disease did not substantially alter the molecular specificity of surfactant PtdCho synthesis. Instead, the almost total absence of PtdGro and PtdIns in both BALF and purified surfactant strongly suggests enhanced degradation of surfactant phospholipid in ARDS, probably by the increased

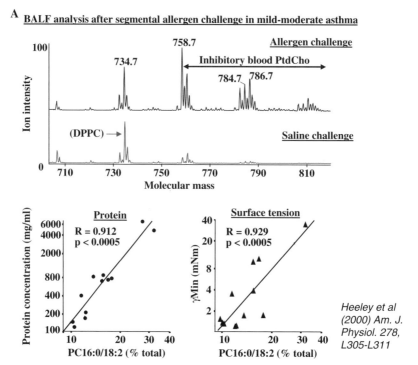

A BALF analysis after segmental allergen challenge in mild-moderate asthma

Heeley et al (2000) Am. J. Physiol. 278, L305-L311

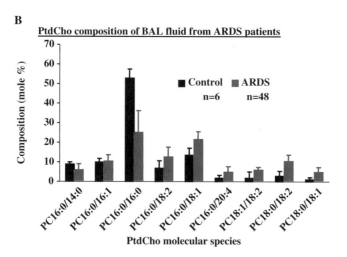

B

Figure 4.7 Alterations of lung lavage phospholipid compositions in lung disease. (a) Accumulation of plasma lipoprotein PtdCho, characterised by a major component at *m/z* 758.7 (PC16:0/18:2) correlates with protein infiltration into the airways and impaired surfactant function.[17] (b) A very different pattern of lung lavage PtdCho in patients with respiratory failure (acute respiratory distress syndrome – ARDS) that appears to be due to accumulation of membrane fragments derived from inflammatory cells.

activity of phospholipase A_2 reported in BALF from patients with this condition.

4.5.2.2 Comparative Physiology and Developmental Aspects of Lung Surfactant Phospholipid

Despite the central role of PC16:0/16:0 in human lung surfactant, ESI-MS analysis of a large range of different animal species, with vastly differing body weights, respiratory rates and metabolic rates, has clearly shown that such a high content of PC16:0/16:0 is not essential to generate a functional surfactant.[19] Indeed, for the marsupial Tasmanian devil, PC16:0/16:0 comprises less than 2% of total surfactant PtdCho which instead is enriched in monounsaturated alkyl acyl plasmanyl species (Figure 4.8a). Comparable, if somewhat less extreme, compositional divergence is shown for surfactant from other animals (Figure 4.8b) but, even for animals with low lung surfactant PC16:0/16:0, the composition of the remaining PtdCho was far from random. Instead, such surfactants contain high concentrations of short-chain disaturated and monounsaturated PtdCho species, principally palmitoyl myristoyl PtdCho (PC16:0/14:0) and palmitoyl palmitoleoyl PtdCho (PC16:0/16:1). Indeed, for a wide range of animal lung surfactants the concentration of PC16:0/16:0 correlated inversely with the sum of PC16:0/14:0 and PC16:0/16:1 and not at all with the sum of longer chain species such as PC16:0/18:1 or PC18:0/18:1.

Many animals undergo periods of hibernation or torpor, for periods lasting from hours to many months, which are generally characterised by decreased body temperature and respiratory rate. Given the generally accepted link between phospholipid transition temperature and body temperature as a central contributor to optimal surfactant function, we were interested to see how surfactant composition, structure and function adapted to torpor. We addressed the intriguing possibility that the molecular species composition of surfactant phospholipid might vary with body temperature to maintain a constant phase structure. However, studies with the fat-tailed dunnart, a small marsupial from Western Australia, clearly showed that this was not the case. With environmental temperature fluctuating over a 50 °C range from day to night, the dunnart responds to this stress by entering torpor for short periods and lowers its body temperature from 37.5 °C to less than 20 °C. Decreased body temperature is accompanied by increases to both lung surfactant concentration and cholesterol content[20,21] but not by any alteration to the molecular species compositions of PtdCho, PtdGro or PtdIns.[19] Instead, the dunnart maintains a lung surfactant based on PC16:0/16:1 that exhibits superior surface tension lowering properties at 23 °C but still works adequately at 37.5 °C.[20] This adaptation makes sense energetically, with modulation of the cholesterol content removing the need for extensive remodelling of surfactant phospholipids during torpor.

The progressive increase to lung surfactant concentration and phospholipid saturation during fetal and postnatal human development is very well documented and has been widely employed as a diagnostic test for fetal lung maturity.[22] However, the largest fractional increment to disaturated PtdCho

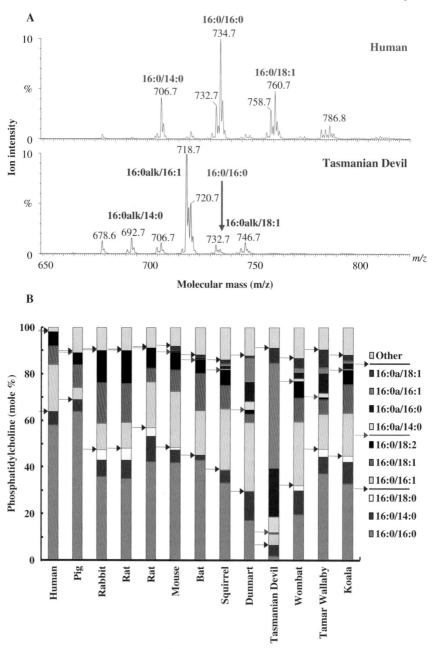

Figure 4.8   Dipalmitoyl PtdCho is not the major component of lung surfactant for all animal species. (a) Comparison of human lung surfactant with that of the marsupial Tasmanian Devil, which is comprised predominantly of monounsaturated plasmanyl PtdCho species. (b) The high variability of lung surfactant PtdCho composition across a wide range of animals, highlighting that composition must be closely tailored for specific lung mechanics for each animal species.

between 15 weeks gestation and birth at 40 weeks, measured both in tissue[23] and lung fluid,[24] was to PC16:0/14:0 rather than to PC16:0/16:0. Subsequent analysis by ESI-MS showed a dramatic modulation to rat surfactant PtdCho composition between 7 and 14 days after birth, with decreased PC16:0/16:0 and corresponding increased PC16:0/14:0 concentration.[25] These adaptations were related to alveolar formation, which is predominately postnatal in the rat, with the suggestion that the phospholipid with shorter acyl groups would facilitate formation of membrane structures with tighter radii of curvature. An alternative explanation comes from the strong inverse correlation between PC16:0/16:0 and respiratory rate over a wide range of animal species, from people (100 breaths/min) to the pygmy shrew (900 breaths/min).[24] Very rapid breathing with proportionally low tidal volumes would require a surfactant that can spread and adsorb very rapidly to the air:liquid interface, and indeed reconstituted surfactant based on a combination of PC16:0/14:0 and PC16:0/16:1 achieved low surface tension values in fewer compression cycles than one based predominantly on PC16:0/16:0.[26]

Finally, exactly the same short-chain PtdCho species are packaged and secreted by fetal human lung epithelial cells that have been hormonally differentiated in culture to the type II cell phenotype.[26] These cells undergo in a few days the normal maturation process that takes many months *in utero* and express many of the proteins characteristic of mature differentiated type II alveolar epithelial cells.[27] The PtdCho composition in these cells, but not that of cultured fibroblasts from the same lungs, is enriched with the same distribution of short-chain molecular species also observed in the developing rat in fast-breathing animals, namely PC16:0/14:0 and PC16:0/16:1 in addition to PC16:0/16:0. Moreover, the composition of PtdCho secreted from these differentiated epithelial cells was further enriched in PC16:0/14:0 and PC16:0/16:1 but not in PC16:0/16:0.[26]

4.6 Conclusions

The analysis of surfactant phospholipid summarised above is just one example of how the increased lipid compositional detail provided by ESI-MS has significantly altered concepts of basic science that have considerable clinical implications. Comparable studies have shown pathology-related changes to phospholipid molecular species compositions in, for example, Alzheimer's disease, diabetes and a range of cancers. There is now considerable interest in the potential for ESI-MS of lipids to provide novel biomarkers of disease. A major challenge for the future application of ESI-MS to phospholipid profiling will be to extend the experimental approach to be able to relate such compositional changes to mechanisms of disease.

References

1. X. L. Han and R. W. Gross, *J. Am. Soc. Mass Spectrom.*, 1995, **6**, 1202–1210.
2. B. Brugger, G. Erben, R. Sandhoff, F. T. Wieland and W. D. Lehmann, *Proc. Natl. Acad. Sci.*, 1997, **94**, 2339–2344.

3. C. J. DeLong, Y. J. Shen, M. J. Thomas and Z. Cui, *J. Biol. Chem.*, 1999, **274**, 29683–29688.

4. A. N. Hunt, G. T. Clark, G. S. Attard and A. D. Postle, *J. Biol. Chem.*, 2001, **276**, 8492–8499.

5. W. Bernhard, C. J. Pynn, A. Jaworski, G. A. Rau, J. M. Hohlfeld, J. Freihorst, C. F. Poets, D. Stoll and A. D. Postle, *Am. J. Respir. Crit. Care Med.*, 2004, **170**, 54–58.

6. K. Ekroos, I. V. Chernushevich, K. Simons and A. Shevchenko, *Anal. Chem.*, 2002, **74**, 941–949.

7. M. Ochs, J. R. Nyengaard, A. Jung, L. Knudsen, M. Voigt, T. Wahlers, J. Richter and H. J. Gundersen, *Am. J. Respir. Crit. Care Med.*, 2004, **169**, 120–124.

8. S. M. Wright, P. M. Hockey, G. Enhorning, P. Strong, K. B. Reid, S. T. Holgate, R. Djukanovic and A. D. Postle, *J. Appl. Physiol.*, 2000, **89**, 1283–1292.

9. A. Mander, S. Langton-Hewer, W. Bernhard, J. O. Warner and A. D. Postle, *Am. J. Respir. Cell Mol. Biol.*, 2002, **27**, 714–721.

10. M. E. Avery, *ACP-Applied Cardiopulmonary Pathophysiology*, 1995, **5**(Suppl. 1), 19–22.

11. R. J. Mason, J. Nellenbogen and J. A. Clements, *J. Lipid Res.*, 1976, **17**, 281–284.

12. S. B. Hall, A. R. Venkitaraman, J. A. Whitsett, B. A. Holm and R. H. Notter, *Am. Rev. Respir. Dis.*, 1992, **145**, 24–30.

13. A. D. Bangham, *Lung*, 1987, **165**, 17–25.

14. A. Gulik, P. Tchoreloff and J. Proust, *Biophys. J.*, 1994, **67**, 1107–1112.

15. Z. Wang, S. B. Hall and R. H. Notter, *J. Lipid Res.*, 1995, **36**, 1283–1293.

16. W. Bernhard, J. Mottaghian, A. Gebert, G. A. Rau, H. A. R. D. von Der and C. F. Poets, *Am. J. Respir. Crit. Care Med.*, 2000, **162**, 1524–1533.

17. E. L. Heeley, J. M. Hohlfeld, N. Krug and A. D. Postle, *Am. J. Physiol. Lung Cell. Mol. Physiol.*, 2000, **278**, L305–L311.

18. A. Mander, S. Langton-Hewer, W. Bernhard, J. O. Warner and A. D. Postle, *Am. J. Respir. Cell Mol. Biol.*, 2002, **27**, 714–721.

19. C. J. Lang, A. D. Postle, S. Orgeig, F. Possmayer, W. Bernhard, A. K. Panda, K. D. Jurgens, W. K. Milsom, K. Nag and C. B. Daniels, *Am. J. Physiol. Regul. Integr. Comp. Physiol.*, 2005, **289**, R1426–R1439.

20. O. V. Lopatko, S. Orgeig, C. B. Daniels and D. Palmer, *J. Appl. Physiol.*, 1998, **84**, 146–156.

21. O. V. Lopatko, S. Orgeig, D. Palmer, S. Schurch and C. B. Daniels, *J. Appl. Physiol.*, 1999, **86**, 1959–1970.

22. L. Gluck, *Prog. Clin. Biol. Res.*, 1980, **44**, 189–201.

23. A. N. Hunt, F. J. Kelly and A. D. Postle, *Early Human Dev.*, 1991, **25**, 157–171.

24. W. Bernhard, S. Hoffmann, H. Dombrowsky, G. A. Rau, A. Kamlage, M. Kappler, J. J. Haitsma, J. Freihorst, H. Von der Hardt and C. F. Poets, *Am. J. Respir. Cell Mol. Biol.*, 2001, **25**, 725–731.

25. R. Ridsdale, M. Roth-Kleiner, F. D'Ovidio, S. Unger, M. Yi, S. Keshavjee, A. K. Tanswell and M. Post, *Am. J. Respir. Crit. Care Med.*, 2005, **172**, 225–232.

26. A. D. Postle, L. W. Gonzales, W. Bernhard, G. T. Clark, M. H. Godinez, R. I. Godinez and P. L. Ballard, *J. Lipid Res.*, 2006, **47**, 1322–1331.

27. L. W. Gonzales, S. H. Guttentag, K. C. Wade, A. D. Postle and P. L. Ballard, *Am. J. Physiol. Lung Cell Mol. Physiol.*, 2002, **283**, L940–L951.

CHAPTER 5

New Developments in Multi-dimensional Mass Spectrometry Based Shotgun Lipidomics

XIANLIN HAN[1,2] AND RICHARD W. GROSS[1-4]

[1] Division of Bioorganic Chemistry and Molecular Pharmacology, Washington University School of Medicine, St. Louis, Missouri 63110, USA

[2] Departments of Medicine, and Molecular Biology & Pharmacology, Washington University School of Medicine, St. Louis, Missouri 63110, USA

[3] Department of Molecular Biology & Pharmacology, Washington University School of Medicine, St. Louis, Missouri 63110, USA

[4] Department of Chemistry, Washington University, St. Louis, MO 63130, USA

5.1 Introduction

Metabolomics is a newly emerging research field that is rapidly growing, following the tremendous progress and similar growth profiles of genomics and proteomics. Since lipids are a major part of the products of cellular metabolism in many cell types, then it follows that lipidomics (defined as the large-scale study of lipids) is a central part of the metabolomics profiles of biological cells and organisms. Accordingly, lipidomics plays a key role in metabolomics research.[1,2] Through the complexity and the unique metabolic signature of the lipidome in metabolomics, many modern technologies (including mass spectrometry, NMR, fluorescence spectroscopy, and microfluidic devices) have been developed to identify, quantify, and understand the structure and function of key metabolic nodes in the lipidome.[3,4]

It is commonly implied that the lipidome can be distinguished on one level through the fact that the large majority of its components are extractable with organic solvents. Thus, sample preparation can be simplified by taking advantage of certain physical properties (*e.g.*, volatility, phase separation) of suitable and selected organic solvents sometimes used in tandem or in sequence. Although this simplifies isolation, it is well known that lipids form aggregates in organic solvents and thus while solutions can appear clear and extraction can be nearly quantitative, great care must be placed in working with a given analytic technique in the appropriate concentration regime to avoid artifacts resulting from lipid aggregation. This problem can become challenging and sometime even a difficult issue for the analysis of different classes and individual molecular species of lipids as their tendency is to aggregate. The analytical results can be greatly affected by changes in analyte mass content or even solvent composition. For example, lipids form aggregates in a manner depending on the physical properties (*e.g.*, head group polarity, aliphatic chain length, and double bond numbers) of each individual lipid molecular species as well as the concentration and solvent employed during a study. Therefore, extremely careful sample preparation and great attention to the selected conditions for lipid analysis are required and are of much greater concern than those for most other metabolites.

Cellular lipidomes are highly complex. It has been estimated that cellular lipidomes contain thousands to tens of thousands of individual molecular species of lipids at the level of attomole to nanomole concentration of lipids per mg of protein.[5] These individual molecular species belong to the very different lipid classes and subclasses, and are comprised of different lengths, degrees of unsaturation, different locations of double bonds, and potential branching in aliphatic chains. Moreover, multiple other factors make the study of this already complex and diverse system even more difficult. These include the facts that, among other factors: (1) cellular lipid molecular species and composition are quite different amongst different species, cell types, cellular organelles, membranes and membrane microdomains (*e.g.*, caveola and/or rafts); and (2) the cellular lipidome is dynamic, depending on nutritional conditions, hormonal influences, health status, exercise levels, and many other factors.

In recent times, the importance of individual constituents in a cellular lipidome in biological functions has become clearer. This is due to the previously known essential role of lipid second messengers of signal transduction, the identification of new signaling molecules, and the important roles that lipids play in facilitating the assembly of signaling constituents, modulating their interactions, and contributing to the regulation of the downstream flow of biological information. In addition, most recently the role of lipids in many epidemic diseases (*e.g.*, obesity, atherosclerosis, stroke, hypertension, and diabetes, which now are collectively referred to as the "metabolic syndrome")[6] has been recognized and substantial attention is being given to the role of lipids and lipid metabolism in energy supply and demand. During the studies, investigators have clearly recognized that the metabolism of individual lipid molecular species or individual lipid classes is interwoven. To conduct research on lipid metabolism only from an isolated system or only being focused on one molecular species or one lipid class has substantial limitation. The metabolism of the entire lipidome of the organelle, the cell type, the organ, the system, or the species should be investigated in a systems biology approach.[7,8] Therefore, the need for such a comprehensive approach for studies of lipid metabolism greatly catalyzes the emerging of lipidomics and accelerates its development.

Among the development of lipidomics, mass spectrometric techniques play a leading role in lipid characterization, identification, and quantitation (see refs 1 and 7–13 for recent reviews). Particularly, electrospray ionization MS (ESI/MS) is the most prominent and has enjoyed the most success. The advantages of ESI/MS are multiple.

- First, its ion source could act as a separation device to selectively ionize a certain category of lipid molecular species based on the electrical propensity of lipid classes. Thus, it is feasible to analyze different lipid classes and individual molecular species with high efficiency without prior chromatographic separation.
- Second, the ionization efficiency of ESI/MS for lipid analysis is incomparably higher in comparison to other traditional mass spectrometric approaches. A detection limitation at a concentration of low fmol/μL and even at amol/μL is frequently achievable and is still improving as the instruments become more and more sensitive.
- Third, ionization efficiency or instrument response factor of individual molecular species in a polar lipid class is essentially identical within experimental error after ^{13}C deisotoping when the experiment is performed in a low lipid concentration region. This is largely due to the minimal source fragmentation and the selective ionization depending on the electrical property largely possessed by the polar head group. Therefore, it is feasible to quantitate individual molecular species of a polar lipid class through direct ratiometric comparison with a selected internal standard or through the peak area measurement from the reconstructed total ion current chromatograph in comparison to a minimal set of external calibration curves.
- Fourth, a nearly linear relationship between an ion peak intensity (or area) of a polar lipid molecular species and the concentration of the compound is present over a wide dynamic range in the low concentration region. For lipids without large dipoles, correction factors or calibration curves for each individual molecular species have to be pre-determined.
- Finally, the reproducibility of a quantitative analysis from an identical sample preparation is excellent ($<5\%$ of experimental error). Thus, it is evident that ESI/MS-based lipid analysis has become an essential tool for measuring cellular lipidomes during cellular perturbations and disease states (see refs 1, 7–9, 12, 14 and 15 for recent reviews).

Three main ESI/MS-based approaches have been independently developed in lipidomics, each of which possesses different advantages and some limitations – which have been extensively discussed previously.[12] Therefore, further developments for each of the approaches are necessary as more and more advanced mass spectrometers are yearly evolving. These approaches are the HPLC-coupled ESI/MS method, tandem mass spectrometry-based techniques, and multi-dimensional MS and array analysis-based technology after ion-source separation. Both of the latter approaches are developed after direct infusion and potentially both can be used as global analyses of individual lipid molecular species directly from a lipid extract of a biological sample without pre-chromatographic separation. Therefore, both of the approaches have now been termed as "shotgun lipidomics". In this article, we will discuss only the most recent developments and applications of shotgun lipidomics, which exploit the synergy between the proximal separation of lipid classes in the ion source and subsequent multi-dimensional mass spectrometry. Shotgun lipidomics is still a growing technology with many improvements anticipated in multiple aspects of this field. The applications of this technology have already been used to identify individual gene functions and provide new insights into the biological mechanisms of multiple disease states. Collectively, further developments of these approaches in lipidomics will lead us to a new level of understanding in lipid-related diseases through a systems biology analysis of disease-related alterations in the lipidome.

5.2 The Principles of Multi-dimensional Mass Spectrometry Based Shotgun Lipidomics

Development of the multi-dimensional MS-based shotgun lipidomics approach was initiated at a very early stage when ESI/MS was employed for lipid analysis.[16,17] The penetration into low abundance regime with quantitation of extremely low abundance species was dramatically improved through the development of two-dimensional MS[18,19] and has now been developed into a mature technology in lipidomics[15] and further improvements in some aspects of this technology are anticipated to improve both the penetrance and accuracy still further. This technology includes multiple simple steps suitable for robotic implementation, including multiplexed extractions, intrasource separation and selective ionization, identification of individual lipid molecular species using multi-dimensional MS and array analyses, quantitation of the identified lipid molecular species using a two-step procedure in conjunction with bioinformatics, and data processing.

5.2.1 Multiplexed Extractions

The first key step in this shotgun lipidomics approach is the multiplexed extractions. This step is vital for mass spectrometric analysis and careful attention to sample preparation is a prerequisite for success. In our experience, the lipids of each biological sample (commonly containing 50 to 500 µg of protein mass content from the cell, tissue, or membrane samples, or a small volume of biological fluid) can be extracted by a modified Bligh and Dyer procedure or other extraction methods (*e.g.*, Folch procedure) under acidic, basic, and/or neutral conditions depending on the lipids of interest to be analyzed, as previously discussed.[12] The pre-selected internal standards (at least one internal standard for each lipid class) are added to the extraction mixture prior to the conduction of extraction. Selection of an internal standard for a lipid class has been discussed in detail.[12] The lipid extract should be further properly diluted to a concentration of less than 100 pmol/μL of total lipids in chloroform/methanol (1:1, v/v) or other concentration depending on the solvents and analytical method employed to warrant the availability of a linear dynamic range prior to direct infusion of the diluted lipid solution to an ESI ion source.

It should be noted that we typically use quite a large volume of chloroform and repeat at least twice (*i.e.*, 2 mL × 2) to extract the small amount of lipids (less than 1 mg of total lipids from the aforementioned biological samples) during the sample preparation. Therefore, a good recovery is facilitated. Moreover, differential recoveries of different lipid classes are eliminated by addition of at least one internal standard for each individual lipid class prior to the extraction. By these two measures, the differential recoveries of different individual molecular species in a lipid class are minimized and typically represent only a secondary factor and most often can be neglected after assurance that they represent < 3% errors.

5.2.2 Intrasource Separation

As is well known, an ESI ion source behaves like an electrophoretic cell which could be utilized to selectively separate different charged moieties and generate discrete class-selective ions under the influence of a high electrical potential (typically ~ 4 kV).[20–22] On the other hand, although there are tens of thousands of individual lipid molecular species present in the cellular lipidome, these species naturally belong to a much smaller number of lipid classes. Different lipid classes possess different electrical properties, largely depending on the nature of the polar head groups, and they can be re-classified into three main categories (*i.e.*, anionic lipids, weak anionic lipids, and lipid classes that are electrically neutral).[1,12] Given the physical factors affecting the process of selective ionization of analytes in the electrospray ion source and the different electrical properties of each of the lipid classes, we recognized that electrospray ion sources can be used to resolve lipid categories in a crude lipid extract into different categories based on the intrinsic electrical properties of each lipid category. Such a separation has now been referred to as intrasource separation of lipids[12,15,19] and is analogous to the use of an ion-exchange column to separate individual lipid classes.[23] However, in comparison to ion-exchange chromatography, intrasource separation is rapid, direct, and reproducible and avoids artifacts inherent in chromatography-based systems.[24] A practical strategy for separation of these categories of lipids based on their differential intrinsic electrical properties has been discussed in detail[1,12,15,19] and has been

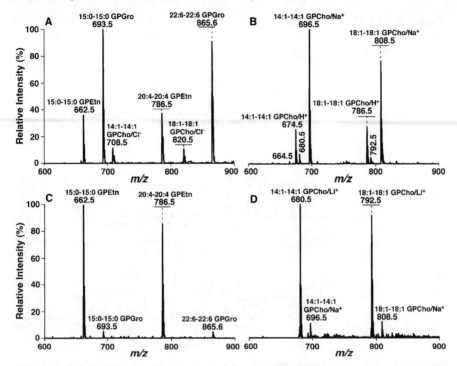

Figure 5.1 Intrasource separation of a model mixture of phospholipids comprising 15:0-15:0 and 22:6-22:6 GPGro (1 pmol/µL each), 14:1-14:1 and 18:1-18:1 GPCho (10 pmol/µL each), and 15:0-15:0 and 20:4-20:4 GPEtn (15 pmol/µL each) molecular species in 1:1 CHCl₃/MeOH. (A) and (C) show mass spectra acquired in the negative-ion mode and (B) and (D) show mass spectra acquired in the positive-ion mode in the absence ((A) and (B)) or the presence ((C) and (D)) of LiOH. The horizontal bars indicate the ion peak intensities after ^{13}C de-isotoping and normalization of molecular species in each class to the one with lower molecular weight.

demonstrated through a model mixture of phospholipids that represent three categories of lipids (Figure 5.1).[25] Through this approach, a comprehensive series of mass spectra can be obtained from a typical lipid extract of any biological samples (*e.g.*, mouse myocardium) (Figure 5.2).

5.2.3 Identification of Lipid Molecular Species by Multi-dimensional Mass Spectrometry and Array Analysis

We recognized that each ion peak in the mass spectra of lipid extracts of biological samples potentially represents at least one and very often more than one lipid molecular species, particularly in those mass spectra acquired by mass spectrometers with a low mass resolution. At this stage, one could perform product ion analyses of individual ion peaks to identify the molecular species

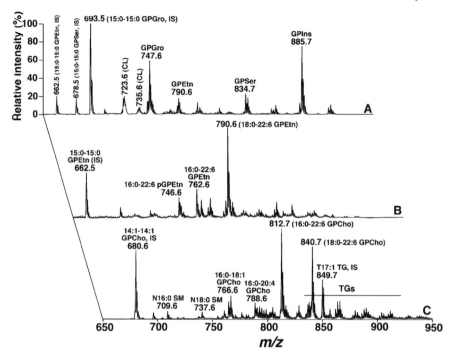

Figure 5.2 Intrasource separation of a lipid extract of mouse myocardium. ESI mass spectra were acquired: (A) in the negative-ion mode after dilution to a total lipid concentration of ∼ 50 pmol/μL with 1:1 chloroform/methanol (v/v); (B) in the negative-ion mode from the diluted lipid solution after addition of 50 nmol LiOH/mg of protein; and (C) in the positive-ion mode from the diluted lipid solution after addition of 50 nmol LiOH/mg of protein. "IS" denotes internal standard; "CL" represents doubly charged cardiolipin; "SM" is sphingomyelin; and "TG" stands for triacylglycerols. All mass spectral traces are displayed after normalization to the base peak in each individual spectrum.

underneath each ion peak. However, we recognized that most of these biological lipid species are linear combinations of aliphatic chains, backbones, and/or head groups, each of which represents a building block of the molecular species under consideration. For example, three moieties linked to the hydroxy groups of glycerol can be recognized as three building blocks and if we can identify these building blocks, then we are able to identify each individual glycerol-derived lipid molecular species in a given sample.[15] An analogous approach can also be used to define other lipid classes (*e.g.*, sphingomyelin in which the phosphocholine head group, the sphingoids (long-chain bases), and the fatty acyl amides represent the three building blocks of each molecular species).[15] Identification of these building blocks can be accomplished by two other powerful tandem mass spectrometric techniques (*i.e.*, neutral loss scanning and precursor ion scanning) by monitoring the specific loss of a neutral fragment or the yield of a fragmental ion in the tandem MS modes. Therefore, all of these building blocks of lipid classes constitute an additional dimension to the molecular ions present in the original mass spectrum termed the first dimension. The crossing peaks of a given primary molecular ion in the first dimension are determined from the building blocks in the second dimension and represent the fragments of this given molecular ions under the conditions employed. Analysis of these crossing peaks (*i.e.*, the fragments) thereby determines the structure of the given molecular ion as well as its isobaric constituents.[15]

This basic two-dimensional mass spectrum can be changed in intensities and profiles by employing various instrumental conditions such as changes in ionization conditions (*e.g.*, source temperature and spray voltage) and in fragmentation conditions (*e.g.*, collision gas pressure, collision energy, and collision gas), among others.[15] Therefore, besides the basic 2D mass spectral unit in which the second dimension is constructed with specific lipid building blocks, each series of ramped changes in instrumental condition facilitates the generation of an additional dimension and can potentially be used for lipid analyses.[15] All these dimensions form a family of multi-dimensional mass spectrometry.

5.2.4 Quantitation of Lipid Molecular Species Using a Two-step Procedure

After identification of each individual molecular ion of a lipid class, quantitation in shotgun lipidomics is performed by a two-step procedure.[15,26,27] First, the abundant and non-overlapping molecular species of a class are quantified by ratiometric comparisons with a pre-selected internal standard of the class after ^{13}C de-isotoping. Next, either all of the determined molecular species of the class plus the pre-selected internal standard or some of the selected ones of these determined species are used as standards to determine the mass content of other low-abundance or overlapping molecular species using one or multiple tandem mass traces (each of which represents a specific building block of the class of interest) by 2D MS. Through use of this second step in the quantitation process, the linear dynamic range of quantitation can be dramatically extended by eliminating background noise and by filtering the overlapping molecular species through a multi-dimensional approach.[12] Of course, additional standards can be added if necessary, but in most cases the naturally occurring distribution of acyl chain length and unsaturation is usually sufficient. It is anticipated that successful applications of bioinformatics and machine learning can correct the small differences in ionization efficiency and/or detector sensitivity (typically less than 7%) in the next generation of applications of computational lipidomics to the shotgun lipidomics approach.

5.2.5 Analysis in the Low Concentration Regime by Shotgun Lipidomics

Unlike other analytes, lipids form aggregates as the lipid concentration increases that depend on the chemical properties of each individual molecular species including the polarity of the head group, the length of aliphatic chains, and the degree of unsaturation. Moreover, formation of lipid aggregation is also dependent on the solvent employed. The higher the hydrophobicity of a lipid molecular species and the more polar the solvent system employed, the lower the concentration at which the lipids will aggregate. It is very difficult to ionize the lipid at its aggregate state and most of the aggregated particles will end up in waste. Thus, as the lipid concentration increases, the ionization efficiency of lipid molecular species containing long and saturated fatty acyl chains dramatically decreases as previously demonstrated.[28] Therefore, it is critical that the lipid concentration in shotgun lipidomics as well as other ESI/MS techniques must be lower than that leading to aggregate formation. The upper limit of the total lipid concentrations of a biological extract is approximately 100 pmol/μL in 2:1, 50 pmol/μL in 1:1, and 10 pmol/μL in 1:2 of chloroform/methanol. In this low concentration region, ionization efficiency of different individual molecular species from the identical class is essentially identical (*i.e.*, the commonly called "ion suppression" is minimal) as demonstrated (Figure 5.3).

5.3   The Chemical Mechanisms Underlying Intrasource Separation

Recently, we employed model mixtures of synthetic glycerophospholipids to examine the factors underlying the intrasource separation and selective ionization of different lipid classes and identified the critical physical and/or chemical interactions which influence the degree of these processes.[25] The model mixtures were comprised of three representative categories of phospholipid classes with distinct electrical propensities, *i.e.*, anionic lipids, weak anionic lipids, and electrically neutral but zwitterionic lipids represented by phosphatidylglycerol (GPGro), phosphatidylethanolamine (GPEtn), and phosphatidylcholine (GPCho) molecular species, respectively (see Figure 5.1). Two individual molecular species at equimolar concentrations from each of these phospholipid classes were selected to represent the variations in chain length and unsaturation typically found in biological phospholipids. It was demonstrated that intrasource separation and selective ionization of phospholipid classes was independent of the lipid concentration and the infusion flow rate within a wide range of experimental conditions (Figure 5.4). The results indicated that GPGro molecular species are selectively ionized relative to GPEtn and GPCho molecular species in the negative-ion mode.

This augmentation in ionization selectivity is consistent with a donor–acceptor model (*e.g.*, conjugate acid–conjugate base model) in which GPGro molecular species can generate their conjugate base by losing a proton (or a small cation, *e.g.*, Li^+, Na^+, *etc.*), whereas both GPCho and GPEtn molecular species behave as a conjugate acid relative to GPGro to accept the proton (or small cation). Although GPEtn and GPCho species (as well as other small ions present in the media) in addition to anionic phospholipids are present in the

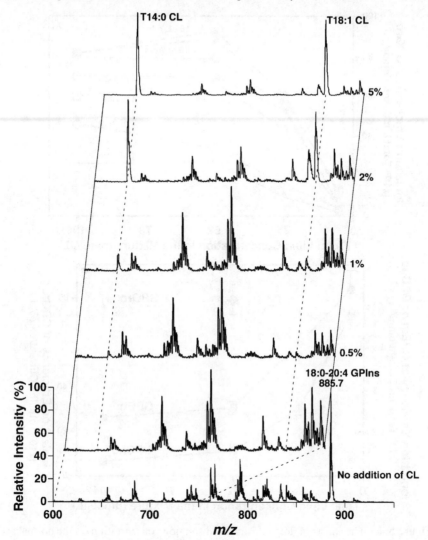

Figure 5.3 The effects of co-existing large amounts of mouse hepatic lipids on quantitative analyses of equimolar mixtures of cardiolipin molecular species at different concentrations. A small amount of an equimolar mixture of cardiolipin molecular species was added into the lipid extracts of mouse liver prior to dilution of the total lipid solution to less than 100 pmol/μL and direct infusion. Mass spectrometric analyses were performed in the negative-ion mode after spiking 0.5, 1, 2, and 5 mol% of total cardiolipin into the mouse liver extracts. The insets represent the expanded spectra of the region containing doubly charged cardiolipin molecular ions as indicated. "CL" represents doubly charged cardiolipin and "GPIns" denotes phosphatidylinositol. (Reprinted from ref. 42 with permission from The American Society for Biochemistry and Molecular Biology, Inc., Copyright 2006.)

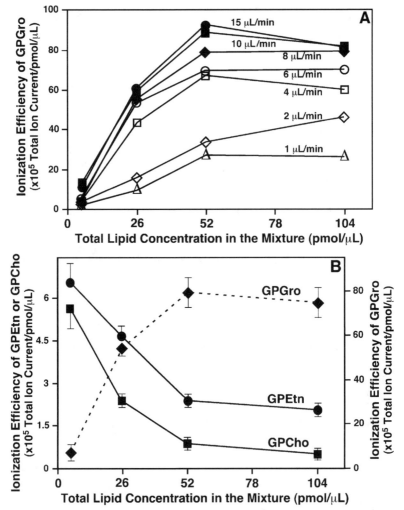

Figure 5.4   The effects of flow rate and lipid solution concentration on the normalized ion count density (*i.e.*, ionization efficiency) of each examined phospholipid class in the negative-ion mode. A phospholipid mixture of 15:0-15:0 and 22:6-22:6 GPGro, 14:1-14:1 and 18:1-18:1 GPCho, and 15:0-15:0 and 20:4-20:4 GPEtn molecular species in a molar ratio of 1:1:10:10:15:15 in 1:1 (v/v) CHCl₃/MeOH was analyzed in the negative-ion mode using a QqQ ESI mass spectrometer. The ionization efficiencies of GPGro (part A and the broken line in part B), GPCho (part B), and GPEtn (part B) were calculated from the normalization of the total ion current of a class to a unit concentration (*i.e.*, pmol/μL) of each class at the indicated lipid concentration and flow rate. The ionization efficiency of GPEtn or GPCho represents the mean ± SD of the ionization efficiencies of the class at the flow rates of 1, 2, 4, 6, 8, 10, and 15 μL/min. The ionization efficiency of PtdGro in Panel B represents the mean ± SD of the ionization efficiencies in Panel A at the flow rates of 4, 6, 8, 10, and 15 μL/min. (Reprinted from ref. 25 with permission from the American Society for Mass Spectrometry, Copyright 2006.)

initial sprayed droplets in the negative-ion mode under (very) weak acidic conditions, anionic lipids are present in greater abundance since anionic lipid molecular species form their conjugate bases more readily than either GPCho or GPEtn molecular species. This preferred selectivity of anionic species over GPCho and GPEtn species is also likely present during the iterative processing of desolvation and reformation of smaller droplets. This selectivity is likely achieved through a small cation transferring from a neutral anionic phospholipid molecular species to GPCho or GPEtn molecular species. This transfer leads to a charge redistribution, preventing GPEtn and GPCho molecular species entering the plate orifice (or other ion inlets). The augmentation of ionization selectivity of GPGro over GPCho and GPEtn represents the cumulative effects resulting from this preferred selectivity during both charged droplet formation as well as electric field dependent selectivity at each step of the droplet desolvation and reformation process.

To examine this hypothesis, we determined the ionization efficiency of individual GPGro or GPEtn classes. We found that the ionization efficiency of GPGro class alone was constant (*i.e.*, $(7.9 \pm 0.6) \times 10^5$ ion current/pmol/µL) over a 1000-fold concentration range examined (0.01 to 10 pmol/µL). This ionization efficiency of GPGro alone was comparable to that of GPGro at its very low concentration in the mixture as aforementioned and was approximately 10 times lower than that at its high concentration in the mixture examined (Figure 5.4b). The ionization efficiency of GPEtn alone at its very low concentration is comparable to that in the mixture. In contrast to GPGro, as the concentration of GPEtn increases, the ionization efficiency of GPEtn becomes smaller. The ionization efficiency of GPEtn averaged $(4.5 \pm 0.5) \times 10^5$ ion current/pmol/µL from a concentration range between 1 and 30 pmol/µL and this value was approximately twice that obtained from the measurement of the phospholipid mixture (Figure 5.4b). Similarly, an average ionization efficiency of $(3.4 \pm 1.2) \times 10^5$ ion current/pmol/µL from GPCho in this concentration range was also obtained. Again, this value was much higher that than obtained from the measurement of the phospholipid mixture (Figure 5.4b). These results suggest that the augmentation of GPGro ionization in the phospholipid mixture is related to a dramatic reduction of GPEtn and/or GPCho ionization due to the favorable association of small cations with GPEtn or GPCho molecular species in comparison to GPGro species in each step of the droplet processing.

5.4 An Enhanced Shotgun Lipidomics for Cardiolipin Analysis

Cardiolipin is localized predominantly in the mitochondrial inner membrane, where it facilitates mitochondrial function through a variety of mechanisms largely related to its unique physical and chemical properties (*i.e.*, highly anionic character, large aliphatic chain to polar head group volume, and specific binding to proteins in the electron transport chain including cytochrome c oxidase) that promote its role as an effector of multiple highly specific biological functions (see refs 29–32 for recent reviews). The essential role of cardiolipin in mito-chondrial function has recently been underscored through identification of a genetic disorder (*i.e.*, Barth syndrome) in which altered cardiolipin metabolism results in cardiolipin depletion and molecular species changes,[33–37] and demon-stration of a substantial cardiolipin depletion in diabetic mouse myocardium[38] in which mitochondrial dysfunction is manifest.

Cardiolipin is an unusual phospholipid comprised of a dimer of phosphati-date molecules linked through a glycerol backbone. LC-MS based techniques for quantification of cardiolipin molecular species have been developed by using either singly charged or doubly charged molecular ions.[39,40] Although shotgun lipidomics approach was also employed to estimate the non-overlap-ping molecular species of cardiolipin on several occasions,[38] the full power of this approach for accurate quantitation of individual low-abundance molecular species present in the entire cardiolipin class has not yet been realized. Recently, we have further extended our shotgun lipidomics approach to this specific lipid class to identify and quantitate low abundance cardiolipin molecular species directly from lipid extracts of biological samples with unprecedented speed, depth of penetration, and accuracy.

This new strategy exploits three chemical principles inherent in cardiolipin molecular species for their identification and quantitation. These chemical principles are (1) the doubly charged character of two phosphates in each cardiolipin molecular species; (2) the specific neutral loss of ketenes from doubly charged cardiolipin molecular ions to yield doubly charged triacyl monolysocardiolipins; and (3) the marked enrichment of one fatty acyl chain in cardiolipin from most of the biological samples.

Through the doubly charged nature of cardiolipins, each ion from cardioli-pin individual molecular species (including the overlapping and low-abundance molecular species) can be directly recognized through searching for the pres-ence of the isotopologue peaks (*i.e.*, $[M-2H + 1]^{2-}$ or $[M-2H + 3]^{2-}$) of doubly charged cardiolipin molecular species since, as is well known, doubly charged molecular ions of other lipid classes are rarely present in the *m/z* region of 600 and 800 of organic extracts at neutral pH. Although the molecular ion peak and the $[M-2H + 2]^{2-}$ ion peak of an individual cardiolipin molecular species may overlap with molecular ion peaks in other lipid classes, the $[M-2H + 1]^{2-}$ or $[M-2H + 3]^{2-}$ isotopologue peaks of a doubly charged cardiolipin molecular species are highly specific for cardiolipin molecular species in chloroform extracts in this mass range under these conditions.

Therefore, any mass spectrometer with a mass resolving power high enough to detect the doubly charged ions can use this approach for shotgun lipidomics of cardiolipin molecular species. For example, Figures 5.5 and 5.6 display the mass spectra of a lipid extract of mouse heart acquired in the negative-ion mode by both a triple quadrupole (*i.e.*, QqQ) mass spectrometer with a mass reso-lution setting of 0.3 Th and a high mass resolution hybrid pulsed instrument (*i.e.*, quadrupole time-of-flight mass spectrometer, QqTOF), respectively. The doubly charged cardiolipin molecular ion peaks are well displayed (see insets of Figures 5.5 and 5.6).

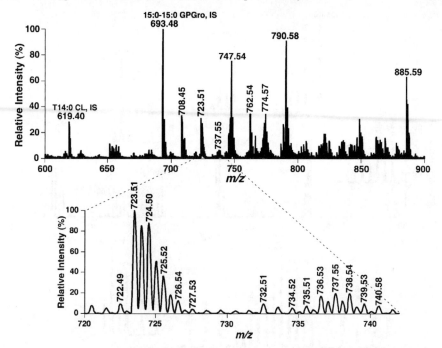

Figure 5.5 Negative-ion ESI mass spectrometric analysis of doubly charged cardiolipin molecular species in a lipid extract of mouse myocardium by a QqQ-type mass spectrometer with a high mass resolution setting of FHMW 0.3 Th. "CL" represents doubly charged cardiolipin; "GPGro" denotes phosphatidylglycerol; and "IS" stands for internal standard. (Reprinted from ref. 42 with permission from The American Society for Biochemistry and Molecular Biology, Inc., Copyright 2006.)

After identification of these unique cardiolipin molecular ion peaks, production analyses of the plus-one isotopologue peaks can be performed to confirm the identities of acyl chains of each individual cardiolipin molecular species. However, 2D MS analyses of all cardiolipin molecular species can be readily conducted based on the second chemical principle, *i.e.*, neutral loss analysis of the ketenes from doubly charged cardiolipin molecular species to yield doubly charged triacyl monolysocardiolipin (*i.e.*, the specific building blocks of cardiolipin molecular species) in the region of interest. For example, Figure 5.7 shows a representative building block (*i.e.*, NL 131.1, neutral loss of C18:2 ketene from doubly charged cardiolipin molecular species) along with two other non-specific building blocks (*i.e.*, precursor-ion (PI) scanning of glycerophosphate and linoleate, *i.e.*, PI 153.2, and PI 279.2, respectively) in a 2D MS analysis format. This 2D MS analysis clearly identifies all the cardiolipin molecular species that contain at least one linoleate.

Once the cardiolipin molecular ion is recognized by searching the $M+1$ isotopologue peak of doubly charged ions and identified through either 2D MS analyses or product-ion analyses, quantitation of cardiolipin molecular species by

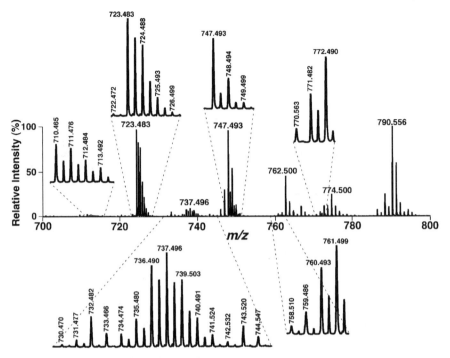

Figure 5.6 Negative-ion ESI mass spectrometric analysis of doubly charged cardiolipin molecular species in a lipid extract of mouse myocardium by a QqTOF-type mass spectrometer. (Reprinted from ref. 42 with permission from The American Society for Biochemistry and Molecular Biology, Inc., Copyright 2006.)

a QqQ-type instrument can be readily performed by ratiometric comparison of the de-isotoped ion peak intensity of the cardiolipin molecular species. Specifically, if we set the monoisotopic peak intensity as 1 and only consider the ^{13}C isotopic distribution in cardiolipin, the peak intensities of the $M + 1$, $M + 2$, $M + 3$, ... isotopologue are $0.01082n$, $0.01082^2 n(n - 1)/2$, $0.01082^3 n(n - 1)(n - 2)/6$, ... , respectively. Therefore, the peak intensity of each of the isotopologue including the mono-isotopologue can be calculated from the peak intensity of the $M + 1$ isotopologue and the calculation of the de-isotoped intensity from the intensity of the $M + 1$ isotopologue peak can be made as follows:

$$I_{\text{total}} = I_1 \times (92.42/n + 1 + 5.41 \times 10^{-3}(n - 1) + 1.95 \times 10^{-5}(n - 1) \\ \times (n - 2) + 5.3 \times 10^{-8}(n - 1)(n - 2)(n - 3) + \ldots \ldots) \tag{1}$$

where I_{total} is the de-isotoped ion intensity of an individual cardiolipin molecular species of interest; I_1 is the peak intensity of its $M + 1$ isotopologue; and n is the total carbon numbers in the species.

It should be emphasized that since the abundance of cardiolipin molecular species is usually quite low relative to other anionic lipids, the presence of baseline noise and/or baseline drift in a QqQ-type instrument could introduce a

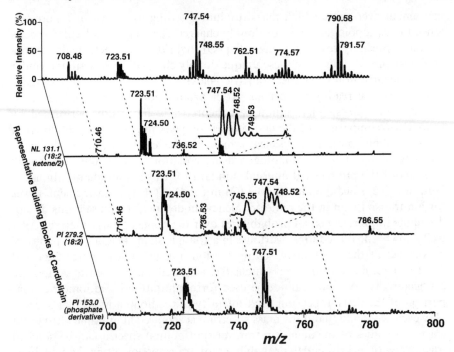

Figure 5.7 Two-dimensional MS analysis of some representative building blocks of doubly charged cardiolipin molecular species in a lipid extract of mouse myocardium by a QqQ-type mass spectrometer with a high mass resolution setting of FHMW 0.4 Th. NL denotes neutral loss scanning and PI represents precursor ion scanning.

large error during quantitation of low-abundance cardiolipin molecular species by use of equation (1) alone. To minimize errors for low abundance species, the third chemical principle can be employed in the second step of the quantitation procedure in shotgun lipidomics. We exploited the marked enrichment of linoleate in cardiolipin compared to that present in other lipid classes and the fact that at least one linoleate chain is present in almost all cardiolipin molecular species[40,41] to generate a strategy that maximally enhanced the signal to noise ratios of most cardiolipins by using precursor-ion scanning of linoleate. Since the linoleate fragment is much more intense than glycerophosphate and triacyl lysocardiolipin ions (Figure 5.7) and the linoleate ion intensity signal is amplified by the presence of multiple linoleoyl chains in most cardiolipin species, the sensitivity and accuracy utilizing linoleate is much higher than that using glycerophosphate (or other less abundant fatty acids in cardiolipin) or ion peaks resulting from ketene loss for quantitation of minor and/or overlapping cardiolipin molecular species in the second step of the quantitation process.

Similarly, we can also use a high mass resolution hybrid pulsed instrument (*i.e.*, QqTOF) to identify and quantitate cardiolipin molecular species at much lower background noise level (reduction in chemical noise) and at much higher resolving power (see the inset of Figure 5.6) in comparison to the QqQ type instrument even with a high mass resolution setting as shown in Figure 5.5. Since the isotopologue peaks of doubly charged cardiolipin molecular species are more precisely resolved using a QqTOF hybrid mass spectrometer than a QqQ instrument, searching these unique doubly charged cardiolipin molecular ions in negative-ion mass spectrometric analysis of lipid extracts of biological samples can be readily performed. After product-ion analyses of the $M + 1$ isotopologue peaks, quantitation can theoretically be readily performed by ratiometric comparison of the de-isotoped ion peak intensity (or peak area) of the cardiolipin molecular species (which can be calculated from the peak intensity of the $m + 1$ isotopologue using equation 1) with the de-isotoped intensity of the pre-selected internal standard for cardiolipin quantitation.

In practice, since the quadrupole (Q) in the QqTOF-type instrument serves for ion-transmission in the radio frequency-mode only, the pulsed ions do not distribute in precisely equal probabilities over the range of mass analysis. As is well known, the pulsed ions distribute in a triangle shape (Figure 5.8a) and are maximized at the position of setting for ion-transmission (Figures 5.8a and 5.8b). Therefore, the ion peak intensities acquired under such a setting are artificially distorted and cannot be used for quantitation by ratiometric comparison of the de-isotoped ion peak intensity of cardiolipin molecular species. However, we found that if the mass spectrum is acquired at multiple settings (at least two values of maximal ion-transmission points) spectra can be stitched together with nearly equal probabilities of transmission of individual ions. Therefore, quantitation by ratiometric comparison of the de-isotoped ion peak intensity of cardiolipin molecular species can also be performed by using a QqTOF-type mass spectrometer.[42]

These results underscore the importance of appropriate stitching in the quantitative analyses of individual cardiolipin molecular species by ratiometric comparison using a QqTOF type instrument. We believe this conclusion also holds for quantitative analyses of lipid classes other than cardiolipin and the importance of the setting points and the transmission levels for other lipids using a quadrupole for pulsed ion transmission should be recognized. By applying this newly developed technique, accurate quantitation of multiple low-abundance cardiolipin molecular species was accomplished and the analyses of the specific profiles of cardiolipin molecular species in the lipid extracts of mouse heart, liver, and skeletal muscle are achieved in an unprecedented manner.[42] Furthermore, multiple oxidized cardiolipin molecular species are also identified through accurate mass analysis followed by confirmation with product-ion analyses.[42]

5.5 Derivatization as an Additional Dimension of Multi-dimension Mass Spectrometry in Shotgun Lipidomics

Derivatization has been previously employed in the ESI/MS analysis of intact lipids[43–45] and other analytes.[46] One purpose of the derivatization is to convert

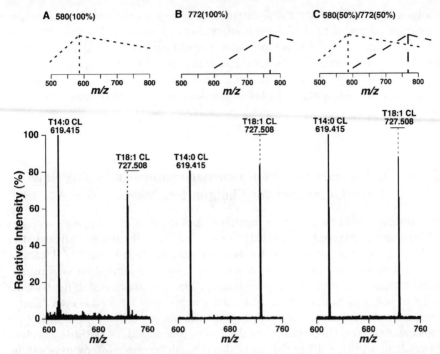

Figure 5.8 The effects of a setting point of ion-transmission on ion peak intensities. The top part of part A schematically illustrates the profile of different ion transmitted when the ion transmission point is set at *m/z* 580 on QqTOF mass spectrometer and the bottom part of part A shows a mass spectrum of the equimolar mixture of tetra 14:0 cardiolipin (T14:0 CL) and T18:1 CL (100 fmol/μL each) acquired on the instrument under the experimental condition, indicating the differential transmission of these two CL ions. Part B shows the same as part A, except that the point of ion transmission is set at *m/z* 772. Part C shows the same as part A, except that the points of ion transmission are hopped at *m/z* 580 and 772 in the equal duty cycle.

non- or less polar lipids into polar lipids which carry an inherent charge, *e.g.*, adding a sulfate group to cholesterol[43] or converting oxosteroids into their oximes.[44] Through this approach, one can dramatically improve the sensitivity of mass spectrometric analysis of these compounds. A detection limitation at the concentration of the low pg/μL for the analysis of these aforementioned less polar lipids can be readily achieved after derivatization as previously described.[43,44] Another purpose of derivatization is to tag a specific group of lipids. Through derivatization, these tagged lipid molecular species can be shifted to a new mass region in which these lipids do not overlap with any other lipid molecular species while overlapping with other lipids occurs in the original mass region. Furthermore, all of these tagged lipid molecular species can be readily distinguished through the facile loss of the derivatized moiety which can be detected by either neutral loss of the tagged probe or precursor-ion monitoring the tagged moiety. Therefore, derivatization provides a new tool for shotgun lipidomics to specifically analyze a class or a group of lipid molecular species of interest directly from a lipid extract of a biological sample. Accordingly, derivatization is a new dimension of multi-dimension mass spectrometry for shotgun lipidomics. The following sections discuss the two newly developed derivatization approaches in shotgun lipidomics for analyses of phospho-ethanolamine-containing phospholipid molecular species and cholesterol mass content.

5.5.1 Derivatization of Phosphoethanolamine-containing Lipid Molecular Species for Shotgun Lipidomics

In shotgun lipidomics, phosphoethanolamine-containing lipids such as ethanolamine glycerophospholipid (GPEtn) and lyso-GPEtn are analyzed after addition of a small amount of LiOH in the negative-ion mode.[12,19] Unfortunately, under these experimental conditions, many low abundant GPEtn molecular species and almost the entire lyso-GPEtn class are buried in the baseline noise or overlapped with some molecular species of other lipid classes. Therefore, it is difficult to identify these low-abundant GPEtn and lyso-GPEtn molecular species. Quantitation of these very low abundant GPEtn molecular species and all lyso-GPEtn species is also difficult by shotgun lipidomics using the aforementioned two-step procedure due to a lack of a highly sensitive and representative tandem mass spectrometric method which can be used to profile very low abundant GPEtn and lyso-GPEtn molecular species under the experimental conditions without derivatization.

We recognized that 9-fluorenylmethoxylcarbonyl chloride (Fmoc chloride, Fmoc-Cl) can specifically tag phosphoethanolamine-containing lipid molecular species in a lipid extract of a biological sample to yield Fmoc-GPEtn and Fmoc-lyso-GPEtn.[47] After one-step *in situ* simple derivatization, Fmoc-GPEtn and Fmoc-lyso-GPEtn molecular species are rendered anionic and can be analyzed directly in the negative-ion mode with enhanced sensitivity. Moreover, derivatization with Fmoc shifts GPEtn molecular species out of the region where GPEtn molecular species potentially overlap with other lipid classes such as phosphatidylserine and phosphatidylinositol in shotgun lipidomics analysis. Furthermore, product-ion analyses of Fmoc-GPEtn molecular species demonstrate a very abundant fragment corresponding to the facile neutral loss of Fmoc (Figure 5.9). The neutral loss of Fmoc from the derivatized phosphoethanolamine-containing lipid molecular species is so facile that over a linear dynamic range of 15 000-fold for quantitation of these lipids can be achieved with a detection limitation at the concentration of amol/μL level and even lyso-GPEtn molecular species in a lipid extract can also be directly profiled.[47] Thus, mass spectrometric analysis by neutral loss of the Fmoc moiety from Fmoc-GPEtn and Fmoc-lyso-GPEtn species can be readily used to quantify lyso-GPEtn and very low abundant or overlapping GPEtn molecular species in the second step of quantitation in shotgun lipidomics.

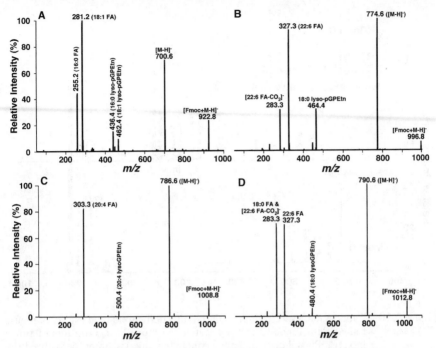

Figure 5.9 Representative negative-ion ESI product-ion analyses of Fmoc-GPEtn molecular species in the lipid extracts of mouse retinas. Fmoc chloride in anhydrous chloroform was added to a small amount of mouse retina lipid extract in a ratio of 1:1 (Fmoc-Cl *vs.* GPEtn content in the extract). The mixture was incubated at room temperature for 5 min and directly diluted with 1:1 of chloroform/methanol to a concentration of approximately 50 pmol/μL of total lipids. Part A shows product ion ESI-MS/MS analyses of Fmoc-derivatized pseudomolecular ions at m/z 922.8 show an abundant ion at m/z 700.6 (representing the neutral loss of Fmoc moiety), two modest fragments at m/z 436.4 and 462.4 (representing lysoplasmenylethanolamine (lyso-pGPEtn) derivatives), and fatty acyl carboxylates of oleate (m/z 281.2) and palmitate (m/z 255.2). These product ions indicate the presence of two Fmoc-derivatized, isobaric ions (*i.e.*, 16:0-18:1 and 18:1-16:0 Fmoc-pGPEtns) in the ion peak of m/z 922.8. Similarly, B indicates the presence of 18:0-22:6 Fmoc-pGPEtn molecular species at m/z 996.8. Part C indicates the presence of 20:4-20:4 Fmoc-GPEtn molecular species at m/z 1008.8, which is an internal standard for quantitation of mouse retina GPEtn molecular species. Part D indicates the presence of 18:0-22:6 Fmoc-GPEtn molecular species at m/z 1012.8.

For example, this technique has been used to identify and quantify GPEtn and lyso-GPEtn molecular species in a lipid extract of mouse retina.[47] Derivatization is carried out at room temperature by mixing each lipid extract in 100 μL of 1:1 (v/v) chloroform/methanol containing approximately 10 nmol of GPEtn with freshly prepared Fmoc-Cl solution in anhydrous chloroform (100 μL) to make a 1:1 molar ratio of GPEtn to Fmoc-Cl and incubating for 5 min. The reaction solution can then be diluted 20-fold with 1:1 (v/v)

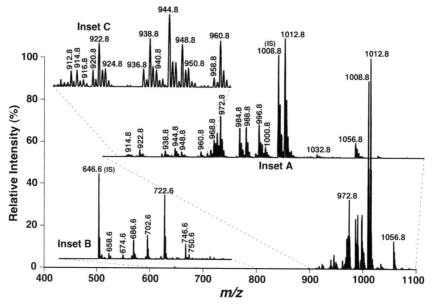

Figure 5.10 Tandem mass spectrometric analysis of a derivatized lipid extract of mouse retinas with Fmoc-Cl in the negative-ion mode by neutral loss of Fmoc moiety. Phosphoethanolamine-containing species were derivatized by addition of equimolar amount of Fmoc chloride. Neutral loss analysis of Fmoc-GPEtn (Inset a) and Fmoc-lyso-GPEtn (Inset b) was performed through coordinately scanning both the first and third quadrupoles with a mass difference (*i.e.*, neutral loss) of 222.2 u, corresponding to the neutral loss of a Fmoc moiety while collision activation was performed in the second quadrupole at collision energy of 30 eV and collision gas pressure of 1 mTorr. Inset c indicates the presence of many very low abundant GPEtn molecular species in the region. (Reprinted from Ref. 42 with permission from The American Society for Biochemistry and Molecular Biology, Inc., Copyright 2006.)

chloroform/methanol and directly infused into the ESI ion source for lipid analysis. Tandem mass spectrometric analysis of the derivatized lipid extract of mouse retinas through neutral loss of 222.2 u (corresponding to a Fmoc moiety) demonstrates over 40 ion peaks in the mass region over *m/z* 900 for Fmoc-GPEtn molecular ions (see the inset A of Figure 5.10) while only approximately 10 abundant ion peaks can be easily identified in the mass spectrum of non-derivatized GPEtn (spectrum not shown).[47]

Two-dimensional mass spectrometric analyses are performed to identify the acyl chain constituents of mouse retina Fmoc-GPEtn molecular species through analysis of all potential building blocks. These building blocks include the Fmoc moiety itself, all naturally occurring fatty acyl chains, and lys-oplamenylethanolamine (lyso-pGPEtn) (Figure 5.11). Therefore, the isobaric species of the ions and the regiospecificity of each molecular species are determined from the presence of multiple cross peaks and the analysis of intensity ratios of the related cross peaks as previously described.[48] Two-dimensional

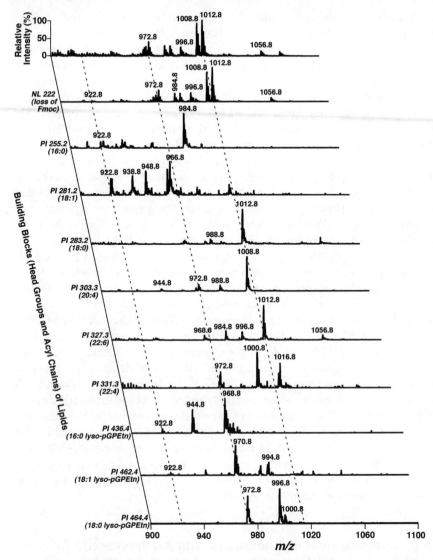

Figure 5.11 Representative two-dimensional ESI mass spectrum of a derivatized lipid extract of mouse retinas with Fmoc-Cl in negative-ion mode. Analyses of Fmoc-GPEtn building blocks including Fmoc moiety, fatty acyl carboxylaytes and lysoplamenylethanolamine derivative ions (lyso-pGPEtn) by precursor-ion (PI) scanning and neutral loss (NL) scanning were performed. "IS" denotes internal standard; (m:n) indicates an acyl chain containing m carbons and n double bonds. All mass spectral traces were displayed after normalization to the base peak in each individual spectrum. (Reprinted from ref. 42 with permission from The American Society for Biochemistry and Molecular Biology, Inc., Copyright 2006.)

mass spectrometric analysis demonstrates the identification of over 50 GPEtn molecular species in lipid extracts of mouse retina (see Table 1 of ref. 47). For example, a very low intensity ion peak at m/z 922.8 is crossed with fragment peaks in the scans of NL222.2, precursor ion (PI) 255.2, PI281.2, PI436.4, and PI462.4 (along the left broken line in Figure 5.11). These fragments indicate that this ion peak at m/z 922.8 contain both 16:0-18:1 and 18:1-16:0 Fmoc-plasmenylethanolamines (Fmoc-pGPEtn) isobaric molecular species. The ratios of fragment intensities of both carboxylates and lyso-pGPEtn were redundantly used to estimate the composition of these isobaric species.

The tandem mass spectrometric analysis through neutral loss of 222.2 u also demonstrates the presence of several very low abundant molecular species in the mass region of m/z 600 to 800, corresponding to Fmoc-lyso-GPEtn molecular species (Inset B of Figure 5.10). This mass spectrum represents the first profile of lyso-GPEtn molecular species directly from a lipid extract of a biological sample. This result also indicates the power of derivatization for lipid analysis by shotgun lipidomics. In addition, by this approach, the GPEtn molecular species are shifted to a higher m/z region that could be selected for by employing the desired derivatization reagent in which overlaps with other endogenous lipid constituents in the lipid extracts are rare.

Derivatization of phosphoethanolamine-containing lipids with Fmoc-Cl also possesses other advantages. For example, this is quite useful for analysis of pGPEtn and lyso-pGPEtn molecular species since these molecular species are very sensitive to oxidation, silica-catalyzed vinyl ether cleavage, or extensive loss during chromatography. The procedure of the derivatization is both simple and effective. Direct quantitation by using tandem mass spectrometric analysis through neutral loss of Fmoc moiety may need multiple internal standards as previously described.[49,50] However, this approach, when used in combination with shotgun lipidomics based on intrasource separation and multi-dimension mass spectrometry, can be used to profile and quantitate the large majority, if not all, of GPEtn and lyso-GPEtn molecular species directly from a lipid extract of a biological sample.

5.5.2 Derivatization of Cholesterol with Methoxyacetic Acid for Quantitation by Shotgun Lipidomics

Cholesterol and cholesterol esters are quite non-polar. Therefore, their ionization efficiencies in an ESI ion source are very low. It has been reported to derive cholesterol to choleterol-3-sulfate by a sulfur trioxide-pyridine complex, thereby increasing in its ionization efficiency.[43] Previously, we determined the cholesterol mass content in a biological sample by using an enzymatic methodology with a cholesterol assay kit.[51] Recently, we have developed an alternative derivatization approach by using methoxyacetic acid, in which cholesterol is converted to cholesteryl methoxyacetate which can be readily detected as [cholesteryl methoxyacetate + MeOH + Li]$^+$ ion in the positive-ion mode in the presence of LiOH.[52] Figure 5.12a shows a mass spectrum acquired from a derivatized

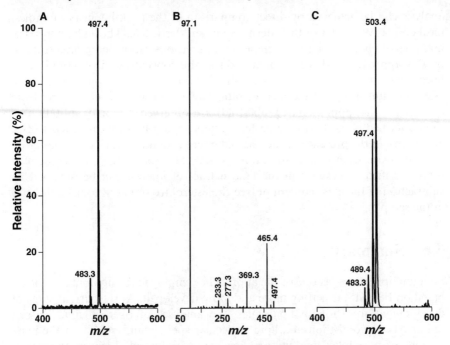

Figure 5.12 Mass spectrometric analysis of cholesterol after derivatization. Part a
shows a mass spectrum of a derivatized cholesterol solution (1 pmol/µl)
with methoxyacetic acid. Part b shows a product-ion ESI spectrum of a
selected ion at m/z 497.4 that is shown in the ESI/MS analysis of the lipid
extract in part a, after derivatization with methoxyacetyl acid identified
it as lithiated cholesteryl methoxyacetate plus a methanol molecule
(see Ref. 52 for proposed fragmentational pathways). Part c shows a
precursor-ion ESI spectrum of m/z 97.1 which was acquired (from the
derivatized human serum lipid extract with methoxyacetic acid in the
presence of LiOH) in the positive-ion mode and used for quantitation of
cholesterol in the lipid extract.

cholesterol solution (1 pmol/µL) with methoxyacetic acid as previously
described,[52] demonstrating an abundant ion at m/z 497.4 (corresponding to
[cholesteryl methoxyacetate + MeOH + Li]$^+$) and a low-abundance ion at
m/z 483.3 (corresponding to [cholesteryl methoxyacetate + H$_2$O + Li]$^+$) Most
importantly, collision-induced dissociation of this molecular ion yields a very
intense fragment at m/z 97.1, corresponding to [methoxyacetate + MeOH + Li]$^+$
(Figure 5.12b). Therefore, this abundant fragment can be used to quantitate the
cholesterol mass content in any lipid extracts after derivatization through
precursor-ion scanning of this fragment as long as an isotope-labeled cholesterol
internal standard co-exists in the samples. Figure 5.12c shows an example of
precursor-ion scanning of a derivatized lipid extract from a human serum
sample. The tandem mass spectrum displays two well-resolved precursor ion
peaks at m/z 497.4 and 503.4 of [cholesteryl methoxyacetate + MeOH + Li]$^+$ and
[d$_6$-cholesteryl methoxyacetate + MeOH + Li]$^+$, respectively. Thus, quantitative

analysis can be achieved by direct comparison of the peak intensity of cholesterol derivative to that of the internal standard derivative. Through analyses of mixtures of cholesterol and d_6-cholesterol at various ratios and concentrations by this approach, a detection limitation at the concentration of 1 fmol/μL, a linear dynamic concentration range from 1 fmol/μL to 100 pmol/μL, and a linear dynamic range of the relative ratio of cholesterol *vs.* d_6-cholesterol from 0.05 to 20 (*i.e.*, approximately 400-fold) are observed.[52] For quantitation of total esterified cholesterols, cholesterol esters should be first converted to free cholesterol in the presence of the selected internal standard (*i.e.*, d_6-cholesterol). After derivatization and quantitation as described above, the mass content of total esterified cholesterol in the lipid extract of interest can be obtained by subtraction of the mass content of free cholesterol from that of total cholesterol in the sample.

5.6   Summary

Shotgun lipidomics, as one of the approaches employed for lipidomics analysis based on ESI/MS, is getting more and more developed. At the current stage of development for shotgun lipidomics, the analyses of over 20 lipid classes, hundreds to over a thousand lipid molecular species, and >95% of the mass content of a cellular lipidome can be readily achieved. Even at this stage, shotgun lipidomics has already showed its powerful and broad applications in biological, pathological, and pathophysiological studies.[38,52–57] It can be anticipated that identification of many biochemical mechanisms underlying lipid metabolism critical to disease states will be increasingly uncovered as the further development of shotgun lipidomics and its penetration into lower and lower abundance regions of mass contents of individual lipid molecular species.

Acknowledgements

This work was supported by National Institute of Health grant P01 HL57278 and National Institute on Aging Grant R01 AG23168. The authors are grateful to Dr Kui Yang for her technical help.

References

1. X. Han and R. W. Gross, *J. Lipid Res.*, 2003, **44**, 1071.
2. M. Lagarde, A. Geloen, M. Record, D. Vance and F. Spener, *Biochim. Biophys. Acta*, 2003, **1634**, 61.
3. L. Feng and G. D. Prestwich (eds), *Functional Lipidomics*, CRC Press, Taylor & Francis Group, Boca Raton, FL, 2006.
4. M. M. Mossoba, J. K. G. Kramer, J. T. Brenna and R. E. Mcdonald (eds), *New Techniques and Applications in Lipid Analysis and Lipidomics*, AOCS Press, Champaign, Illinois, 2006.

5. E. Fahy, S. Subramaniam, H. A. Brown, C. K. Glass, A. H. Merrill, Jr., R. C. Murphy, C. R. Raetz, D. W. Russell, Y. Seyama, W. Shaw, T. Shimizu, F. Spener, G. Van Meer, M. S. Vannieuwenhze, S. H. White, J. L. Witztum and E. A. Dennis, *J. Lipid Res.*, 2005, **46**, 839.
6. S. Kenchaiah, J. C. Evans, D. Levy, P. W. Wilson, E. J. Benjamin, M. G. Larson, W. B. Kannel and R. S. Vasan, *N. Engl. J. Med.*, 2002, **347**, 305.
7. M. Pulfer and R. C. Murphy, *Mass Spectrom. Rev.*, 2003, **22**, 332.
8. W. J. Griffiths, *Mass Spectrom. Rev.*, 2003, **22**, 81.
9. R. Welti and X. Wang, *Curr. Opin. Plant Biol.*, 2004, **7**, 337.
10. W. C. Byrdwell, *Oily Press Lipid Library*, 2003, **16**, 171.
11. J. Schiller, R. Suss, J. Arnhold, B. Fuchs, J. Lessig, M. Muller, M. Petkovic, H. Spalteholz, O. Zschornig and K. Arnold, *Prog. Lipid Res.*, 2004, **43**, 449.
12. X. Han and R. W. Gross, *Mass Spectrom. Rev.*, 2005, **24**, 367.
13. F.-F. Hsu and J. Turk in "Modern Methods for Lipid Analysis by Liquid Chromatography/Mass Spectrometry and Related Techniques", ed. W. C. Byrdwells, AOCS Press, Champaign, Ill, 2005, 61.
14. P. T. Ivanova, S. B. Milne, J. S. Forrester and H. A. Brown, *Mol. Interv.*, 2004, **4**, 86.
15. X. Han and R. W. Gross, *Expert Rev. Proteomics*, 2005, **2**, 253.
16. X. Han and R. W. Gross, *Proc. Natl. Acad. Sci. U. S. A.*, 1994, **91**, 10635.
17. X. Han, R. A. Gubitosi-Klug, B. J. Collins and R. W. Gross, *Biochemistry*, 1996, **35**, 5822.
18. X. Han and R. W. Gross, *Anal. Biochem.*, 2001, **295**, 88.
19. X. Han, J. Yang, H. Cheng, H. Ye and R. W. Gross, *Anal. Biochem.*, 2004, **330**, 317.
20. L. Tang and P. Kebarle, *Anal. Chem.*, 1991, **63**, 2709.
21. M. G. Ikonomou, A. T. Blades and P. Kebarle, *Anal. Chem.*, 1991, **63**, 1989.
22. S. J. Gaskell, *J. Mass Spectrom.*, 1997, **32**, 677.
23. R. W. Gross and B. E. Sobel, *J. Chromatogr.*, 1980, **197**, 79.
24. C. J. Delong, P. R. S. Baker, M. Samuel, Z. Cui and M. J. Thomas, *J. Lipid Res.*, 2001, **42**, 1959.
25. X. Han, K. Yang, J. Yang, K. N. Fikes, H. Cheng and R. W. Gross, *J. Am. Soc. Mass Spectrom.*, 2006, **17**, 264.
26. X. Han, H. Cheng, D. J. Mancuso and R. W. Gross, *Biochemistry*, 2004, **43**, 15584.
27. D. Schwudke, J. Oegema, L. Burton, E. Entchev, J. T. Hannich, C. S. Ejsing, T. Kurzchalia and A. Shevchenko, *Anal. Chem.*, 2006, **78**, 585.
28. M. Koivusalo, P. Haimi, L. Heikinheimo, R. Kostiainen and P. Somerharju, *J. Lipid Res.*, 2001, **42**, 663.
29. M. Schlame, D. Rua and M. L. Greenberg, *Prog. Lipid Res.*, 2000, **39**, 257.
30. J. B. Mcmillin and W. Dowhan, *Biochim. Biophys. Acta*, 2002, **1585**, 97.
31. I. M. Cristea and M. Degli Esposti, *Chem. Phys. Lipids*, 2004, **129**, 133.
32. G. M. Hatch, *Biochem. Cell Biol.*, 2004, **82**, 99.
33. P. Vreken, F. Valianpour, L. G. Nijtmans, L. A. Grivell, B. Plecko, R. J. Wanders and P. G. Barth, *Biochem. Biophys. Res. Commun.*, 2000, **279**, 378.

34. M. Schlame, J. A. Towbin, P. M. Heerdt, R. Jehle, S. Dimauro and T. J. Blanck, *Ann. Neurol.*, 2002, **51**, 634.
35. F. Valianpour, R. J. Wanders, H. Overmars, P. Vreken, A. H. Van Gennip, F. Baas, B. Plecko, R. Santer, K. Becker and P. G. Barth, *J. Pediatr.*, 2002, **141**, 729.
36. P. G. Barth, F. Valianpour, V. M. Bowen, J. Lam, M. Duran, F. M. Vaz and R. J. Wanders, *Am. J. Med. Genet. A*, 2004, **126**, 349.
37. Z. Gu, F. Valianpour, S. Chen, F. M. Vaz, G. A. Hakkaart, R. J. Wanders and M. L. Greenberg, *Mol. Microbiol.*, 2004, **51**, 149.
38. X. Han, J. Yang, H. Cheng, K. Yang, D. R. Abendschein and R. W. Gross, *Biochemistry*, 2005, **44**, 16684.
39. F. Valianpour, R. J. Wanders, P. G. Barth, H. Overmars and A. H. Van Gennip, *Clin. Chem.*, 2002, **48**, 1390.
40. G. C. Sparagna, C. A. Johnson, S. A. Mccune, R. L. Moore and R. C. Murphy, *J. Lipid Res.*, 2005, **46**, 1196.
41. M. Schlame, M. Ren, Y. Xu, M. L. Greenberg and I. Haller, *Chem. Phys. Lipids*, 2005, **138**, 38.
42. X. Han, K. Yang, J. Yang, H. Cheng and R. W. Gross, *J. Lipid Res.*, 2006, **47**, 864.
43. R. Sandhoff, B. Brugger, D. Jeckel, W. D. Lehmann and F. T. Wieland, *J. Lipid Res.*, 1999, **40**, 126.
44. S. Liu, J. Sjövall and W. J. Griffiths, *Rapid Commun. Mass Spectrom.*, 2000, **14**, 390.
45. W. J. Griffiths, S. Liu, G. Alvelius and J. Sjövall, *Rapid Commun. Mass Spectrom.*, 2003, **17**, 924.
46. D. W. Johnson, *J. Mass Spectrom.*, 2001, **36**, 277.
47. X. Han, K. Yang, H. Cheng, K. N. Fikes and R. W. Gross, *J. Lipid Res.*, 2005, **46**, 1548.
48. X. Han and R. W. Gross, *J. Am. Soc. Mass Spectrom.*, 1995, **6**, 1202.
49. B. Brugger, G. Erben, R. Sandhoff, F. T. Wieland and W. D. Lehmann, *Proc. Natl. Acad. Sci. U. S. A.*, 1997, **94**, 2339.
50. R. Welti, W. Li, M. Li, Y. Sang, H. Biesiada, H.-E. Zhou, C. B. Rajashekar, T. D. Williams and X. Wang, *J. Biol. Chem.*, 2002, **277**, 31994.
51. X. Han, H. Cheng, J. D. Fryer, A. M. Fagan and D. M. Holtzman, *J. Biol. Chem.*, 2003, **278**, 8043.
52. H. Cheng, X. Jiang and X. Han, *J. Neurochem.*, 2007, **101**, 57.
53. X. Su, X. Han, D. J. Mancuso, D. R. Abendschein and R. W. Gross, *Biochemistry*, 2005, **44**, 5234.
54. L. J. Pike, X. Han and R. W. Gross, *J. Biol. Chem.*, 2005, **280**, 26796.
55. X. Han, *Curr. Alz. Res.*, 2005, **2**, 65.
56. H. Cheng, S. Guan and X. Han, *J. Neurochem.*, 2006, **97**, 1288.
57. H. Han, J. Yang, K. Yang, Z. Zhao, D. R. Abendschein and R. W. Gross, *Biochemistry*, 2007, **46**, 6417.

CHAPTER 6

Neutral Lipidomics and Mass Spectrometry

ROBERT C. MURPHY, MARK FITZGERALD AND
ROBERT M. BARKLEY

University of Colorado at Denver and Health Sciences Center, Department
of Pharmacology, Mail Stop 8303, 12801 E. 17th Avenue, P.O. Box 6511,
Aurora, CO 80045, USA

6.1 Introduction

There are a host of hydrophobic compounds present within cells that can be
termed neutral lipids. In large part, this designation recognizes lack of ionic
charge on these molecules in aqueous solution as well as biosynthetic pathways
based on carbocation and carbanion condensations that assemble lipid substances. Neutral lipids (NL) by this definition correspond to uncharged lipid
substances which migrate rapidly with relatively nonpolar organic solvent
systems. Such compounds include alcohols, esters, and even ceramides. All of
these neutral lipids can be components of lipidomic studies; however, there
have only been three general classes of neutral esters that have been the focus of
studies of what can be called neutral lipidomics. These include the glyceryl
lipids such as triacylglycerols, the cholesterol esters (CE), and wax esters (WE).
This is not to say that other neutral lipids such as squalene (a hydrocarbon
intermediate in the biosynthesis of steroids) or dolichols (long-chain alcohols)
are not important and in the future these likely will themselves become targets
of lipidomic studies. This chapter will focus on the application of mass
spectrometry to determine the structure of the common neutral lipids described
above as well as techniques that have emerged to study the different molecular
species of neutral lipids present within mammalian cells.

6.1.1 Neutral Lipid Analysis and Lipidomics

The emergence of the strategy to study complexity of cellular biochemistry
through global analysis of components within the cell has resulted in the

development of various ways to assess events taking place within the cell from the standpoint of qualitative analysis (structural complexity) as well as quantitative analysis (relative changes in abundance as related or unrelated processes take place). As new approaches have developed, it became apparent that it was necessary to explore alternative methods to analyze neutral lipidsas individual molecular species if one was to assess their regulation both in synthesis and metabolism as well as the interaction of lipid components of biochemical pathways as a result of intracellular and even extracellular events. Mass spectrometry has emerged as one of the most generally applicable techniques, yet ionization of neutral lipids is a major concern since it is absolutely required to form gas-phase ions for the mass spectrometric experiment.

Neutral lipids constitute a challenging class of lipids in which to engage studies of lipidomics from several standpoints. For example, triacylglycerols are quite abundant in certain cells and fluids, such as plasma, but they can be difficult to analyze because of the enormous complexity of molecular species present within cells. Furthermore, electrospray ionization, used quite successfully for the study of phospholipid molecular species, is a somewhat insensitive ionization strategy and the generally applicable ionization method for neutral lipids, electron ionization, causes extensive decomposition that limits molecular weight determination.

This chapter will focus attention on recent advances in mass spectrometry that have been developed to specifically address lipidomic-like questions for neutral lipids. This chapter will focus on glyceryl lipids, cholesterol esters, and wax esters – which also present some of the most complex mixtures of lipids that are made within cells.

6.2 Glyceryl Lipids

6.2.1 Overview

The esterification of fatty acids to the hydroxyl groups of glycerol results in three major NL species: the triacylglycerols (TAGs), diacylglycerol (DAGs), and monoacylglycerols (MAGs) (Figure 6.1). The TAGs play an important role as a major source of energy reserve stored in a cell when they are metabolized by a lipase to release fatty acids followed by subsequent generation of ATP through β-oxidation. TAGs also are a vehicle by which dietary fatty acids are transported from the intestine to the liver for packaging in lipoproteins and subsequent distribution to all cells for storage. Furthermore, it is now known that virtually all cells contain reservoirs of TAGs in a structure called the adiposome,[1] and a complex series of events lead to the deposition of TAGs within the adiposome as well as turnover of the fatty acyl components. While DAGs can serve as intermediates both in the synthesis and metabolism of TAGs,[2] the diacylglycerols can also be important signaling molecules when they are produced through the hydrolysis of phospholipids by phospholipase C.[3] It is now known that a typical cell can have several hundreds of different

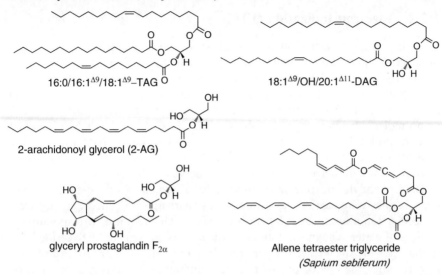

Figure 6.1 Representative structures of triacylglycerols (TAG), diacylglycerol (DAG), and monoacylglycerol (2-AG), the glycerol prostaglandin $F_{2\alpha}$,[4] and a naturally occurring allene tetraester triacylglycerol.[6]

molecular species of both DAGs and TAGs within their lipidome, which challenges the power of all of the strategies developed for their analysis. The monoacylglycerols can also be metabolites and precursors of the DAGs, but recent evidence has shown that a unique MAG, 2-arachidonoyl glycerol, can be a substrate for prostaglandin H synthase-2 to generate a series of rather interesting glycerol prostaglandins (Figure 6.1).[4] Studies of the molecular species of these glyceryl lipids have now emerged as an important goal for studies of lipidomics. In addition to the expected simple glyceryl esters, more complex structures have been discovered, including oxidized species[5] and even a very unexpected allyl tetraester (Figure 6.1).[6]

There have been a large number of diverse methods developed for the analysis of glyceryl lipids found in biological samples in the past 50 years. Very sophisticated techniques of chromatographic separation have emerged which are covered by recent reviews.[7–10] Also, analysis by nuclear magnetic spectroscopy has been explored.[11] However, in the past, most method development has largely dealt with the analysis of edible oils and fats that appear in diets. Advances made in this area have had an important impact on the development of specific lipidomic strategies; however, since they often deal with large quantities of glycerol esters available for analysis, analytical sensitivity has not limited their application. For example, detection of species eluting from HPLC columns by UV absorbance has been employed.[12] Nonetheless, excellent reviews have emerged in the literature which detail many aspects of gas chromatography (GC),[13] unique techniques of high pressure liquid chromatography (HPLC),[14] as well as various means to employ thin-layer chromatography as the final analytical tool.[15]

6.2.2   Electron Ionization (EI)

Mass spectrometry with electron ionization was initially applied to the analysis of glyceryl lipids and a review of this even appeared as early as 1960.[16] Triacylglycerols yield a characteristic EI mass spectrum dominated by ions corresponding to the loss of each fatty acyl substituent as a neutral carboxylic acid yielding one, two, or three diglyceride-type product ions (Figure 6.2). Little or no molecular ion is typically observed for TAGs, DAGs, or MAGs which renders molecular weight determination by EI difficult. Nonetheless, it is possible to discern the complexity of molecular species of TAGs by careful analysis of the numerous fragment ions observed.[17]

The EI behavior of triacylglycerols has been studied in some detail; however, these studies were carried out a number of years ago. In general, a very weak molecular ion abundance [M$^{\cdot}$] is observed as well as an ion corresponding to the loss of water. The most abundant ions have been termed "diglyceride ions" in that they are the charged species retaining two fatty acyl groups esterified to glycerol after the loss of one of the three different fatty acyl groups as a free carboxylic acid or the simple loss of one of the carboxyl radicals. The former ion results in an even mass ion, whereas the latter ion appears at an odd mass-to-charge ratio. There are also ions present which correspond to the carboxylic acid itself, either as an acylium ion or the acylium ion $RCO^+ + 74$ u and $RCO^+ + 128$ u. The structures of these ions have been suggested (Scheme 6.1).[18] In the case of diglycerides and monoglycerides, these are typically studied after formation of the trimethylsilyl ether derivative. While there are some ions corresponding to the fatty acyl substituents, most of the mass spectrum is dominated by fragment ions of the trimethylsilyl group.[19]

The emergence of combined gas chromatography/mass spectrometry (GC/MS), in particular capillary GC/MS, has had an enormous impact on the analysis of triacylglycerides. Even capillary GC could not completely resolve

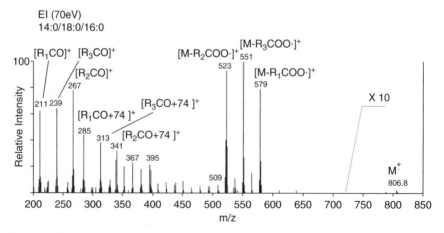

Figure 6.2   Electron ionization mass spectra (70 eV, positive ions) of the triacylglycerol 14:0/18:0/16:0, redrawn and annotated from Ref. 18.

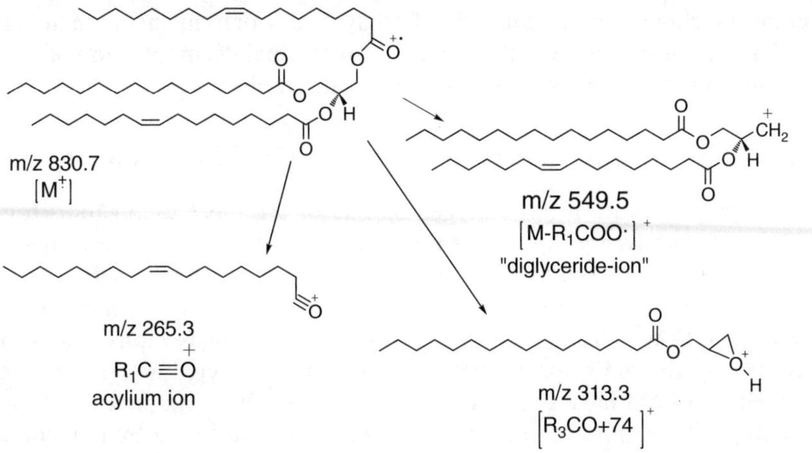

m/z 830.7
[M$^+$]

m/z 549.5
[M-R$_1$COO$\cdot$]$^+$
"diglyceride-ion"

m/z 265.3
R$_1$C≡O
acylium ion

m/z 313.3
[R$_3$CO+74]$^+$

Scheme 6.1

the complex mixture of triglycerides present in cellular TAGs, yet it was possible to identify the major molecular species of components present based upon their diglyceride fragment ions.[20] This strategy remains one of the more useful techniques, largely because of the ultimate sensitivity of modern GC/MS instruments and the extraordinary separation power of capillary GC. It is also possible to couple 2-dimensional chromatography with GC to further enhance separation of individual molecular species.[21] Specific techniques have been developed and summarized by Andrikopoulos[12] in determining regioisomers based upon GC/MS techniques.

6.2.3 Negative-ion Chemical Ionization (NCI)

The emergence of chemical ionization, and in particular chemical ionization using ammonia as reagent gas (NH$_2^-$) and negative ion analysis, added considerably to the analytical power of GC/MS and gas-phase techniques for the analysis of TAGs. The power of capillary GC, now coupled with a considerably less energetic ionization technique, resulted in observable abundant molecular ion species as well as fragment ions such as carboxylate anions, revealing the fatty acyl substituent esterified to a glycerol backbone. Computer aided interpretation of the data has been recently described,[22] with some suggestion of the gas-phase ion chemistry responsible for the observed ions. While the capillary GC has substantial resolving power because of its high number of theoretical plates, it does require very high temperatures to pass some of the higher molecular weight triacylglycerols through the capillary GC column. This results in decomposition of some molecular species; most prone are the polyunsaturated fatty acyl substituted TAGs. While the GC/MS approach, especially with negative-ion CI, has been a very valuable tool for the analysis of vegetable oils and animal fats, it is not likely to be the first method of choice for the analysis of triglycerides present in the mammalian cellular lipidome because of the potential for thermal decomposition of specific (polyunsaturated) species during gas chromatography.

6.2.4   Electrospray Ionization/Tandem Mass Spectrometry

Electrospray ionization has emerged as one of the most useful tools for the analysis of the glyceryl lipids in spite of the fact that these neutral lipids are neither positive nor negative ions in solution and must be charged by an ionic species present in the electrospray mobile phase. Various attachment ions with TAGs and DAGs have been used to generate gas-phase ions that can be analyzed by the mass spectrometer. In general, two separate strategies have emerged and both have been used in lipidomic studies. The first involved the formation of lithium ion adducts of glycerol lipids followed by tandem mass spectrometric analysis by collision-induced decomposition. The second strategy has been to form the ammonium ion adduct. Both of these adducts have both strengths and weaknesses in the mass spectrometric experiment, which should be considered in the choice of a method for neutral lipid, lipidomic studies.

6.2.4.1   Lithium Adducts

Detailed studies of the mass spectrometry of alkali metal adducts of TAGs after collision-induced decomposition (CID) have appeared. Chen and Gross[23] studied the high-energy collisional activation of both $[M + NH_4]^+$ and $[M + Na]^+$ adducts of triacylglycerols and found that when one pure molecular species was in the ion source of the mass spectrometer, there was sufficient information to determine positions of fatty acyl groups on the glycerol back-bone as well as positions of double bonds within the fatty acyl group by remote site fragmentation mechanisms.[23] Further studies of the lithiated adducts of triacylglycerols were carried out in the tandem quadrupole instrument and detailed mechanisms of decomposition of the CID ions were elucidated.[24] In general, the lithiated molecular ions of triacylglycerols undergo an abundant loss of each of the fatty acids as a neutral carboxylic acid (Scheme 6.2) and loss of each fatty acyl group as a lithium carboxylate salt to form the most abundant product ions observed in the tandem mass spectrometer (Figure 6.3). Again, the relative abundances of the resultant diglyceride-like ions permitted assignment of the fatty acid substituents and even some information about double bond locations in polyunsaturated fatty acids.[24]

In one of the first lipidomic studies of neutral lipids, lithiated adducts of triacylglycerols were used as the means by which one could use electrospray ionization to directly analyze the complex mixture of TAGs isolated from a biological sample.[25] Since there was no chromatographic separation of the lipid classes, let alone molecular species, the mass spectrometer was used as a means by which one could separate molecular ion species of TAGs and obtain quantitative information through the use of a single internal standard. In this approach, developed by Han and Gross,[25,26] neutral loss scans for individual

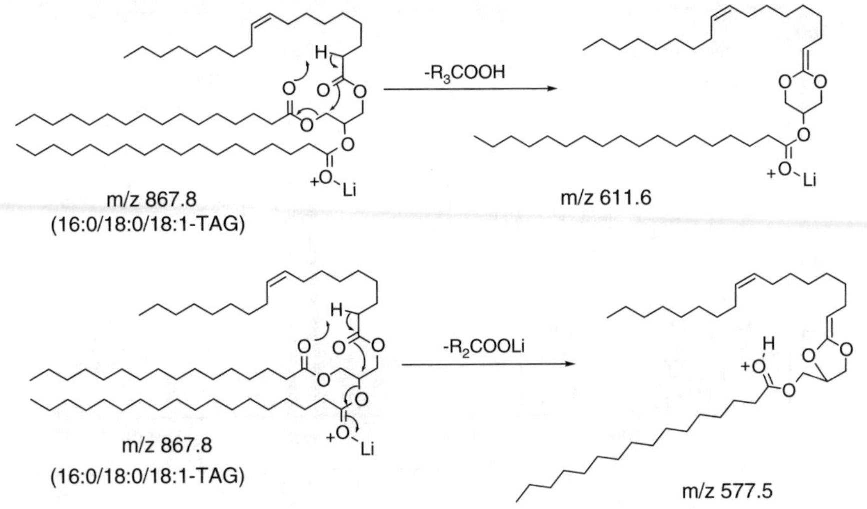

m/z 867.8
(16:0/18:0/18:1-TAG)

-R₃COOH → m/z 611.6

m/z 867.8
(16:0/18:0/18:1-TAG)

-R₂COOLi →

m/z 577.5

Scheme 6.2

fatty acids were used to assign fatty acyl components present within each observed molecular ion species. For example, in Figure 6.4 the neutral loss of 16:0, 18:1, and 18:0 can be seen as unique product ions following the neutral loss of 256 u (NL of 16:0), 282 u (NL of 18:1), and 284 u (NL of 18:0) from the ion at m/z 840 after collisional activation. After correction for the carbon-13 isotopes present in each molecular species, it was possible to assess a rather constant response for each molecular species over the range of fatty acids expected in the mammalian cells. In addition, a correction for the stable isotope of lithium-6 needs to be considered since lithium is a mixture of lithium-6 (7.5%) and lithium-7 (92.5%).

A refinement of this technique was embodied in the "shotgun" strategy where a crude lipid extract of cells or tissue was carried out followed by the addition of LiOH to the sample.[27,28] This addition of base serves several uses. The first was to reduce signals from acidic lipids in the crude extract because they would become negatively charged and not form positive ions. The electrically neutral lipids in this basicified extract could be observed, which included phosphatidylcholine as well as the glyceryl lipids as [M + Li]⁺. An example of this technique applied to a crude biological extract is shown in Figure 6.5. While clear successes with this approach have been obtained, there has to be some concern about the well-known suppression effect in electrospray ionization and whether or not the complex mixture of minor molecular species can in fact be observed. However, for major species clearly this is a very attractive approach.

6.2.4.2 Ammonium Ion Adducts

The second charging species which has been employed for lipidomics studies is that of the ammonium ion for TAGs and DAGs.[29] In order to drive adduct

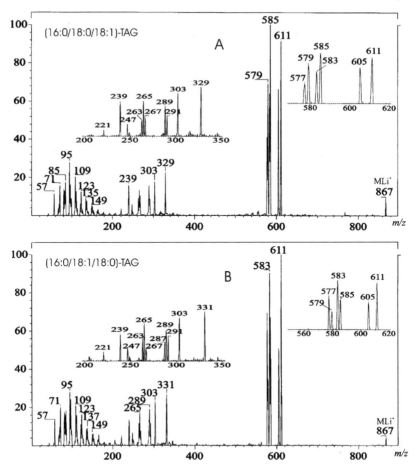

Figure 6.3 Tandem mass spectra of lithiated adducts of two positionally isomeric TAG species that contain three distinct fatty acid substituents (a) 16:0/18:0/18:1-TAG and (b) 16:0/18:1/18:0-TAG. (Reproduced with permission of the American Society for Mass Spectrometry.[24])

formation and suppress favorable alkali metal ion attachment, the electrospray mobile phase has 5–10 mM concentration of ammonium salts such as ammonium acetate or ammonium formate present so that the ammonium ion dominates as a possible charging species. In this case, an abundant ion corresponding to the $[M+NH_4]^+$ adduct can be observed. Collisional activation of the $[M+NH_4]^+$ results in abundant formation of diglyceride ions (Scheme 6.3) corresponding to the loss of each of the fatty acyl groups as a protonated carboxylic acid as well as the loss of gaseous ammonia (Figure 6.6a).

Recently, the complexity of mixtures of TAGs isolated from mammalian cells (RAW 264.7 cells) was demonstrated using tandem mass spectrometry of the $[M+NH_4]^+$ adduct ion.[30] In this case, multiple species were found at each

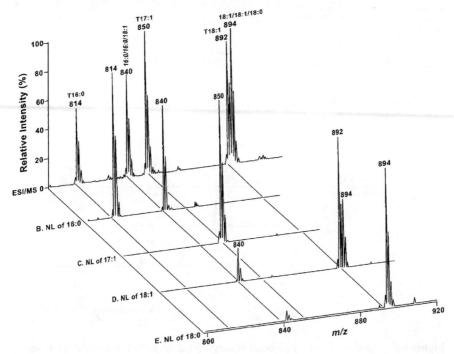

Figure 6.4 Neutral loss scanning of specific fatty acyl groups while equal molar of 16:0/16:0/16:0, 18:1/16:0/16:0, 17:1/17:1/17:1, 18:1/18:1/18:1, and 18:0/18:1/18:0-TAGs (2 pmol each and 200 µL containing 50 mM LiOH) were infused into a tandem quadrupole mass spectrometer. (a) Summed total ion current (positive ions), (b) neutral loss of 256 u (16:0), (c) neutral loss of 268 u (17:1), (d) neutral loss of 282 u (18:1), and (e) neutral loss of 284 u (18:0). All NL mass spectra were displayed after normalization in the individual spectrum. (Reproduced with permission from Academic Press.[25])

and every observed even mass-to-charge ratio. These ions revealed the abundance of isobaric TAG and DAG molecular species corresponding to the same nominal number of carbon fatty acyl atoms and total double bonds in the total fatty acyl groups. For example, collisional activation of the ion at m/z 824.7 (from the complex mixture of triglycerides from RAW cells) resulted in the formation of at least five diglyceride-type ions (Figure 6.6b), corresponding to the loss of five unique fatty acids and NH_3. This meant that multiple compounds must be present within a single mass-to-charge peak since any individual trisubstituted TAG would yield only three, but not five, diglyceride-type ions.

Further studies using MS^3 of the resulting diglyceride ions revealed the exact nature of each individual component present.[31] In this strategy, the resultant diglyceride ion, which arises following loss of one of the fatty acids, was collisionally activated in an ion trap mass spectrometer and the resulting ion product corresponded to either an acylium ion ($RC\equiv O^+$) or the fatty acid

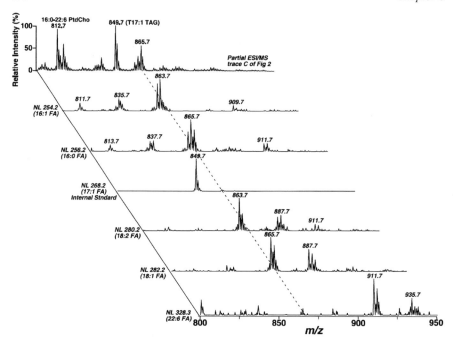

Figure 6.5   Analysis of triacylglycerol molecular species by 2-dimensional electrospray
ionization of a mouse liver chloroform extract. After the addition of LiOH
to the sample, a first dimension spectrum was obtained (top trace) in a
positive ion mode. Neutral loss scanning (NL) of all naturally occurring
fatty acyl chains was utilized to confirm the molecular species assignments,
identify isobaric molecular species, and quantify triacylglycerol molecular
species by comparison with a selected internal standard (T17:1 TAG).
(Reproduced with permission from Elsevier, Inc.[26])

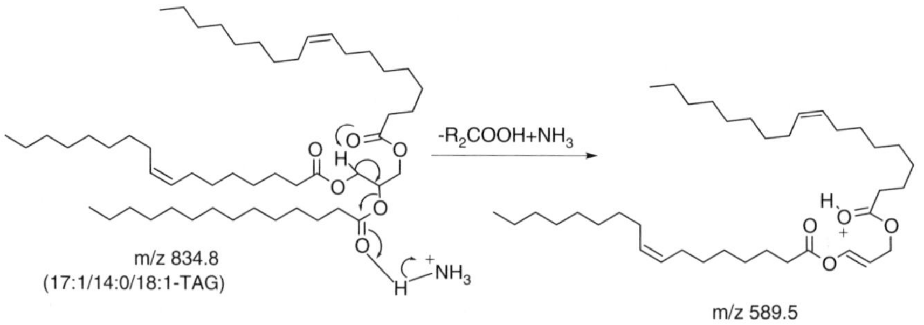

Scheme 6.3

containing a portion of the glycerol backbone (RCO + 74). Detailed studies
using isotope labeled TAGs were used to reveal the exact mechanism for the
appearance of these MS^3 product ions (Scheme 6.4). With identification of each
acylium ion, which had a mass that corresponded to the known fatty acyl

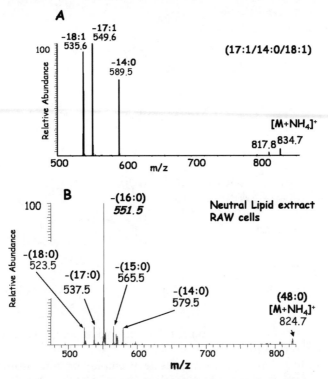

Figure 6.6 Tandem mass spectra of triacylglycerol molecular species after electro-spray ionization as the ammonium adduct $[M+NH_4]^+$. (a) Synthetic triacylglycerol, 17:1/14:0/18:1, and collisional decomposition of m/z 834.7. (b) Product ions from m/z 824.7 present in the neutral lipid extract of RAW 264.7 cells using a tandem quadrupole mass spectrometer.

components present in the mixture (*e.g.*, m/z 265 for oleoyl), a corresponding neutral loss took place that in turn identified the third and final fatty acyl substituent in the triglyceride. In summary, the first MS/MS revealed one of the fatty acyl groups and the second stage of tandem mass spectrometry (MS^3) of the product ion revealed the second fatty acyl group from the observed ions and the third fatty acyl group from the corresponding neutral loss from the diglyceride ion. In this way, identification of each triglyceride molecular species present in complex mixtures could be achieved through sequential study of each $[M+NH_4]^+$ and its diglyceride-like product ion.

The unique finding of such MS^3 studies was that for mammalian TAGs, each observed mass-to-charge ratio defines the total sum of the fatty acyl carbons with their double bonds, but as many as 50 different components could be isobaric.[30] Thus, any method measuring a single ion (*e.g.* methods that measure only $[M+Li]^+$, $[M+NH_4]^+$ or $M^{\cdot}$) without some other means of separation does not uniquely measure one component, but a mixture of several components if the glyceryl lipids are isolated from cellular adiposomes. Therefore, to get an idea of what fatty acyl groups are actually present, one needs to have

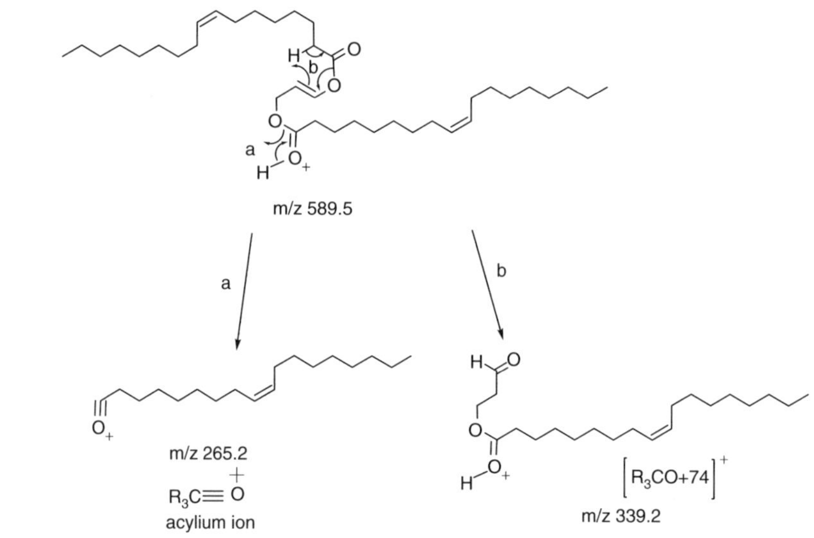

Scheme 6.4

an alternative strategy to gain some insight into what components are present within each observed ion.

Our laboratory has developed a strategy employing nine isotope-labeled internal standards where each deuterium atom is attached to the glycerol backbone (D_5-TAGs) with 18 different fatty acyl groups esterified (Table 6.1). With a series of neutral loss scans for each target fatty acyl group, one can examine the complex mixture of TAGs present in an extract and reveal those $[M + NH_4]^+$ ions that contain specific fatty acids. An example of this is shown in Figure 6.7a, which reports the profile of neutral losses of 18 different fatty acids using 9 different TAG internal standards. Precursor molecular ions (Figure 6.7b) for these neutral losses reveal the abundance of molecular ion species containing a specific fatty acyl group (in this case containing esterified palmitic acid). Since deuterium-labeled internal standards were employed in this strategy, the abundance for each molecular species could be compared to the observed abundance for the same neutral loss from one of the 9 internal standards. While this method is not intended to provide absolute quantitative numbers, it is an accurate and precise method by which one can assess relative changes in the TAG and DAG molecular species as one alters biochemical events taking place in the same cell population. The only caveat is that equal amounts of the internal standard must be added to each and every sample that is a part of the comparative study. The weakness is that the abundance ratios cannot be compared between different fatty acyl containing molecular species.

Detailed studies of the behavior of TAGs during electrospray ionization and molecular ion response as the $[M + NH_4]^+$ adducts have been carried out in a series of studies by Evans and co-workers.[31,32] In these studies a large number of TAG standards were synthesized and the collision-induced decomposition

Table 6.1 Deuterium labeled internal standards available for quantitative mass spectrometric analysis of glycerol lipids. In each case the isotope label is on the glycerol carbon atoms (D_5-glycerol).

		m/z	*Fatty acyl*		*Fatty acyl*	
D_5-DAG	$[M+H]^+$	$[M+NH_4]^+$	(*sn*-1,3)	NL (amu)	(*sn*-2)	NL (amu)
14:0/14:0	518.5	535.5	14:0	245.3	–	–
15:0/15:0	546.5	563.5	15:0	259.3	–	–
16:0/16:0	574.5	591.6	16:0	273.3	–	–
17:0/17:0	602.6	619.6	17:0	287.3	–	–
19:0/19:0	658.6	675.7	19:0	315.4	–	–
20:0/20:0	686.7	703.7	20:0	329.3	–	–
D_5-TAG						
14:0/16:1/14:0	754.7	771.7	14:0	245.3	16:1	271.2
15:0/18:1/15:0	810.8	827.8	15:0	259.3	18:1	299.3
16:0/18:0/16:0	840.8	857.8	16:0	273.3	18:0	301.3
17:0/17:1/17:0	852.8	869.8	17:0	287.3	17:1	285.3
19:0/12:0/19:0	840.8	857.8	19:0	315.4	12:0	217.2
20:0/20:1/20:0	978.9	996.0	20:0	329.4	20:1	327.4
20:2/18:3/20:2	938.8	955.8	20:2	325.3	18:3	295.3
20:4/18:2/20:4	932.8	949.8	20:4	321.3	18:2	297.3
20:5/22:6/20:5	976.7	993.8	20:5	319.3	22:6	345.3

behavior studied in an attempt to assess response factors during CID. In general, it was found that there was variation in abundance of diglyceride ions related to not only the chain length, which was somewhat modest, but also with the position of the esterified fatty acid on the glycerol backbone. The most abundant fatty acid loss always was from position *sn*-1,3, whereas the least abundant diglyceride ion (loss of protonated fatty acid + NH$_3$) was from the *sn*-2 position. Perhaps more important was that as the number of double bonds increased, there was a significant increase in the proportion of the $[M+NH_4]^+$ which decomposed to $[M+H]^+$ by the loss of neutral NH$_3$. Thus it has become clear that there are many structurally sensitive factors which lead to the observed abundance of diglyceride-type ions in the tandem mass spectra of $[M+NH_4]^+$ adducts of TAGs. One would have to completely separate the TAGs into pure molecular species in order to assess information such as position of fatty acid esterification and utilize quantitative response factors. However, this investigator suggested it would be possible to calculate the yield of product ions based upon the chemical structure of a pure TAG.[32]

6.2.5 Atmospheric Pressure Chemical Ionization

An alternative technique for neutral lipid ionization has been atmospheric pressure chemical ionization (APCI) where the nebulized flowing liquid stream, such as that from the HPLC, is directed towards a corona discharge that effects ionization of molecules within the plasma.[33] This technique can lead to the production of $[M+H]^+$ ions from TAGs due to the mobility of protons within

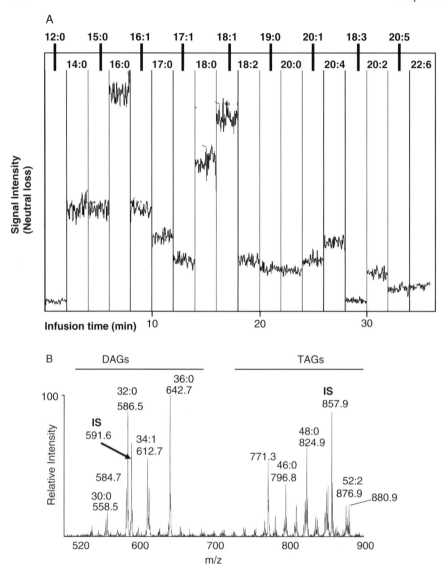

Figure 6.7 (a) Infusion of diacylglycerols and triacylglycerols isolated from RAW 264.7 into a tandem quadrupole mass spectrometer using nanoelectrospray (Nanomate). Recording of ion currents for neutral loss scans for each of the fatty acyl groups listed at the top of the figure. These different neutral loss transitions were recorded over a 30 min period. (b) Summed mass spectra corresponding to period 4 for the neutral loss for 16:0 (273 u) from each $[M+NH_4]^+$ adduct ion. The diacylglycerol mass ranges from m/z 500 to 700 while the triacylglycerols typically appear between m/z 700 and 900. The deuterium labeled internal standards correspond to 16:0/16:0 d_5-DAG (m/z 591.6) and 16:0/18:0/16:0-d_5-TAG (m/z 857.9) that were added to the RAW cell Bligh/Dyer extract prior to isolation of these neutral lipids and electrospray mass spectrometry.

the plasma and the basicity of the three ester groups within TAGs. This ionization technique leads to considerably more activation of the molecular ion species and hence it is possible to observe fragment ions formed directly during the ionization process. This is the case for TAGs with the formation of diglyceride type ions during ionization (likely by Scheme 6.3). A comparison of the various atmospheric pressure ionization techniques from flowing liquid streams was recently reported for the saturated TAG, trielaidin, with APCI and electrospray ionization with and without the presence of 10 mM ammonium formate in the mobile phase (Figure 6.8). In this case, the abundance of the $[M + NH_4]^+$ ions from electrospray ionization with ammonium formate was significantly higher than that observed for the $[M + H]^+$ from APCI as well as over 10-fold greater without the presence of ammonium formate. The third ionization technique to be employed in this study was atmospheric pressure photoionization, a process very similar in terms of energetics to that of APCI, only using 10.6 eV photons to effect ionization.[34] The formation of the abundant diglyceride-type ion, typified in the mass spectrum in Figure 6.8, revealed that APCI itself leads to the formation of product ions that can be used to assist in the structural characterization of glyceryl lipids. Furthermore, there is some abundance of the molecular ion, although it has been stated that one of the weaknesses of the APCI ionization approach is that saturated TAGs have a very low abundance of the molecular ion species which limits the ability to determine the molecular weight of the TAGs themselves.[35] One particular

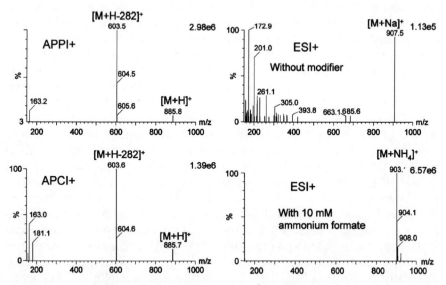

Figure 6.8   Mass spectrometry of the all-*trans* (Δ^9) triacylglycerol 18:1/18:1/18:1 using atmospheric pressure photoionization (APPI+), atmospheric pressure chemical ionization (APCI), positive ion electrospray ionization without buffer (ESI+), and positive ion electrospray ionization in the presence of ammonium formate. (Reproduced with permission from the American Chemical Society.[34])

solution to the reduced abundance of the molecular ion is to split the HPLC effluent into two separate mass spectrometers, one carrying out APCI as the ionization technique and the other carrying out electrospray ionization.[35] The electrospray technique would clearly reveal the molecular ion species as $[M + NH_4]^+$, or even the alkali metal adduct ion depending upon electrolytes present in the HPLC mobile phase, while the APCI mass spectrometer would provide useful diglyceride-type fragment ion composition.

An important aspect of APCI is therefore the appearance of the diglyceride-type ions which reveal the fatty acyl substituents present in the TAGs. The usefulness of these abundant ions can be illustrated with four different TAG molecular species (Figure 6.9). The abundance of the $[M + H]^+$ ion appears to be related to the total number of double bonds; however, ions corresponding to the loss of each of the fatty acids can be observed.[33] In general, the loss of the fatty acid from the *sn*-2 position leads to a less abundant ion than that corresponding to the loss of a fatty acid from either *sn*-1 or *sn*-3. This is illustrated in Figure 6.9c with the appearance of the ion $[SO]^+$ at m/z 605 which resulted from the loss of linoleic acid from *sn*-2, whereas the ions at m/z 603 and 601 were significantly more abundant and these correspond to the loss of either the *sn*-1 (S) or *sn*-3 (O) from the diglyceride ion (see the figure caption for the abbreviations used by the authors). A large number of studies have now been reported in the use of APCI to characterize different components in complex mixtures of TAGs, in particular from vegetable oils.[36–38]

In order to unambiguously define the composition of the TAGs using APCI and diglyceride ion formation, pure molecular species must be present in the ion source which requires considerable purification of naturally occurring glyceryl lipids. This may require sophisticated HPLC separations or, alternatively, analysis of somewhat less complex mixtures of glyceryl lipid species than that seen in mammalian cells as described above.[33] The reason for this is that all the diglyceride ions would appear in the same mass spectrum, with ion intensities largely related to the abundance of different molecular species rather than revealing the fatty acyl position unambiguously. Successful use of APCI has largely involved HPLC/MS applied to rather simple mixtures. In fact, one of the strategies developed for molecular species analysis with APCI, which has been termed the "bottom up solution", has been to use HPLC/MS to separate individual molecular species into relatively pure components prior to APCI analysis.[39] Thus, one can assess the relative contribution of individual molecular species. A complex mathematical approach has been described to provide information about the individual TAG components based upon the ratio of $[M + H]^+$ to the sum of the diglyceride ions, called the critical ion ratio.[39] However, the procedure has not been rigorously tested in a very complex mixture and if multiple TAG molecular species eluted in the same HPLC peak, it is likely that a serious compromise of this mathematical approach would result. Nonetheless, with improved separation strategies of triacylglycerides such as 2D-HPLC or even 3D-HPLC, it is possible that direct analysis of the triglycerides with APCI-LC/MS could become a competitive technique for molecular species analysis of

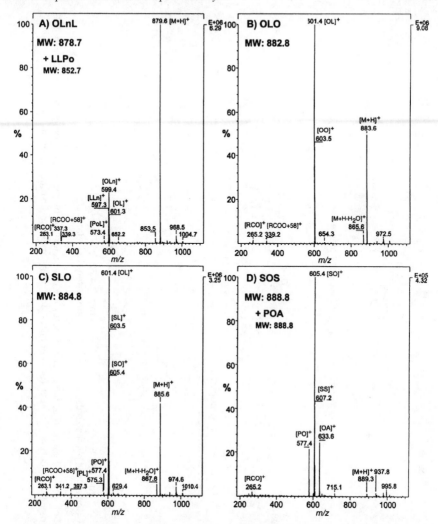

Figure 6.9 Atmospheric pressure chemical ionization (APCI-mass spectra) of triacyl-glycerols isolated from soy bean oil. (A) OLnL, (B) OLO, (C) SLO, and (D) SOS. Abbreviations for the fatty acyl groups are as follows: O, oleoyl; Ln, linolenoyl; S, stearoyl; L, linoleoyl; P, palmitoyl; A, arachidoyl. (Reproduced with permission from the American Oil Chemists' Society.[33])

very complex mixtures of TAGs and DAGs such as those found in mammalian cells.

6.3 Cholesterol Esters

The steryl esters and specifically cholesterol esters (CE) represent neutral lipids where there is a modest degree of molecular species complexity (Figure 6.10).

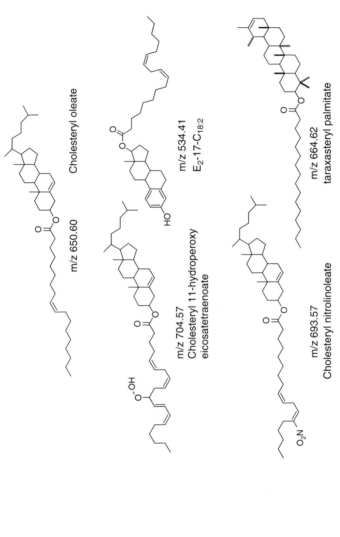

Figure 6.10   Representative structures of cholesterol esters that have been structurally characterized by mass spectrometry. The corresponding M^+ ions are indicated for each structure. See text for references describing mass spectrometric behavior.

CEs are neutral lipids thought to represent storage molecules for energy and cholesterol required by the cell. When present within mammalian cells, they typically are within lipid bodies which are neutral lipid droplets bounded by a phospholipid monolayer that have the hydrophobic tails of the phospholipids pointed towards the neutral lipid droplets. In addition to this, CEs are transported in circulating plasma by various lipoproteins because of their very poor water solubility. While methods to measure total cholesterol lipids are well established, considerably less emphasis has been placed upon identifying and describing the variety of molecular species of cholesterol esters present within the cell. It is now clear that specific molecular species of cholesterol esters reflect unique biochemical events taking place within the cell and reflect the availability of specific fatty acyl CoAs for the synthetic enzyme acyltransferase CoA (ACAT). Cholesterol can also be esterified with fatty acyl groups present in phospholipids, through action of the acyl CoA independent lecithin-cholesterol acyltransferase (LCAT).

Most CEs are simple esters of the reaction of cholesterol with fatty acids (Figure 6.10). However, several variations of the general structure have been reported. Recent structural studies have characterized oxidized forms of cholesterol which might be present in oxidized LDL, which is known to play an important role in atherosclerosis.[40] Various aldehydes and carboxylic acids have been found to be products of free radical oxidation of cholesteryl linoleate and some evidence has been gathered to suggest they are not as easily hydrolyzed to cholesterol and linoleic acid as non-oxidized esters.[41] One unique CE was identified as cholesteryl nitrolinoleate.[42] Esters of steroid hormones have also been described, such as esterdiol fatty acid esters (Figure 6.10), and they had been found to be some of the most potent steroidal estrogens present in mammalian cells.[43] Interestingly, the fatty acid composition of these steroidal esters is different from those synthesized by LCAT. Other steroidal esters have been described including the formation of interesting phytosteryl esters such as taraxasteryl palmitate (Figure 6.10) present in the smoke of burning leaves.[44]

6.3.1 Electron Ionization

The first studies of cholesterol esters and other steryl esters involved electron ionization and our understanding of the mass spectrometric behavior of these molecules came from the study of short chain ester analogs.[45] Chemical ionization has also been useful for characterizing steroidal esters.[46] Since many of the strategies involved separating molecules by gas chromatography prior to entering into the mass spectrometer, relatively high temperatures were employed to elute these somewhat nonvolatile compounds from the GC column. Thus, they entered the ion source with excess thermal energy and rapidly underwent the loss of the fatty acid by an expected ester "ene" reaction (Scheme 6.5). Typically a molecular ion is of very low abundance (or perhaps even absent) and what is observed is one abundant ion corresponding to the

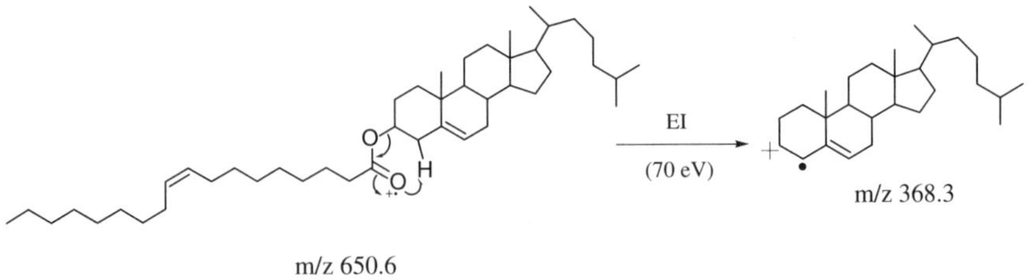

Scheme 6.5

molecular ion of cholestadiene (at m/z 368). The elution of components at various GC retention times characteristic of the fatty acyl ester, as well as the appearance of virtually a single ion, has been used as a method to identify cholesterol ester molecular species.

6.3.2 Chemical Ionization

Negative-ion chemical ionization has been found to yield much more structural information in that abundant ions derived from both the cholesterol portion of the molecule as well as the fatty acyl portion of the molecule are observed (Figure 6.11). In this case, the negative ion at m/z 367 is thought to also be deprotonated cholestadiene, perhaps derived from nucleophilic attack of the ester moiety by the species NH_2^- present in the chemical ionization plasma.[47] The ion at m/z 385 (deprotonated cholesterol) readily loses water to form deprotonated cholestadiene at m/z 367. The carboxylate anion derived from the fatty acyl substituent is revealed by the ion at m/z 283 and subsequent loss of water to form a ketene-type ion at m/z 265 is observed. These ion structures were established using a series of synthetic cholesterol esters under negative ammonia CI mass spectrometric conditions.[48]

6.3.3 Electrospray Ionization

Cholesterol esters readily form ammonium ion adducts and generate $[M + NH_4]^+$ during electrospray ionization. Collisional activation of this adduct ion results in the formation of virtually a single product ion, that being m/z 369, which is the protonated cholestadiene. Nonetheless, the ESI tandem mass spectrometric method can be used to quantitate CEs present in biological samples. A method has been developed using stable isotope dilution mass spectrometry to measure CEs in plasma and 13 different molecular species were quantitated ranging from 14:0 to 22:4 CE fatty acyl groups esterified to cholesterol.[49] Electrospray ionization has also been employed as a sensitive means to detect unstable oxidized cholesterol esters.[50] In these studies, Ag^+ salts were added to the mobile phase and ionization during electrospray resulted in formation of $[M + Ag]^+$ in a process called ion coordination electrospray.[50] In this case silver cations avidly formed π-complexes with the

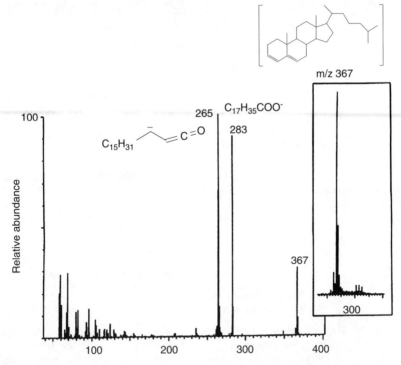

Figure 6.11 Negative ion ammonia chemical ionization mass spectrum of cholesteryl stearate. The carboxylate anion and corresponding ketene are the most abundant ions as well as deprotonated cholestadiene, which is enlarged in the inset. Redrawn and annotated from Ref. 47.

double bonds present in cholesterol esters. Since cholesterol itself has at least one double bond, this served as a useful means to detect these molecules.

6.4 Wax Esters

Wax esters are a surprisingly complex mixture of lipid molecular species in spite of their rather simple chemical structure, being long-chain fatty acids esterified to long-chain fatty alcohols. These compounds are found in the bacterial, animal, and plant kingdoms since they serve an important role in preventing dehydration of living organisms. For example, wax esters on the cuticles of insects reduce loss of water through evaporation.[51] Many birds generate a complex mixture of wax esters within preen glands that are used to repel water on feathers. The meibomian gland in the human eye synthesizes wax esters that coat the outer surface of the eye to prevent the loss of water from this critical organ.[52] The lipid-secreting cells in the human skin, sebaceous glands, synthesize wax esters complex in number and structure.[53]

The complexity of wax ester molecular species arises from the variety of fatty acids and fatty alcohols that constitute these simple esters. Most abundant

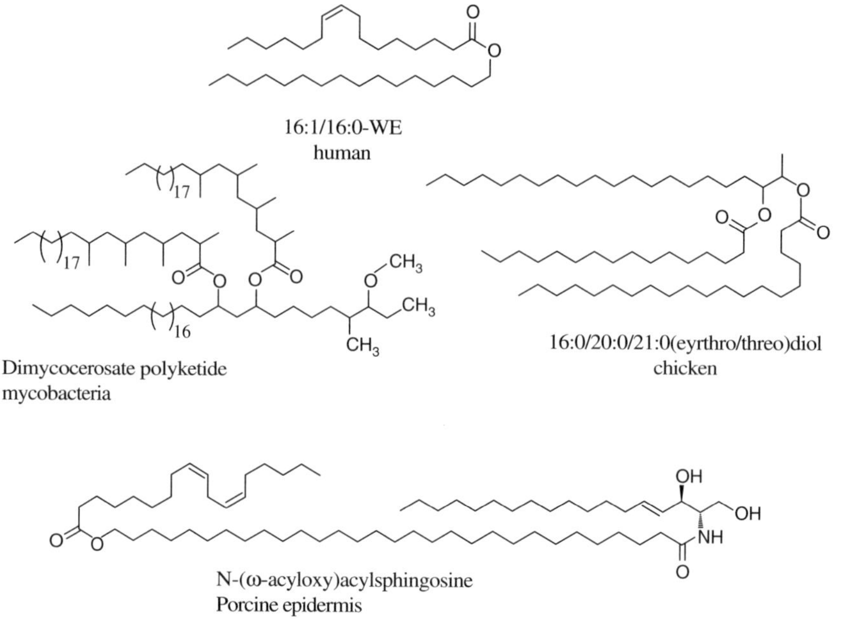

16:1/16:0-WE
human

Dimycocerosate polyketide
mycobacteria

16:0/20:0/21:0(eyrthro/threo)diol
chicken

N-(ω-acyloxy)acylsphingosine
Porcine epidermis

Figure 6.12 Representative structures of wax esters that correspond to the esterification of long-chain alcohols to long-chain acids. See text for references
to individual mass spectrometric identification of the more complex
structures.

human wax esters are composed of either straight-chain monounsaturated or
saturated fatty acids that are esterified to either straight-chain or methyl
branched saturated alcohols (Figure 6.12). Much more complex wax esters
are known in nature. For example, a complex wax ester has been described in
porcine epidermis (Figure 6.12).[54] This wax ester is derived from the long-chain
unsaturated fatty acid linoleic acid, esterified to the ω-hydroxyl group of a
very long-chain fatty acid which is itself acylated to the free amino group of
the long-chain base, forming a ceramide-like molecule. Bacterial wax esters
are often complex, as illustrated with the dimycocerosate polyketide found
in mycobacteria.[55] Even a wax triester has also been found.[56] Perhaps not
surprising, wax esters are likely the least studied of the complex lipidomes
present in living organisms, including human subjects.

Since the late 1950s, electron ionization mass spectrometry has been employed to analyze wax esters and still remains the most successful technique to
structurally characterize both the simple and complex wax esters that are
isolated from biological sources.[57] However, considerable use has been made
of gas chromatography, most recently capillary gas chromatography, in separating many of the wax esters which have sufficient volatility to elute from gas
chromatographic columns stable at high temperatures.[58] Nonetheless, certain
wax esters, in particular the diesters and triesters,[7] have insufficient volatility
to be separated and analyzed as intact molecules by gas-phase techniques.

Therefore, an alternative strategy has been widely employed to breakdown wax esters to their constituent fatty acids and fatty alcohols and to analyze these components independently, since these degradation products are more amenable to both gas chromatography and electron ionization mass spectrometry. Newer techniques of HPLC are beginning to emerge to separate wax esters; however, desorption and spray ionization techniques are not sufficiently sensitive at the present time to afford viable strategies for structural characterization of separated molecular species of wax esters eluting from the HPLC column. This remains an important challenge for the future.

6.4.1 Electron Ionization

Structural characterization of wax esters has relied heavily on the use of electron ionization and single stage mass spectrometry for analysis. However, that is not to say that this is an ideal technique to gain useful information about these interesting esters. The behavior of wax esters after electron ionization can be illustrated by the mass spectrum of a normal chain wax ester n-hexadecyl-n-hexadecanoate (Figure 6.13). For saturated wax esters a molecular ion is observed and this radical cation is used to characterize the total number of carbon atoms and double bonds present in the species. During the electron ionization process, the wax ester yields an abundant array of fragment ions with a typical base peak corresponding to the protonated carboxylic acid, in this case observed at m/z 257 (Scheme 6.6).[59,60] With short-chain esters such as methyl esters, this type of ion is not typically observed, but the additional carbon atoms in the alcohol chain permit a hydrogen atom from the carbon atom three carbons removed from the alkoxyl group to be abstracted followed by radical-driven cleavage of the carbon–oxygen bond. There is also an ion corresponding to the acylium ion from the carboxylic acid moiety $CH_3(CH_2)_{14}$-CO^+ at m/z 239. The alcohol moiety is revealed by the abundant ion corresponding to $C_{16}H_{32}$ (m/z 224), which is likely a McLafferty-type rearrangement ion corresponding to cleavage of the alkyl oxygen bond (Scheme 6.6).[60] A similar hydrogen atom rearrangement two methylene groups down the alcohol chain could explain another abundant odd-electron ion species observed at m/z 196. A hydrocarbon series of ions usually dominates at low mass starting at m/z 43 and with the sequential addition of 14 u, which clearly shows the saturated nature of this molecule.

When a single double bond is present in the fatty acid portion of the ester, the observed fragment ions are derived from alternative pathways of decomposition that become predominant. Unfortunately, there are no specific ions that can be used to assign positions of double bonds in the fatty acid carbon chain; however, there is information that can be employed to confirm the position of unsaturation as being in the fatty acyl group as opposed to the alcohol. One of the more abundant ions is a ketene ion, observed at m/z 236 in the example illustrated (Scheme 6.7 and Figure 6.13). This ion would correspond to the loss of methanol from the fatty acid methyl ester and is typically quite abundant in

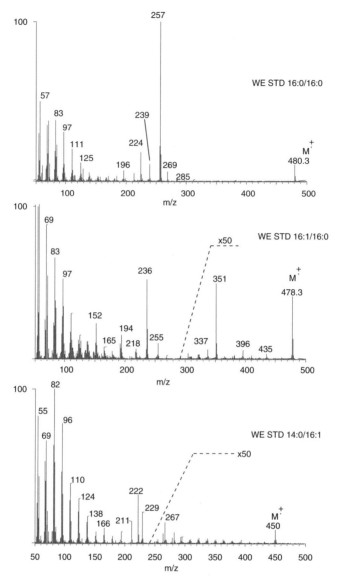

Figure 6.13 Electron ionization (70 eV) mass spectra of synthetic wax esters corresponding to (a) palmitoyl palmitate (acid/alcohol; 16:0/16:0), (b) palmitoyl palmitoleate (acid/alcohol; 16:1/16:0), and palmitoleoyl tetradecanoate (acid/alcohol; 14:0/16:1).

unsaturated fatty acid methyl esters.[61] There is an ion corresponding to the protonated carboxylic acid observed in this case at m/z 255. The abundance ratio between the ketene ion and protonated carboxylic acid ion clearly indicate, as well as the mass of these ions, that the unsaturation is in the fatty acid portion of the wax ester.

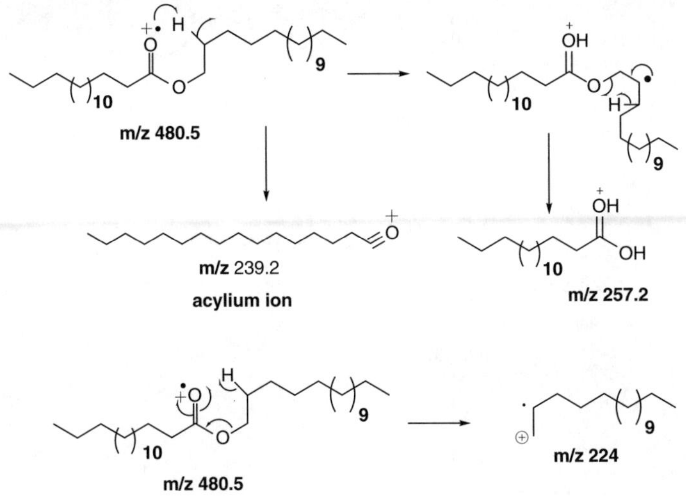

Scheme 6.6

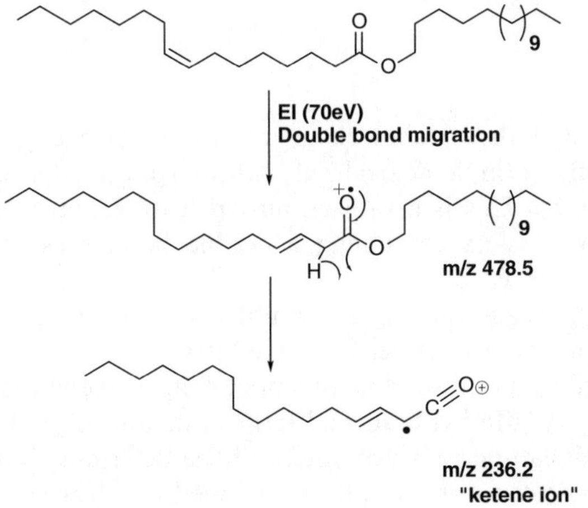

Scheme 6.7

A double bond present in the alcohol portion of the wax ester drastically alters the fragmentation pattern of the saturated species (Figure 6.13). In this case, the most abundant ions are derived from the alcohol portion of a wax ester. An ion corresponding to the protonated carboxylic acid is observed, but a more abundant odd-electron ion species corresponding to cleavage of the carbon–oxygen bond of the alcohol is observed (m/z 222) (Scheme 6.8). In addition to the protonated carboxylic acid, an acylium ion is observed at m/z 211. Perhaps most striking is the large number of odd-electron ion species in a series starting at m/z 82 proceeding by 14 u intervals to m/z 222. These would correspond to specific cleavages likely driven by the double bond

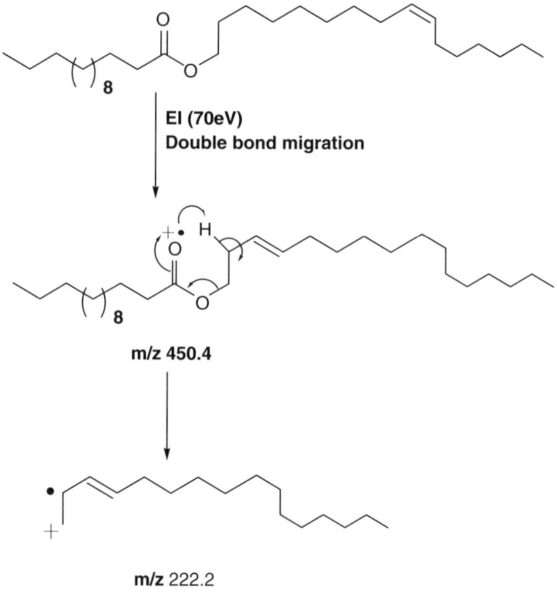

Scheme 6.8

moving throughout the alcohol chain due to excess energy during electron ionization, resulting in a delocalized radical cation adjacent to a double bond (Scheme 6.8). This is a rather unusual feature that is characteristic of normal-chain wax esters containing a double bond present in the alcohol portion of the molecule.

Perhaps the most challenging structural task is to assign the alkyl group position of the intact ester either from the fatty acid or fatty alcohol perspective. This is a critical point since methyl branching is one of the major motifs in wax ester diversity. Methyl-branched fatty acids and fatty alcohols are even found in mammalian-derived wax esters.[62] The determination of alkyl group branching of wax esters has typically been done by hydrolysis of the wax esters to their constituent components and using chromatographic separation techniques as well as mass spectrometry to assign methyl group positions. Yet some EI-derived ions from intact wax esters have been suggested to reveal methyl branching as iso- or anteiso-fatty acids.[63] Complex wax esters from vicinal diols have been isolated from various avian preen glands. Recently, the wax ester lipidome from the red knot (*Calidris canutus*) was described to be a mixture of both monoester waxes and diester waxes.[64,65] Electron ionization mass spectrometry of the intact wax esters was used to identify specific components that make up this complex mixture of wax esters based upon the elution of components with specific molecular ions[65] (Figure 6.14). In the case of diester waxes, these did not yield a molecular ion but formed an abundant ion corresponding to loss of 143 u, which was indicative of the molecular weight of these diesters. Intense ions corresponding to cleavage of the carboxylic acid

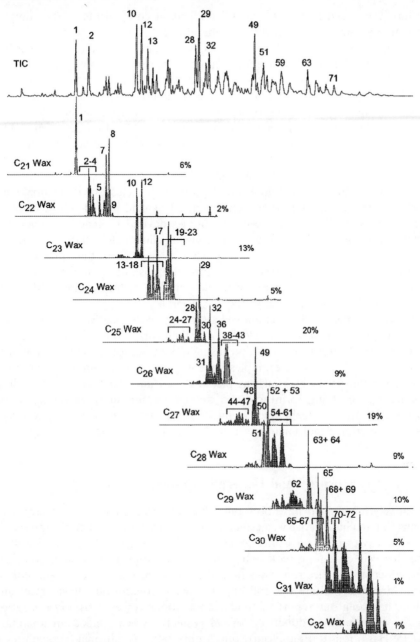

Figure 6.14 Total ion chromatogram (TIC) of the wax esters isolated from *C. canutus*. Partial mass chromatograms of specific molecular ions C21 (*m/z* 340), C22 (*m/z* 354), C23 (*m/z* 368), C24 (*m/z* 382), C25 (*m/z* 396), C26 (*m/z* 410), C27 (*m/z* 424), C27 (*m/z* 438), C28 (*m/z* 452), C29 (*m/z* 466), C30 (*m/z* 480), C31 (*m/z* 494), and C32 (*m/z* 500). See original reference for identification of specific wax ester molecular species.[65] (Reproduced with permission from the American Oil Chemists' Society.)

groups were used in part to identify individual components in the complex mixture by using a mass chromatogram approach.[64]

Another interesting polyunsaturated wax ester[66] was recently described as a product of squalene metabolism by a marine microbe (*Marinobacter squalenivorans*). Electron ionization mass spectrometry did reveal a molecular ion for this product at m/z 496; but none of the other typical ions seen during electron ionization were observed. However, after hydrogenation, the doubly protonated carboxylate species was observed at m/z 271. In order to unambiguously assign the methyl branching, it was necessary to degrade this molecule to the representative alcohols and fatty acids and compare the mass spectra to that of authentic isoprene derivatives.[66]

The determination of double bond positions by electron ionization mass spectrometry has always been a challenging task and indeed this remains so with wax esters. There have been techniques devised by reacting intact wax esters with dimethyldisulfide to form a vicinal dimethyldisulfide derivative.[67] These derivatized wax esters yielded characteristic ions that underwent a neutral loss of 48 u (CH_3SH) that defined the position of the double bond. Such approaches have been useful for determining olefin positions only when one double bond was present in the wax ester.

Other investigators have employed a strategy to degrade wax esters to their constituent fatty acid or fatty alcohol and then carry double bond determination or methyl substitution positions using more traditional techniques of fatty acid analysis.[61,68] Specific examples have involved making unique remote site fragmentation derivatives such as picolinyl and nicotinate esters[69] as well as comparing the EI fingerprint to that of authentic compounds. The combination of these EI and GC techniques of intact and hydrolyzed wax esters has been used to determine molecular species in human hair[70] and follicular cells.[71]

6.4.2   Electrospray and Desorption Ionization

Relatively few reports have been published concerning the use of electrospray ionization to characterize wax esters. In part, this has been due to the inherently low sensitivity of electrospray ionization for these neutral lipids, as well as the absence of unique collision induced decomposition product ions. As was observed for electron ionization, the appearance of even a single double bond drastically alters the collision-induced decomposition mass spectrum. There has been one report of fundamental studies of electrospray ionization of the ammoniated adduct ion of wax esters[72] which yielded an abundant $[M + NH_4]^+$ ion during electrospray ionization. Collisional activation of $[M + NH_4]^+$ yielded doubly protonated carboxylate cation, as well as a few other ions.

Electrospray ionization was employed to structurally characterize an aromatic wax ester isolated from *Rhodococcus* bacteria grown on phenyldecane. The unique wax ester was purified by thin-layer chromatography then characterized by electrospray ionization as the sodium adduct with a molecular ion of *m/z* 47 that could be collisionally activated to yield the sodiated carboxylate species at *m/z* 271.[73] One report used ESI to analyze lithiated adduct neutral lipid ions from human tear fluid but could not detect the expected major wax ester.[74]

6.5 Conclusions

The analysis of complex mixtures of lipids as closely related molecular species and as unique elements of the cellular lipidome is emerging as an analytical possibility. Techniques such as electrospray ionization of cationized neutral lipids followed by collision-induced decomposition is emerging as a very general technique by which one can assess not only the structural variants of molecular species, but also attempt to understand the quantitative changes that can take place within a cell. Many challenges remain because of the extraordinary large number of molecular species that are present in major and minor components and challenges of quantitative analysis. In addition, techniques such as electron ionization and gas chromatography/mass spectrometry remain useful approaches to study the complexity of neutral lipidomes with species such as wax esters. The only limitation of this latter technique is the thermostability of such compounds. As sensitivity in mass spectrometry improves, other neutral lipids which are of lower abundance in cells (*e.g.* dolichols) may become the target of lipidomic studies.

Acknowledgement

This work was supported in part by the Large Scale Collaborative Grant, Lipid MAPS (GM069338) from the National Institutes of Health.

References

1. P. Liu, Y. Ying, Y. Zhao, D. I. Mundy, M. Zhu and R. G. Anderson, Chinese hamster ovary K2 cell lipid droplets appear to be metabolic organelles involved in membrane traffic, *J. Biol. Chem.*, 2004, **279**, 3787–3792.
2. K. Athenstaedt and G. Daum, The life cycle of neutral lipids: synthesis, storage and degradation, *Cell. Mol. Life Sci.*, 2006, **63**, 1355–1369.
3. Y. J. Sigal, M. I. McDermott and A. J. Morris, Integral membrane lipid phosphatases/phosphotransferases: common structure and diverse functions, *Biochem. J.*, 2005, **387**, 281–293.
4. C. A. Rouzer and L. J. Marnett, Glycerylprostaglandin synthesis by resident peritoneal macrophages in response to a zymosan stimulus, *J. Biol. Chem.*, 2005, **280**, 26690–26700.
5. J. P. Suomela, M. Ahotupa, O. Sjoval, J. P. Kurvinen and H. Kallio, New approach to the analysis of oxidized triacylglycerols in lipoproteins, *Lipids*, 2004, **39**, 507–512.

6. H. W. Sprecher, R. Maier, M. Barber and R. T. Holman, Structure of an optically active allene-containing tetraester triglyceride isolated from the seed oil of *Sapium sebiferum*, *Biochemistry*, 1965, **4**, 1856–1863.

7. J. Cvacka, O. Hovorka, P. Jiros, J. Kindl, K. Stransky and I. Valterova, Analysis of triacylglycerols in fat body of bumblebees by chromatographic methods, *J. Chromatogr. A*, 2006, **1101**, 226–237.

8. W. W. Christie, *Lipid Analysis*, 3rd edn, The Oily Press: Bridgwater, England, 2003.

9. M. Buchgraber, F. Ulberth, H. Emons and E. Anklam, Triacylglycerol profiling by using chromatographic techniques, *Eur. J. Lipid Sci. Technol.*, 2004, **106**, 621–648.

10. V. Ruiz-Gutiérrez and L. J. R. Barron, Methods for the analysis of triacylglycerols, *J. Chromatogr. B*, 1995, **671**, 133–168.

11. M. S. F. Lie Ken Jie and J. Mustafa, High-resolution nuclear magnetic resonance spectroscopy – applications to fatty acids and triacylglycerols, *Lipids*, 1997, **32**, 1019–1034.

12. N. K. Andrikopoulos, Chromatographic and spectroscopic methods in the analysis of triacylglycerol species and regiospecific isomers of oils and fats, *Crit. Rev. Food Sci. Nutr.*, 2002, **42**, 473–505.

13. H.-G. Janssen, W. Boers, H. Steenbergen, R. Horsten and E. Flöter, Comprehensive two-dimensional liquid chromatography X gas chromatography: evaluation of the applicability for the analysis of edible oils and fats, *J. Chromatogr. A*, 2003, **1000**, 385–400.

14. P. Dugo, T. Kumm, M. L. Crupi, A. Cotroneo and L. Mondello, Comprehensive two-dimensional liquid chromatography combined with mass spectrometric detection in the analyses of triacylglycerols in natural lipidic matrixes, *J. Chromatogr. A*, 2006, **1112**, 269–275.

15. B. Nikolova-Damyanova, Quantitative thin-layer chromatography of triacylglycerols. Principles and application, *J. Liq. Chromatogr. Rel. Technol.*, 1999, **22**, 1513–1537.

16. R. Ryhage and E. Stenhagen, Mass spectrometry in lipid research, *J. Lipid Res.*, 1960, **1**, 361–390.

17. R. A. Hites, Mass spectrometry of triglycerides, *Methods Enzymol.*, 1975, **35**, 348–359.

18. M. Barber, T. O. Merren and W. Kelley, The mass spectrometry of large molecule I. The triglycerides of straight chain fatty acids, *Tetrahedron Lett.*, 1964, **18**, 1063–1067.

19. M. Barber, J. R. Chapman and W. A. Wolstenholme, Lipid analysis by coupled mass spectrometry-gas chromatography, *Int. J. Mass Spectrom. Ion Phys.*, 1968, **1**, 98–101.

20. P. Laakso, Mass spectrometry of triacylglycerols, *Eur. J. Lipid Sci. Technol.*, 2002, **104**, 43–49.

21. L. Mondello, A. Casilli, P. Q. Tranchida, M. Furukawa, K. Komori, K. Miseki, P. Dugo and G. Dugo, Fast enantiomeric analysis of a complex essential oil with an innovative multidimensional gas chromatographic system, *J. Chromatogr. A*, 2006, **1105**, 11–16.

22. J.-P. Kurvinen, P. Rua, O. Sjövall and H. Kallio, Software (MSPECTRA) for automatic interpretation of triacylglycerol molecular mass distribution spectra and collision induced dissociation product ion spectra obtained by ammonia negative ion chemical ionization mass spectrometry, *Rapid Commun. Mass Spectrom.*, 2001, **15**, 1084–1091.

23. C. Cheng, M. L. Gross and E. Pittenauer, Complete structural elucidation of triacylglycerols by tandem sector mass spectrometry, *Anal. Chem.*, 1998, **70**, 4417–4426.

24. F. F. Hsu and J. Turk, Structural characterization of triacylglycerols as lithiated adducts by electrospray ionization mass spectrometry using low-energy collisionally activated dissociation on a triple stage quadrupole instrument, *J. Am. Soc. Mass Spectrom.*, 1999, **10**, 587–599.

25. X. Han and R. W. Gross, Quantitative analysis and molecular species fingerprinting of triacylglyceride molecular species directly from lipid extracts of biological samples by electrospray ionization tandem mass spectrometry, *Anal. Biochem.*, 2001, **295**, 88–100.

26. X. Han, J. Yang, H. Chen, H. Ye and R. W. Gross, Toward fingerprinting cellular lipidomics directly from biological samples by two-dimensional electrospray ionization mass spectrometry, *Anal. Biochem.*, 2004, **330**, 317–331.

27. X. Han and R. W. Gross, Global analyses of cellular lipidomes directly from crude extracts of biological samples by ESI mass spectrometry: a bridge to lipidomics, *J. Lipid Res.*, 2003, **44**, 1071–1079.

28. X. Han and R. W. Gross, Shotgun lipidomics: electrospray ionization mass spectrometric analysis and quantitation of cellular lipidomes directly from crude extracts of biological samples, *Mass Spectrom. Rev.*, 2005, **24**, 367–412.

29. K. L. Duffin, J. D. Henion and J. J. Shieh, Electrospray and tandem mass spectrometric characterization of acylglycerol mixtures that are dissolved in nonpolar solvents, *Anal. Chem.*, 1991, **63**, 1781–1788.

30. A. M. McAnoy, C. C. Wu and R. C. Murphy, Direct qualitative analysis of triacylglycerols by electrospray mass spectrometry using a linear ion trap, *J. Am. Soc. Mass Spectrom.*, 2005, **16**, 1498–1509.

31. X. Li and J. J. Evans, Examining the collision-induced decomposition spectra of ammoniated triglycerides as a function of fatty acid chain length and degree of unsaturation. I. The OXO/YOY series, *Rapid Commun. Mass Spectrom.*, 2005, **19**, 2528–2538.

32. X. Li, E. J. Collins and J. J. Evans, Examining the collision-induced decomposition spectra of ammoniated triglycerides as a function of fatty acid chain length and degree of unsaturation. II. The PXP/YPY series, *Rapid Commun. Mass Spectrom.*, 2006, **20**, 171–177.

33. W. C. Byrdwell, Atmospheric pressure chemical ionization mass spectrometry for analysis of lipids, *Lipids*, 2001, **36**, 327–346.

34. S.-S. Cai and J. A. Syage, Comparison of atmospheric pressure photoionization, atmospheric pressure chemical ionization, and electrospray

ionization mass spectrometry for analysis of lipids, *Anal. Chem.*, 2006, **78**, 1191–1199.

35. W. C. Byrdwell and W. E. Neff, Dual parallel electrospray ionization and atmospheric pressure chemical ionization mass spectrometry (MS), MS/MS and MS/MS/MS for the analysis of triacylglycerols and triacylglycerol oxidation products, *Rapid Commun. Mass Spectrom.*, 2002, **16**, 300–319.

36. J. Parcerisa, I. Casals, J. Boatella, R. Codony and M. Rafecas, Analysis of olive and hazelnut oil mixtures by high-performance liquid chromatography-atmospheric pressure chemical ionization mass spectrometry of triacylglycerols and gas-liquid chromatography of non-saponifiable compounds (tocopherols and sterols), *J. Chromatogr. A*, 2000, **881**, 149–158.

37. K. Nagy, D. Bongiorno, G. Avellone, P. Agozzino, L. Ceraulo and K. Vèkey, High performance liquid chromatography-mass spectrometry based chemometric characterization of olive oils, *J. Chromatogr. A*, 2005, **1087**, 90–97.

38. L. Fauconnot, J. Hau, J. M. Aeschlimann, L.-B. Fay and F. Dionisi, Quantitative analysis of triacylglycerol regioisomers in fats and oils using reversed-phase high-performance liquid chromatgoraphy and atmospheric pressure chemical ionization mass spectrometry, *Rapid Commun. Mass Spectrom.*, 2004, **18**, 218–224.

39. W. C. Byrdwell, The bottom-up solution to the triacylglycerol lipidome using atmospheric pressure chemical ionization mass spectrometry, *Lipids*, 2005, **40**, 383–417.

40. G. Hoppe, A. Ravandi, D. Herrera, A. Kuksis and H. F. Hoff, Oxidation products of cholesteryl linoleate are resistant to hydrolysis in macrophages, form complexes with proteins, and are present in human atherosclerotic lesions, *J. Lipid Res.*, 1997, **38**, 1347–1360.

41. C. Suarna, R. T. Dean, P. T. Southwell-Keeley, D. E. Moore and R. Stocker, Separation and characterization of cholesteryl oxo- and hydroxy-linoleate isolated from human atherosclerotic plaque, *Free Radic. Res.*, 1997, **27**, 397–408.

42. E. S. Lima, P. D. Mascio and D. S. P. Abdalla, Cholesteryl nitrolinoleate, a nitrated lipid present in human blood plasma and lipoproteins, *J. Lipid Res.*, 2003, **44**, 1660–1666.

43. J. M. Larner, S. L. Pahuja, C. H. Shackleton, W. J. McMurray, G. Giordano and R. B. Hochberg, The isolation and characterization of estradiol-fatty acid esters in human ovarian follicular fluid, *J. Biol. Chem.*, 1993, **268**, 13893–13899.

44. V. O. Elias, B. R. T. Simoneit, A. S. Pereira and J. N. Cardoso, Mass spectra of triterpenyl alkanoates, novel natural products, *J. Mass Spectrom.*, 1997, **32**, 1356–1361.

45. L. G. Partridge and C. Djerassi, Mass spectrometry in structural and stereochemical problems. 250. Characteristic fragmentation of cholesterol acetate, *J. Org. Chem.*, 1977, **42**, 2799–2805.

46. D. H. Albert, L. Ponticorvo and S. Lieberman, Identification of fatty acid esters of pregnenolone and allopregnanolone from bovine corpora lutea, *J. Biol. Chem.*, 1980, **255**, 10618–10623.

47. R. P. Evershed and L. J. Goad, Capillary gas chromatography/mass spectrometry of cholesteryl esters with negative ammonia chemical ionization, *Biomed. Environ. Mass Spectrom.*, 1987, **14**, 131–140.

48. R. P. Evershed, M. C. Prescott, N. Spooner and L. J. Goad, Negative ion ammonia chemical ionization and electron impact ionization mass spectrometric analysis of steryl fatty acyl esters, *Steroids*, 1989, **53**, 285–309.

49. G. Liebisch, M. Binder, R. Schifferer, T. Langmann, B. Schulz and G. Schmitz, High throughput quantification of cholesterol and cholesteryl ester by electrospray ionization tandem mass spectrometry (ESI-MS/MS), *Biochim. Biophys. Acta*, 2006, **1761**, 121–128.

50. H. Yin, C. M. Havrilla, J. D. Morrow and N. A. Porter, Formation of isoprostane bicyclic endoperoxides from the autoxidation of cholesteryl arachidonate, *J. Am. Chem. Soc.*, 2002, **124**, 7745–7754.

51. S. Patel, D. R Nelson and A. G. Gibbs, Chemical and physical analyses of wax ester properties, *J. Insect Sci.*, 2001, **1**, 1–7.

52. N. Nicolaides, J. K. Kaitaranta, T. N. Rawdah, I. Macy, F. M. I. Boswell and R. E. Smith, Meibomian gland studies: comparison of steer and human lipids, *Invest. Ophthalmol. Vis. Sci.*, 1981, **20**, 522–536.

53. M. E. Stewart, Sebaceous gland lipids, *Semin. Dermatol.*, 1992, **11**, 100–105.

54. P. W. Wertz and D. T. Downing, Ceramides of pig epidermis: structure determination, *J. Lipid Res.*, 1983, **24**, 759–765.

55. K. C. Onwueme, C. J. Vos, J. Zurita, J. A. Ferreras and L. E. N. Quadri, The dimycocerosate ester polyketide virulence factors of mycobacteria, *Prog. Lipid Res.*, 2005, **44**, 259–302.

56. M. K. Logani, D. B. Nhari and R. E. Davies, Triester wax: a novel neutral lipid from the skin of the rhino mutant mouse, *Biochim. Biophys. Acta*, 1975, **388**, 291–300.

57. R. Ryhage and E. Stenhagen, Mass spectrometric studies. II. Saturated normal long-chain esters of ethanol and higher alcohols, *Ark. Kemi*, 1959, **14**, 483–495.

58. K. Stransky, M. Zarevucka, I. Valterova and Z. Wimmer, Gas chromatographic retention data of wax esters, *J. Chromatogr. A*, 2006, **1128**, 208–219.

59. C. Djerassi and C. Fenselau, Mass spectrometry in structural and stereochemical problems. LXXXVI. The hydrogen-transfer reactions in butyl propionate, benzoate, and phthalate, *J. Am. Chem. Soc.*, 1965, **87**, 5756–5762.

60. A. J. Aasen, H. H. Hofstetter, B. T. R. Iyengar and R. T. Holman, Identification and analysis of wax esters by mass spectrometry, *Lipids*, 1971, **6**, 502–507.

61. R. C. Murphy, Mass Spectrometry of Lipids. *The Handbook of Lipid Research*, ed. F. Snyder, Vol. 7, Plenum Press, New York, 1993.

62. D. J. Harvey, J. M. Tiffany, J. M. Duerden, K. S. Pandher and L. S. Mengher, Identification by combined gas chromatography-mass spectrometry of constituent long-chain fatty acids and alcohols from the meibomian glands of the rat and a comparison with human meibomian lipids, *J. Chromatogr.*, 1987, **414**, 253–263.

63. D. J. Harvey and J. M. Tiffany, Identification of meibomian gland lipids by gas chromatography-mass spectrometry: Application to the meibomian lipids of the mouse, *J. Chromatogr.*, 2006, **301**, 173–187.

64. J. S. S. Damste, M. Dekker, B. E. van Dongen, S. Schouten and T. Piersma, Structural identification of the diester preen-gland waxes of the red knot (*Calidris canutus*), *J. Nat. Prod.*, 2000, **63**, 381–384.

65. M. H. A. Dekker, T. Piersma and J. S. S. Damste, Molecular analysis of intact preen waxes of *Calidris canutus* (Aves: Scolopacidae) by gas chromatography/mass spectrometry, *Lipids*, 2000, **35**, 533–541.

66. J. F. Rontani, A. Mouzdahir, V. Michotey, P. Caumette and P. Bonin, Production of a polyunsaturated isoprenoid wax ester during aerobic metabolism of squalene by marinobacter squalenivorans sp. nov, *Appl. Environ. Mcirobiol.*, 2003, **69**, 4167–4176.

67. C. Pepe, P. Scribe and A. Saliot, Double bond location in mono-unsaturated wax esters by gas chromatography/mass spectrometry of their dimethyl disulphide derivatives, *Org. Mass Spectrom.*, 1993, **28**, 1365–1367.

68. W. Francke, G. Lubke, W. Schroder, A. Reckziegel, V. Imperatriz-Fonseca, A. Kleinert, E. Engels, K. Hartfelder and R. Radtke W. Engels, Identification of oxygen containing volatiles in cephalic secretions of workers of Brazilian stingless bees, *J. Braz. Chem. Soc.*, 2000, **11**, 562–571.

69. D. J. Harvey, Identification by gas chromatography and mass spectrometry of lipids from the rat Harderian gland, *J. Chromatogr.*, 1991, **565**, 27–34.

70. Y. Masukawa, H. Tsujimura and G. Imokawa, A systematic method for the sensitive and specific determination of hair lipids in combination with chromatography, *J. Chromatogr. B, Analyt. Technol. Biomed. Life Sci.*, 2005, **823**, 131–142.

71. K. M. Nordstrom, J. N. Labows, K. J. McGinley and J. J. Leyden, Characterization of wax esters, triglycerides, and free fatty acids of follicular casts, *J. Invest. Dermatol.*, 1986, **86**, 700–705.

72. J. L. Krank and R. C. Murphy, Analysis of wax ester molecular species by liquid chromatography tandem mass spectrometry, *Proc. Am. Soc. Mass Spectrom.*, 2005, **53**, A051893.

73. H. M. Alvarez, H. Luftmann, R. A. Silva, A. C. Cesari, A. Viale, M. Waltermann and A. Steinbuchel, Identification of phenyldecanoic acid as a constituent of triacylglycerols and wax ester produced by *Rhodococcus opacus* PD630, *Microbiology*, 2002, **148**, 1407–1412.

74. B. M. Ham, J. T. Jacob, M. M. Keese and R. B. Cole, Identification, quantification and comparison of major non-polar lipids in normal and dry eye tear lipidomes by electrospray tandem mass spectrometry, *J. Mass Spectrom.*, 2004, **39**, 1321–1336.

CHAPTER 7

Bioinformatics of Lipids

EOIN FAHY

San Diego Supercomputer Center, University of California, San Diego, 9500
Gilman Drive, La Jolla, CA 92093-0505, USA

7.1 Introduction

In the past decade, bioinformatics has become an integral part of research and
development in the biomedical sciences, propelled by the sequencing of the
human genome and advances in high-performance computing. The latter has
motivated research for the elucidation of the macromolecular parts lists in cells.
Unlike in genomics and proteomics, where advanced methodologies have
yielded parts lists of genes and proteins in cells, there has been little effort to
systematically obtain lists of cellular lipids and their changes. Only recently
organized efforts have begun to emerge in identifying and cataloguing lipids. In
particular, lipid research or "lipidomics" is emerging as a rapidly expanding
field where these molecules are generating great interest with regard to their
roles in membrane structure and organization, energy production, signaling,
metabolism, inflammation, apoptosis and gene regulation, and other processes
in cells, as well being implicated in major pathologies such as diabetes, heart
disease, and cancer. When efforts to sequence genomes and to measure
proteomes were initiated, concomitantly significant investments were made in
bioinformatics infrastructures that could accommodate the avalanche of data.
Structured vocabularies were developed to classify these macromolecules,
database schemas were developed to house the data in relational tables, and
numerous algorithms were developed to carry out extensive gene and protein
sequence analysis.[1] However, when efforts to measure mammalian lipids
were initiated there was great paucity in informatics resources concerning
lipids and lipid data. Even the lipid nomenclature organized by IUPAC was
decades old[2–5] and there was no infrastructure that tied lipids to the genes and
proteins associated with them. This void was soon recognized and efforts
are underway to create a comprehensive bioinformatics infrastructure for
lipidomics.[6] In this chapter, we will provide an overview of the current status

of lipid bioinformatics and discuss the emerging developments in this field. We present the new classification and structural representation schemes for lipids that is beginning to be widely accepted, organized database efforts for lipidomic data spearheaded by the LIPID MAPS and LipidBank projects, schemes for integrating lipidomic, genomic and proteomic data, methods for quantitative analysis of lipids in cells, and early efforts to map lipid-associated pathways in normal and pathological conditions. We will also present significant challenges that remain in the development of lipid bioinformatics.

7.2 Lipid Classification, Nomenclature, and Structure Representation

7.2.1 Classification

The first step towards classification of lipids is the establishment of an ontology that is extensible, flexible, and scalable. One must be able to classify, name, and represent these structures in a logical manner which is amenable to databasing and computational manipulation. Lipids have been loosely defined as biological substances that are generally hydrophobic in nature and in many cases soluble in organic solvents.[7] These chemical features are present in a broad range of molecules such as fatty acids, phospholipids, sterols, sphingolipids, terpenes, and others.[8] There are a number of online resources (Table 7.1) which outline comprehensive classification schemes for lipids. In view of the fact that lipids comprise an extremely heterogeneous collection of molecules from a structural and functional standpoint, it is not surprising that there are significant differences with regard to the scope and organization of current classification schemes. The Lipid Library site (http://www.lipidlibrary.co.uk) defines lipids as "fatty acids and their derivatives, and substances related biosynthetically or functionally to these compounds". The Cyberlipids website (http://www.cyberlipid.org) has a broader approach and includes isoprenoid-derived molecules such as steroids and terpenes. Both of these resources classify lipids into "simple" and "complex" groups, with simple lipids being those yielding at most two types of distinct entities on hydrolysis (e.g. acylglycerols: fatty acids and glycerol) and complex lipids (e.g. glycerophospholipids: fatty acids, glycerol, and headgroup) yielding three or more products on hydrolysis. The LipidBank database (http://lipidbank.jp) in Japan defines 17 top-level categories in their classification scheme covering a wide variety of animal and plant sources. The LIPID MAPS group (http://www.lipdmaps.org) has taken a more chemistry based approach (Figure 7.1) and defines lipids as hydrophobic or amphipathic small molecules that may originate entirely or in part by carbanion-based condensations of thioesters (fatty acids, polyketides, etc.), and/or by carbocation-based condensations of isoprene units (prenols, sterols, etc.). This classification scheme[6] (Table 7.2) organizes lipids into well-defined categories that cover eukaryotic and prokaryotic sources, and which is equally applicable to archaea and synthetic (man-made) lipids. Biosynthetically related

Table 7.1   Online Resources for Lipid Classification and Databasing.

Resource	URL	Country	Comments
Lipid Library	http://www.lipidlibrary.co.uk	U.K.	Multiple reference topics and examples, MS and NMR libraries, literature service
Cyberlipid Center	http://www.cyberlipid.org	France	Descriptions of lipid classes with examples and literature references
LipidBank	http://lipidbank.jp	Japan	Lipid database with a wide variety of categories, including plant lipids
LipidBase	http://lipidbase.jp	Japan	An updated online version containing LipidBank records classified according to LIPID MAPS scheme, search interfaces
LIPIDAT	http://www.lipidat.chemistry.ohio-state.edu	U.S.	Database composed mostly of phospholipids and associated thermodynamic data
LIPID MAPS	http://www.lipidmaps.org	U.S.	Classification scheme, lipid and protein databases, MS libraries, search tools

compounds that are not technically lipids due to their water solubility are included for completeness in this classification scheme. Lipids are divided into eight categories (Fatty Acyls, Glycerolipids, Glycerophospholipids, Sphingolipids, Sterol Lipids, Prenol Lipids, Saccharolipids, and Polyketides) containing distinct classes and subclasses of molecules and provide a 12-digit identifier for each unique lipid molecule. An important database field is the LIPID ID, a unique 12-character identifier based on this classification scheme. The format of the LIPID ID, outlined in Table 7.3, provides a systematic means of assigning unique IDs to lipid molecules and allows for the addition of large numbers of new categories, classes, and subclasses in the future. The last four characters of the ID comprise a unique identifier within a particular subclass and are randomly assigned. By initially using numeric characters this allows 9999 unique IDs per subclass, but with the additional use of 26 uppercase

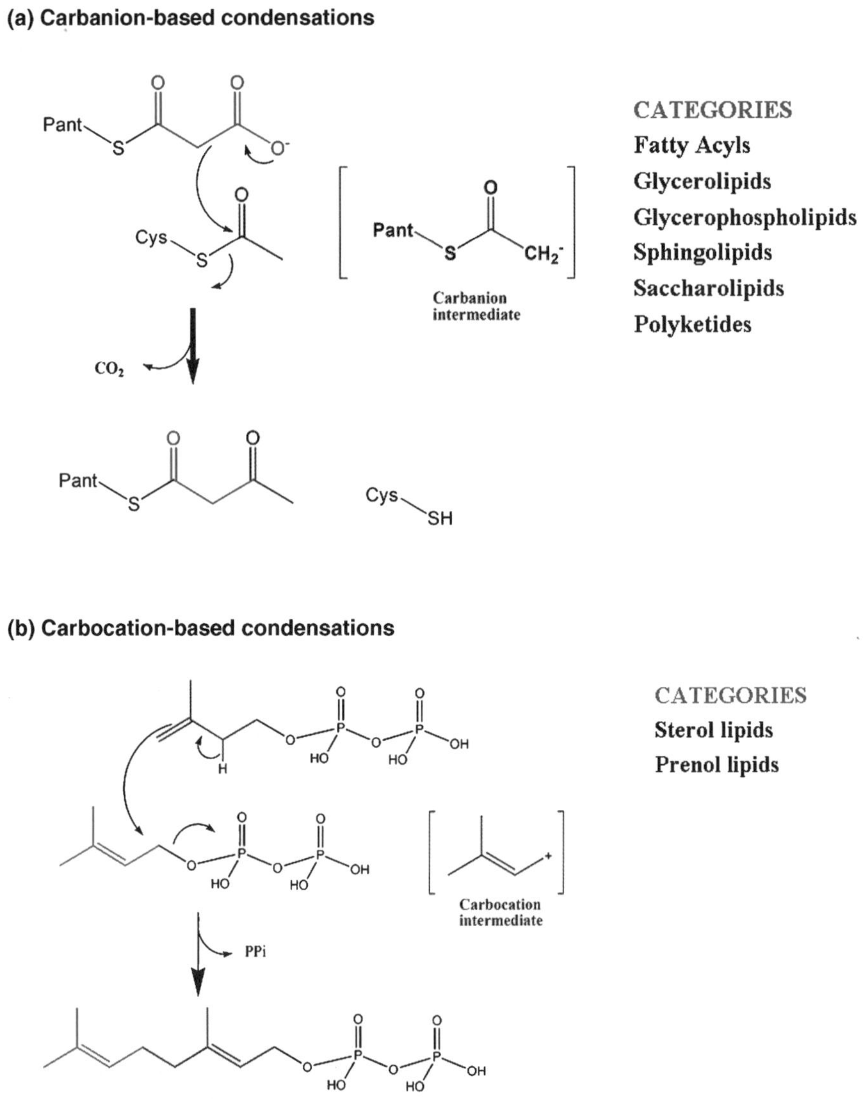

(a) Carbanion-based condensations

CATEGORIES
Fatty Acyls
Glycerolipids
Glycerophospholipids
Sphingolipids
Saccharolipids
Polyketides

(b) Carbocation-based condensations

CATEGORIES
Sterol lipids
Prenol lipids

Figure 7.1 The LIPID MAPS chemistry-based approach defines lipids as molecules that may originate entirely or in part by (a) carbanion-based condensations of thioesters and/or by (b) carbocation-based condensations of isoprene units.

alphabetic characters, a total of 1.68 million possible combinations can be generated, providing ample scalability within each subclass. In cases where lipid structures were obtained from other sources such as LipidBank or LIPIDAT, the corresponding IDs for those databases are included to enable cross-referencing. The LIPID MAPS classification scheme has been adopted

Table 7.2   LIPID MAPS Lipid Categories and Examples.

Category	Abbrev.	Example
Fatty acyls	FA	Dodecanoic acid
Glycerolipids	GL	1-Hexadecanoyl-2-(9Z-octadecenoyl)-*sn*-glycerol
Glycerophospholipids	GP	1-Hexadecanoyl-2-(9Z-octadecenoyl)-*sn*-glycero-3-phosphocholine
Sphingolipids	SP	*N*-(Tetradecanoyl)-sphing-4-enine
Sterol lipids	ST	Cholest-5-en-3β-ol
Prenol lipids	PR	2*E*,6*E*-Farnesol
Saccharolipids	SL	UDP-3-*O*-(3*R*-Hydroxy-tetradecanoyl)-αD-*N*-acetylglucosamine
Polyketides	PK	Aflatoxin B_1

Table 7.3   Format of 12-Character LIPID ID.

Characters	Description	Example
1–2	Fixed database designation	LM
3–4	2-letter category code	FA
5–6	2-digit class code	03
7–8	2-digit subclass code	02
9–12	Unique 4-character identifier within subclass	AG12

by the LipidBank (http://lipidbank.jp) consortium in Japan. The LIPID MAPS classification scheme has also been adopted by KEGG (Kyoto Encyclopedia of genes and genomes) where functional hierarchies involving lipids, reactions, and pathways have been constructed (http://www.genome.ad.jp/brite/). The development of this system has been enriched by interaction with lipidologists across the world, anticipating that this system will be internationally accepted and utilized.

7.2.2   Nomenclature

Nomenclature of lipids falls into two main categories: systematic names and common or trivial names. The latter includes abbreviations which are a convenient way to define acyl/alkyl chains in acylglycerols, sphingolipids, and glycerophospholipids. The generally accepted guidelines for systematic names have been defined by the International Union of Pure and Applied Chemists and the International Union of Biochemistry and Molecular Biology (IUPAC-IUBMB) Commission on Biochemical Nomenclature (http://www.chem.qmul.ac.uk/iupac/).[2–5] The use of core structures such as prostanoic acid, cholestane, or phosphocholine is strongly recommended as a way of simplifying systematic nomenclature; commercially available software packages that perform structure-to-name conversions generally create overly complicated names for many categories of lipids.

Many lipids, in particular the glycerolipids, glycerophospholipids, and sphingolipids, may be conveniently described in terms of a shorthand name where abbreviations are used to define backbones, headgroups, and sugar units, and the radyl substituents are defined by a descriptor indicating carbon chain length and number of double bonds. These shorthand names lend themselves to fast, efficient text-based searches and are used widely in lipid research as compact alternatives to systematic names. The glycerophospholipids in the LIPIDAT database (http://www.lipidat.chemistry.ohio-state.edu), for example, may be conveniently searched with a shorthand notation that has been extended to handle side-chains with acyl, ether, branched-chain, and other functional groups.[9] The use of a shorthand notation for selected lipid categories[3] that incorporates a condensed text nomenclature for glycan substituents has been deployed by LIPID MAPS. The abbreviations for the sugar units follow the current IUPAC-IUBMB recommendations.[3]

7.2.3   Structure Representation

In addition to having rules for lipid classification and nomenclature, it is important to establish clear guidelines for drawing lipid structures. This is a prerequisite for any useful lipid molecular database in terms of consistent molecular structure presentation. Large and complex lipids are difficult to draw, which leads to the use of shorthand and unique formats that often generate more confusion than clarity among lipid researchers. The LIPID MAPS consortium has chosen a consistent format for representing lipid structures where, in the simplest case of the fatty acid derivatives, the acid group (or equivalent) is drawn on the right and the hydrophobic hydrocarbon chain is on the left (Scheme 7.1).

7.3   Lipid Molecular Databases

Modern biology has become increasingly sophisticated to permit complex database schemas and approaches which enable efficient data accrual, storage, and dissemination. A large number of repositories such as GenBank, SwissProt, and ENSEMBL (http://www.ensembl.org) support nucleic acid and protein databases; however, there are only a few specialized databases (*e.g.* LIPIDAT[9] and LipidBank[10]) that are dedicated to cataloging lipids. The LIPIDAT database, developed by Martin Caffrey's group at Ohio State University, focuses on the biophysical properties of glycerolipids and sphingolipids and contains over 12 000 unique molecular structures. The LipidBank online database in Japan contains over 6000 structures over a broad range of lipid classes and is a rich resource for associated spectral data, biological properties, and literature references.

Given the importance of these molecules in cellular function and pathology, it is essential to have a well-organized database of lipids with a defined ontology that is extensible, flexible, and scalable. The ontology of lipids must incorporate

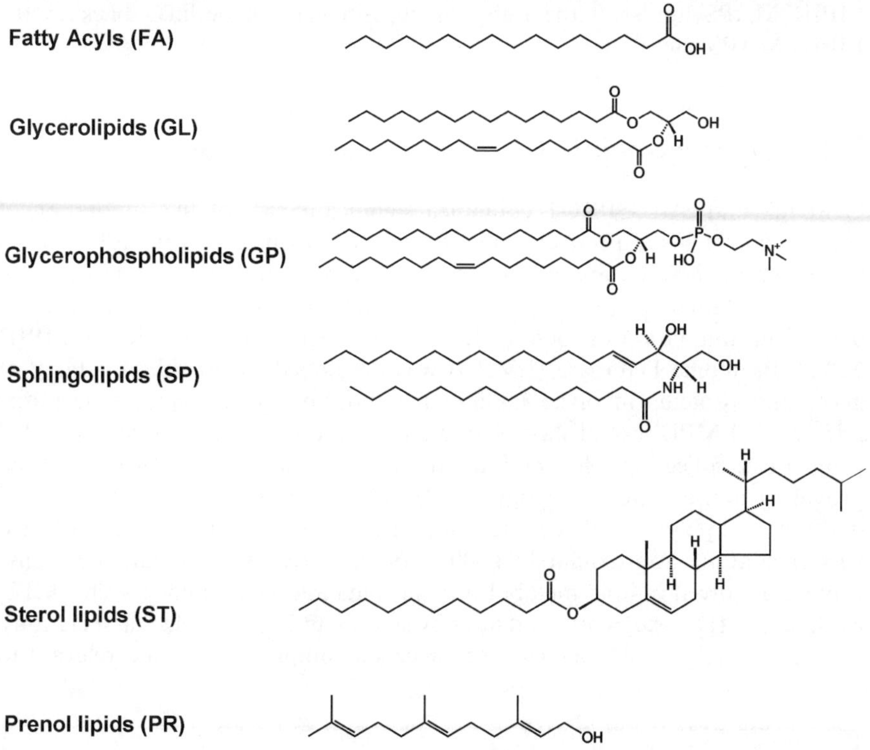

Fatty Acyls (FA)

Glycerolipids (GL)

Glycerophospholipids (GP)

Sphingolipids (SP)

Sterol lipids (ST)

Prenol lipids (PR)

Scheme 7.1 Consistent format for representing lipid structures.

classification, nomenclature, structure representations, definitions, related biological/biophysical properties, cross-references, and structural features (formula, molecular weight, number of carbon atoms, number of various functional groups, *etc.*) of all objects stored in the database. This ontology is then transformed into a well-defined schema that forms the foundation for a relational database of lipids. An object-relational database of lipids, based on the above classification scheme, and containing structural, biophysical, and biochemical characteristics is available on the LIPID MAPS website with browsing and searching capabilities. The database currently contains over 10,000 structures which have been obtained from the LIPID MAPS core facilities as well as the LipidBank and LIPIDAT databases. All structures have been classified and redrawn according to LIPID MAPS guidelines. A number of different molecular viewing formats: GIF image, Chemdraw CDX, and the Java-based Marvin and JMol interfaces are offered. This database stores curated information on lipids in a web-accessible format and will provide a community standard for lipids.

In an effort to increase public availability, LIPID MAPS lipid structures are now available on NCBI's PubChem website (http://pubchem.ncbi.nlm.nih.gov) where they have been assigned PubChem Substance IDs. All the deposited

LIPID MAPS lipids on the Pubchem website have hyperlinks back to the LIPID MAPS site.

7.4   Lipid-associated Protein/Gene Databases

GenBank and SWISSPROT contain a significant part of the annotation of genes and proteins respectively and most of the known lipid proteins have been annotated in these databases. However, there is no unique gene and protein database of "lipid" proteins that contains comprehensive summary and context dependent annotation of these molecules. In order to fill this void the LIPID MAPS Proteome Database (LMPD) was developed to provide a catalog of genes and proteins involved in lipid metabolism and signaling. The initial release of LMPD establishes a framework for creating a lipid-associated protein list, collecting relevant annotations, databasing this information, and providing a user interface (Figure 7.2). The LMPD is an object-relational database of lipid-associated protein sequences and annotations.[11] The current version contains approximately 4,000 records, representing human and mouse proteins involved in lipid metabolism and signaling. Users may search LMPD by database ID or keyword, and filter by species and/or lipid class associations; from the search results, one can then access a compilation of data relevant to

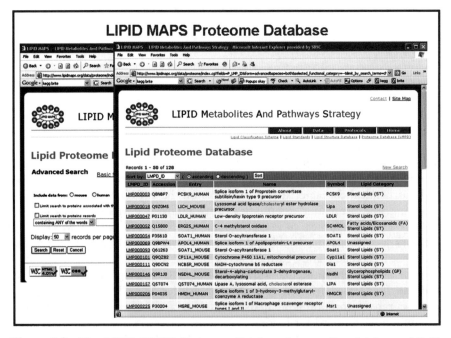

Figure 7.2   The LIPID MAPS Proteome Database is an online database of lipid-associated protein sequences and annotations which may be searched by various criteria. The URL is http://www.lipidmaps.org/data/proteome/.

each protein of interest, cross-linked to external databases. The LIPID MAPS Proteome Database (LMPD) is publicly available from the LIPID MAPS Consortium website (http://www.lipidmaps.org/data/proteome/index.cgi). A list of lipid-related GO (Gene Ontology, http://www.geneontology.org)[12] terms and KEGG[13] pathway data was compiled, using lipid-specific keywords, such as trivial names of classes, subclasses, and individual lipid compounds. The UniProt[14] proteins annotated with those GO and KEGG terms were then collected, and these proteins were classified (based on their substrates/products or interactions) according to the lipid classification scheme previously described. Annotations are organized by category: Record Overview, Gene/ GO/KEGG Information, UniProt Annotations, and Related Proteins. The record overview contains LMPD_ID, species, description, gene symbols, lipid categories, EC number, molecular weight, sequence length, and protein sequence. Gene information includes Entrez Gene ID, chromosome, map location, primary name, primary symbol, and alternate names and symbols; Gene Ontology (GO) IDs and descriptions, and KEGG pathway IDs and descriptions. UniProt annotations include primary accession number, entry name and comments such as catalytic activity, enzyme regulation, function, and similarity.

7.5 Lipid-associated Pathways and Networks

Lipids play central roles in energy storage, cell membrane structure, cellular communication, and regulation of biological processes such as inflammatory response, neuronal signal transmission, and carbohydrate metabolism. Organizing these processes into useful, interactive pathways, and networks represents a great bioinformatics challenge. The KEGG consortium maintains collection of manually drawn pathway maps representing current knowledge on the molecular interaction and reaction networks, several of which pertain to lipids (http://www.genome.jp/kegg/pathway.html), including fatty acid biosynthesis and degradation, sterol metabolism, and phospholipids pathways. Additionally, the KEGG Brite (http://www.genome.jp/kegg/brite.html) collection of hierarchical classifications includes a section devoted to lipids where the user can select a lipid of interest and view reactions and pathways involving that molecule. A number of category-specific lipid pathways have been constructed, notably the SphinGOMAP, a pathway map of approximately 400 different sphingolipid and glycosphingolipid species[15] (http://www.sphingomap.org). LIPID MAPS is developing a comprehensive graphical pathway display, editing, and analysis program (Biochemical Pathways Workbench) which will have the capability to display a variety of lipid-related entities and states such as lipids, enzymes, genes, pathways, activation states, and experimental results. This graphical tool will be tightly integrated with the underlying LIPID MAPS relational databases and will play an important role in the reconstruction of lipid networks using emerging data from LIPID MAPS core facilities and other public sources. The large amount of data emerging from experimental lipid research is continually increasing our knowledge of the complexity of these pathways.

7.5.1   Lipidomic Data

With the advent of sensitive analytical instrumentation such as mass spectrometry, it is now possible to obtain quantitative data on large numbers of lipid species under a variety of experimental conditions, allowing us to investigate time-dependent changes in lipid-associated processes in response to a variety of stimuli. Studies by Brown *et al.*[16,17] have quantified over 400 phospholipids from cell extracts using ESMS and statistical bioinformatics techniques. This ability to simultaneously assess the metabolic dynamics of hundreds of phospholipid species reveals a wealth of information regarding the cellular lipidome. On a more general scale, the LIPID MAPS consortium has embarked on a time-dependent study of a wide range of lipid classes in mouse macrophage cells, in response to stimulation of the toll-like receptor 4 (TLR4) with Kdo_2-Lipid A (http://www.lipidmaps.org/data/kdo2lipidatimecourse/). Quantitative data from these experiments are being used to validate existing lipid networks and elucidate novel interactions. Transcription factors often function as the end points of signal transduction pathways. Changes in the activities of these proteins result in changes in the rate of transcription of specific target genes. As a result, the mRNA composition of the cell is altered. In many cases this transcriptional reprogramming leads to changes in cellular proliferation, differentiation, and other cellular phenotypes. These global changes in mRNA profiles can be evaluated using gene expression microarrays. Gene-array technologies have also been utilized to assess lipid-associated proteins at the transcriptional level. Glass and co-workers[18] have used arrays containing 10 000 mouse sequences to study nuclear–receptor repression in macrophage cells via toll-like receptor (TLR) signaling. These nuclear receptors such as PPARs and LXRs play critical roles in lipid homeostasis. Both lipid mass spectrometric data and gene expression data are also made easily accessible to the research community at the LIPID MAPS website.

7.6   Bioinformatics Tools for Lipidomics Research

7.6.1   Data Collection

Data analysis of lipidomics experiments represents significant challenges both in volume and complexity. Typical experiments yield component lists of lipids with quantitative content data and a catalog of interactions and networks involving lipids. A truly comprehensive "lipidomics" approach must incorporate multiple separation and identification techniques, maintaining sufficient sensitivity to distinguish closely related metabolites while remaining robust enough to cope with the wide variety of heterogeneous classes of cellular lipid species. Electrospray ionization mass spectrometry (ESI-MS), with its extraordinary sensitivity and its capacity for high throughput, has become the technique of choice for the analyses of complex mixtures of lipids from biological samples.[19] From a bioinformatics standpoint, one must record and store all the experimental conditions and protocols pertaining to a particular set of experiments (metadata), as well as the actual experimental measurements (data). A variety of commercial and in-house laboratory information management systems (LIMS) have been deployed to capture and database the metadata. The LIPID MAPS consortium has developed a Java-based LIMS system to enable barcoding and data entry of the various protocols, reagents, and experimental parameters used at LIPID MAPS experimental facilities. The LIMS modules have been customized to capture metadata generated by mass spectrometry procedures and other experiments.

The mass spectral data itself is generated from a range of instruments such as MALDI, ion-trap, and quadrupole time-of-flight (QToF) spectrometers from different manufacturers – consequently the native binary data formats produced by each type of mass spectrometer also differ and are usually proprietary. This impedes the analysis, exchange, comparison, and publication of results from different experiments and laboratories, and prevents the bioinformatics community from accessing data sets required for software development. In an effort to overcome this problem, developers at the Institute for Systems Biology have introduced the "mzXML" format,[20] an open, generic XML (extensible markup language) representation of MS data and have also developed an accompanying suite of supporting programs to covert binary data files such as those created by Sciex/ABI Analyst, Micromass MassLynx, and ThermoFinnigan Xcalibur into the common mzXML format (Figure 7.3).

7.6.2 Standards Libraries

The interpretation of MS and NMR data for lipids can be very challenging due to a number of inherent properties of these molecules. Many lipids contain unsubstituted aliphatic chains or substructures whose ^{1}H NMR chemical shifts are in the upfield spectral region (0–2.5 ppm) and not well dispersed, even at high field strength. The structural heterogeneity of many classes of lipids precludes accurate and reliable prediction of MS fragmentation in tandem spectra, in contrast to the situation for MS interpretation of proteins where facile cleavage of the peptide bonds has led to the development of search algorithms which compare MS/MS fragmentation data to a list of "virtual spectra" computed from a protein database. In this situation, comparison of MS and NMR spectral data with libraries of lipid standards can be very informative. The Lipid Library website (http://www.lipidlibrary.co.uk/masspec.html) contains a large set of over 1600 electron-impact (EIMS) mass spectra of fatty acid derivatives, as well as ^{1}H and ^{13}C NMR spectra obtained from various sources. An online library of lipid standards, including tandem mass spectral data generated by the LIPID MAPS core facilities, has recently been made available to the public. This database currently consists of

Data Analysis Pipeline from heterogeneous MS sources
via mzXML

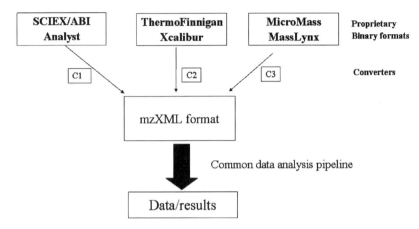

Figure 7.3   Raw MS data files in proprietary binary formats are converted to a common text-based mzXML document format, using vendor-specific conversion programs. The mzXML files may then be processed and analyzed by a single informatics pipeline.

approximately 200 analytes spanning 7 lipid categories. An online search interface allowing end-users to search by precursor or product ion mass has been developed and will be released shortly.

7.6.3   MS Prediction Tools

Certain classes of lipids such as acylglycerols and phospholipids composed of an invariant core (glycerol and headgroups) and one or more acyl/alkyl substituents are good candidates for MS computational analysis since they fragment more predictably in ESMS leading to loss of acyl side-chains, neutral loss of fatty acids, loss of water, and other diagnostic ions depending on the phospholipid headgroup.[19] It is possible to create a virtual database of permutations of the more common side-chains for glycerolipids and glycerophospholipids and calculate "high-probability" product ion candidates in order to compare experimental data with predicted spectra. The LIPID MAPS group has developed a suite of search tools allowing a user to enter an m/z value of interest and view a list of matching structure candidates, along with a list of calculated of neutral-loss ions and "high-probability" product ions. These search interfaces have been integrated with structure drawing and isotopic distribution tools. The LipidBase group at the University of Tokyo has created a "Lipid Search" online interface (http://lipidsearch.jp/manual_search/) for performing similar types of analyses on phospholipids. A novel algorithm and databases has recently been developed to predict the presence of eicosanoids and related polyunsaturated fatty acids (PUFAs) from LC-UV-MS/MS data where emphasis is placed on predicting fragments generated by carbon–carbon bond cleavage.[21]

7.6.4 Automated Structure Drawing

In addition to having rules for lipid classification and nomenclature, it is important to establish clear guidelines for drawing lipid structures. Large and complex lipids are difficult to draw, which leads to the use of many unique formats that often generate more confusion than clarity among lipid research community. For example, the use of SMILES strings to represent lipid structures, while being very compact and accurate in terms of bond connectivity, valence, and chirality, causes problems when the structure is rendered. This is due to the fact that the SMILES format does not include 2-dimensional coordinates – therefore the orientation of the structure is arbitrary, making visual recognition and comparison of related structures much more difficult. Additionally, the structure-drawing step is typically the most time-consuming process in creating molecular databases of lipids. However, many classes of lipids lend themselves well as targets for automated structure drawing, due to

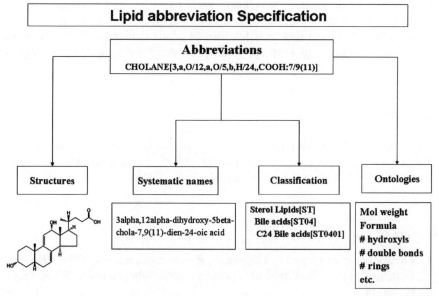

Figure 7.4 Lipid structure, nomenclature, classification, and ontology creation may be performed computationally, using well-defined, unique abbreviations as a starting point. This example demonstrates the conversion of a text abbreviation for a bile acid into a dataset containing structure (molfile), systematic name, classification, and various molecular attributes such as formula, molecular weight, number of functional groups, double bonds, and rings.

their consistent 2-dimensional layout. A suite of structure-drawing tools has been developed and deployed which dramatically increase the efficiency of data entry into lipid-structure databases and permit "on-demand" structure generation in conjunction with a variety of mass spectrometry-based informatics tools. Concurrently, a generalized lipid abbreviation format has been developed which enables structures, systematic names, and ontologies to be generated automatically from a single source format (Figure 7.4).

7.7   Conclusion

Lipids are recognized as key participants in the regulation and control of cellular function and have important roles in signal transduction processes. The diversity in lipid chemical structure presents a challenge for establishing practical methods to generate and manage high volumes of complex data that translate into a snapshot of cellular lipid changes. The need for high-quality bioinformatics to manage and integrate experimental data becomes imperative at several different levels: (a) definition of lipid classification and ontologies, (b) relational database design, (c) capture and automated pipelining of experimental data, (d) efficient management of metadata, (e) development of lipid-centric search tools, (f) analysis and visual display of results, and (g) integration of the lipid knowledge base into biochemical pathways and interactive maps. Whereas many of these requirements are shared by other areas of biology, the unique structural and functional features of lipids demand a high degree of specialization which will keep the emerging lipid bioinformatics community busy for many years to come.

Acknowledgements

The author thanks Dr Shankar Subramaniam for his review of this document. This work was supported by the LIPID MAPS Large-Scale Collaborative Grant GM-069338 from the National Institutes of Health.

References

1. Encyclopedia of Genetics, Genomics, Proteomics and Bioinformatics, eds. L. B. Jorde, P. F. R. Little, M. J. Dunn and S. Subramaniam, Wiley Press, NJ, 2005.
2. *J. Lipid Res.*, 1978, **19**, 114–128. Website: http://www.chem.qmul.ac.uk/iupac/lipid/.
3. M. A. Chester, *Eur. J. Biochem.*, 1998, **257**, 293–8. Website: http://www.chem.qmul.ac.uk/iupac/misc/glylp.html.
4. *Eur. J. Biochem.*, 1987, **167**, 181–184. Website: http://www.chem.qmul.ac.uk/iupac/misc/prenol.html.

5. *Eur. J. Biochem.*, 1989, **186**, 429–458. Website: http://www.chem.qmul.ac.uk/iupac/steroid/.
6. E. Fahy, S. Subramaniam, H. A. Brown, C. K. Glass, A. H. Merrill Jr., R. C. Murphy, C. R. Raetz, D. W. Russell, Y. Seyama, W. Shaw, T. Shimizu, F. Spener, G. Van Meer, M. S. VanNieuwenhze, S. H. White, J. L. Witztum and E. A. Dennis, *J. Lipid Res.*, 2005, **46**, 839–862.
7. A. Smith, Oxford Dictionary of Biochemistry and Molecular Biology, 2nd edn, Oxford University Press, 2000.
8. W. W. Christie, Lipid Analysis, 3rd edn, Oily Press, Bridgewater, UK, 2003.
9. M. Caffrey and J. Hogan, *J. Chem. Phys. Lipids.*, 1992, **61**, 1–109. Website: http://www.lipidat.chemistry.ohio-state.edu.
10. K. Watanabe, E. Yasugi and M. Ohshima, *Trends Gycosci Glycotechnol.*, 2000, **12**, 175–184.
11. D. Cotter, A. M. Maer, C. Guda, B. Saunders and S. Subramaniam, *Nucleic Acids Res.* (Database issue), 2006, **34**, D507-1.
12. M. A Harris, J. Clark, A. Ireland, J. Lomax, M. Ashburner, R. Foulger, K. Eilbeck, S. Lewis, B. Marshall and C. Mungall *et al.*, *Nucleic Acids Res.*, 2004, **32**, D258–D261.
13. M. Kanehisa and S. Goto, *Nucleic Acids Res.*, 2000, **28**, 27–30.
14. R. Apweiler, A. Bairoch, C. H. Wu, W. C. Barker, B. Boeckmann, S. Ferro, E. Gasteiger, H. Huang, R. Lopez and M. Magrane *et al.*, *Nucleic Acids Res.*, 2004, **32**, D115–D119.
15. M. C. Sullards, E. Wang, Q. Peng and A. H. Merrill Jr., *Cell Mol. Biol. (Noisy-le-grand)*, 2003, **49**(5), 789–797.
16. J. S. Forrester, S. B. Milne, P. T. Ivanova and H. A. Brown, *Mol. Pharmacol.*, 2004, **65**, 813–821.
17. P. T. Ivanova, S. B. Milne, J. S. Forrester and H. A. Brown, *Mol. Interventions*, 2004, **4**, 86–96.
18. S. Ogawa, J. Lozach, C. Benner, G. Pascual, R. K. Tangirala, S. Westin, A. Hoffmann, S. Subramaniam, M. David, M. G. Rosenfeld and C. K. Glass, *Cell*, 2005, **122**(5), 707–721.
19. R. C. Murphy, J. Fiedler and J. Hevko, *Chem. Rev.*, 2001, **101**, 479–526.
20. P. G. Pedrioli, J. K. Eng, R. Hubley, M. Vogelzang, E. W. Deutsch, B. Raught, B. Pratt, E. Nilsson, R. H. Angeletti, R. Apweiler, K. Cheung, C. E. Costello, H. Hermjakob, S. Huang, R. K. Julian, E. Kapp, M. E. McComb, S. G. Oliver, G. Omenn, N. W. Paton, R. Simpson, R. Smith, C. F. Taylor, W. Zhu and R. Aebersold, *Nat. Biotechnol.*, 2004, **22**(11), 1459–1466.
21. Y. Lu, S. Hong, E. Tjonahen and C. N. Serhan, *J. Lipid Res.*, 2005, **46**(4), 790–802.

CHAPTER 8

Mass Spectrometry in Glycobiology

JOÃO RODRIGUES, CARLA ANTONIO, SARAH
ROBINSON AND JANE THOMAS-OATES

Department of Chemistry, University of York, Heslington, York
YO10 5DD, UK

8.1 Introduction

The field of glycobiology is considered to encompass the study of glycoconjugates, molecules that contain carbohydrates, in terms of their structure and function and the roles they play in biology. However, glycoproteins and polysaccharides, the glycoconjugates that generally attract most attention from glycobiologists, have little relevance in a metabolomic/metabonomic context, although glycolipids, oligosaccharides, small molecule glycosides and intermediates in carbohydrate metabolism all fall within the scope of metabolite glycobiology.

The analysis of glycoconjugates is particularly challenging, since these compounds are structurally very diverse, and occur in nature in a wide range of isomeric forms, both structural and diastereoisomeric. Many of the monosaccharides of which these molecules consist are epimers (diastereoisomers having opposite configurations at only one stereogenic centre) (Figure 8.1a). These epimers exist as both α and β anomers (stereoisomers deriving from the centre of chirality generated on hemiacetal ring closure) (Figure 8.1b). These different forms of the monosaccharides may be joined to each other or to non-carbohydrate moieties either *via* the anomeric or the other carbons in the molecule (Figures 8.1c and 8.1d), and may be joined to each other in linear or branched chains (Figure 8.1d).

This level of structural complexity places heavy demands on the analytical approaches used to identify and differentiate these isomeric structures. Mass spectrometry offers particular strengths in its compatibility with high resolution separation techniques, especially GC, LC and CE, and its ready applicability

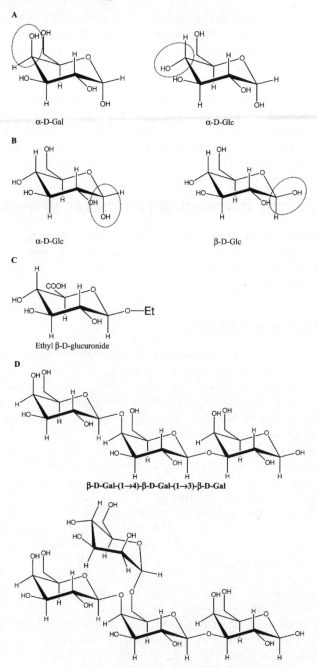

Figure 8.1 (a) Structures of the epimers galactose and glucose; (b) structures of the anomers α- and β-glucose; (c) structure of ethyl glucuronide; (d) structures of a linear trisaccharide and a branching tetrasaccharide.

for determining monosaccharide sequence, branching and patterns of non-carbohydrate substitution. Nonetheless, due to their polarity, structural variety, the instability of some glycoconjugates, and their poor UV absorbance, glycoconjugates are a particularly challenging subclass of the metabolome to analyse.

In this chapter we discuss mass spectrometry-based methods for the analysis of glycoconjugate metabolites, focusing on intermediates of cellular energy metabolism (although we intentionally avoid plant metabolites, as these are covered in Chapter 10), glycoconjugate biomarkers of inborn metabolic disorders, and glucuronide biomarkers of opiate and alcohol abuse.

8.2 Glycolytic Intermediates and Sugar Phosphates

Phosphorylated carbohydrates, such as glycolytic intermediates and sugar phosphates, are key compounds in metabolism. They are important intermediates of cellular energy metabolic pathways, such as glycolysis and the pentose phosphate pathway.

8.2.1 Sampling and Extraction

Sampling of cells and rapid inactivation of metabolic processes, prior to cell separation and extraction, are of the utmost importance when measuring intermediary metabolites with high turnover rates, such as glycolytic intermediates and sugar phosphates. Important requirements when performing an extraction protocol are that: (i) metabolites are completely extracted; (ii) metabolite levels do not change either enzymatically or chemically; and (iii) metabolites are not destroyed.

The importance of fast sampling has been well reported.[1] An extraction protocol based on a cold methanol quenching method was applied for the determination of glycolytic metabolites in yeast cells. The method involves rapid quenching of metabolism by spraying yeast cells with methanol/water (60% v/v) at –40 °C, followed by a neutral extraction of metabolites with chloroform at –40 °C. Quenching time was estimated to be less than 1 second. Cold methanol quenching was shown to be an efficient method to reduce changes in metabolite concentration during sampling, as long as the temperature is kept below –20 °C throughout the procedure; it has been used successfully in a range of different applications.[2–8]

Another useful example of rapid sampling is by quenching in liquid nitrogen.[9,10] Hajjaj *et al.*[9] used both cold methanol and liquid nitrogen to stop metabolism and concluded that both methods were equally efficient. Consequently, metabolites, including glycolytic intermediates, were extracted from filamentous fungi using different protocols. It was shown that extraction using boiling buffered ethanol (75% v/v) followed by solvent evaporation gave the best recoveries of metabolites. One of the advantages of the boiling ethanol method is that the evaporation of ethanol after the extraction results in a concentration of metabolites.[11] Glycolytic intermediates and sugar phosphates can also be extracted from animal tissues using trichloroacetic acid.[12,13] However, the simplicity and universality of the cold methanol approach as described by de Koning and van Dam[1] turns out to be an advantage, as it can be used as a general tool for quantification of metabolites in different organisms, such as bacterial, yeast and mammalian cells.[14] Furthermore, a recent publication by Villas-Bôas *et al.*[15] suggested that quenching yeast cells using cold methanol followed by extraction of intracellular metabolites, including sugar phosphates, with pure methanol presented the most reliable method when compared to other sample preparation methods.

8.2.2 Sample Handling and Derivatisation

In the extraction step, large volumes of solvents are often used and metabolites are thus obtained in dilute solutions. Prior to analysis it is therefore necessary to remove solvents and concentrate the sample. Solvents can be evaporated to dryness using a centrifugal concentrator.[16] Another method commonly used for sample concentration is freeze-drying, where aqueous samples are dehydrated and dried from the frozen solution. Using this process, metabolites are stable and freeze-dried samples can be stored for long periods.[17] Samples can be then stored, preferably at –80 °C, prior to analysis.

Glycolytic intermediates and sugar phosphates are in volatile and very unstable compounds. In order to analyse this class of metabolite, for example by GC-MS, derivatisation is, therefore, required. Silylation is the simplest, quickest and most popular method of derivatisation for GC-MS applications.[18,19] Organic compounds containing reactive functional groups are converted to their silylated derivatives (usually trimethylsilyl, TMS) which are highly volatile, thermally more stable, and therefore, have excellent gas chromatographic properties.[20] Some metabolites (*e.g.*, amino acids, carbohydrates) contain more than one reactive group, and thus, more than one derivatisation product may be formed. The main disadvantage associated with derivatisation protocols is the time and sample losses associated with the additional step in sample handling; in many cases, alternative analytical methods are suggested to avoid the derivatisation step.

8.2.3 Analytical Methods

Analytical methods for the determination of glycolytic intermediates and sugar phosphates must be selective and sensitive. Selectivity in detection helps to overcome interferences from other extracted compounds and matrix interferences, which is a particularly important consideration in biological samples. High sensitivity is necessary to minimise the amount of biological material that is extracted, reduce the sample load on the analytical system, and to detect the low concentrations of metabolites (below 1 mM) present in biological samples. MS, combined with on-line separations, is the most widely applied technology in metabolome analysis. It allows the detection and identification of a wide range of metabolites in a single run. Moreover, the ability to perform multi-stage tandem MS experiments provides specificity to the analysis and enhances sensitivity by improving the signal-to-noise ratio.

8.2.3.1   Direct Infusion

Direct infusion MS is a rapid and high-throughput tool used to provide sample classification in metabolic fingerprinting-based strategies with minimal sample preparation. An interesting example was recently published by Castrillo and co-workers,[21] reporting a direct infusion electrospray MS method for metabolite profiling of intracellular metabolites of the yeast *Saccharomyces cerevisiae*. Sampling and extraction methods were optimised and metabolites were reproducibly analysed by direct infusion positive ion electrospray MS. In this study, tricine was chosen as a non-salt buffer for quenching and extraction of metabolites, and more than 25 metabolites, ranging from glycolytic intermediates and nucleotides to amino acids, could be simultaneously detected. Moreover, this method can be applied for the study of other microorganisms and biological systems.

Targeted analysis using MALDI-TOF MS was reported by Wittmann and Heinzle[22] for the quantification of lysine, alanine and glucose in *Corynebacterium glutamicum*. The method uses 2,5-dihydroxybenzoic acid (DBH) as the matrix and L-α-^{15}N-lysine, L-1-^{13}C-alanine and 1-^{13}C-glucose stable isotope labelled internal standards. Sensitive and reproducible quantification of metabolites was achieved with standard deviations of 3.2% for glucose in positive ion MALDI-TOF MS. These MALDI-TOF MS results showed excellent agreement with those obtained by HPLC (lysine, alanine) and enzyme assay (glucose), highlighting the possibility of analysing low molecular weight metabolites by MALDI-TOF MS with minimal sample amounts (1–3 μL volumes of low mM concentration samples) required.

Advantages of using direct infusion MS analysis are the speed and ease of the analysis compared to conventional chromatographic separations. The choice of buffers in MS analysis is of the utmost importance as salt buffers can cause ion suppression effects and compromise the mass spectrometric analysis.

8.2.3.2   GC-MS

GC-MS-based methods have been extensively applied in metabolomics for the analysis of volatile and thermally stable compounds, providing high resolution separations and subsequent sensitive detection of compounds by mass spectrometry. However, the majority of metabolites analysed require chemical derivatisation to impart volatility and thermal stability, with silylation the most popular method of derivatisation for GC-MS applications. An interesting and recent example of a GC-MS method was reported by Strelkov *et al.*[23] for the comprehensive metabolome analysis of *Corynebacterium glutamicum*. Different classes of metabolites were analysed by GC-MS as their silylated derivatives. Simultaneous detection of approximately 300 metabolites was achieved and a total of 121 metabolites (corresponding to 168 derivatives) were identified, ranging from amino acids to organic acids, sugars and sugar phosphates, fatty acids, and other classes of compound. This approach was also employed to study the effect of different growth conditions on *Corynebacterium glutamicum*. This constitutes, a rapid, reliable method allowing the sensitive and simultaneous analysis of a wide and significant number of metabolites using a GC-MS approach.

8.2.3.3 LC-MS

The column chemistry most commonly used in LC-MS applications is reversed-phase C_{18}. However, polar metabolites, such as glycolytic intermediates and sugar phosphates, show minimal retention and generally elute close to the void volume without chromatographic separation. To overcome this, other column chemistries along with MS-compatible mobile phase compositions are required.

Feurle *et al.*[24] applied negative ion mode HPLC-ESI-MS/MS for the analysis of sugar phosphate standards using a β-cyclodextrin bonded stationary phase and an MS compatible mobile phase composed of acetonitrile/aqueous ammonium acetate. In addition to the successful separation, detection and structural characterisation of the sugar phosphates by MS/MS, identification of isomeric compounds such as glucose-1-phosphate (Glc1P) and glucose-6-phosphate (Glc6P) was achieved based on their product ion spectra, which yield very different and characteristic product ions. Moreover, this on-line ESI-MS/MS system allowed the detection and structural characterisation of D-1-deoxyxylulose-5-phosphate, enzymatically formed from pyruvate and D-glyceraldehyde phosphate by yeast transketolase.

Buchholz *et al.*[25] reported the quantification of intracellular metabolites in *Escherichia coli* K12 under defined growth conditions by means of LC negative-ion mode electrospray MS/MS. Two MS compatible chromatographic methods were used: method 1 used two Nucleodex β-OH columns in series and method 2 used a Hypercarb porous graphitised carbon column. It was shown that, although the Hypercarb column required longer equilibration times, its use led to improved separation of isomers such as Glc6P/fructose-6-phosphate (Fru6P) and 3-phosphoglycerate (3PG)/2-phosphoglycerate (2PG), not resolved using method 1. The LC-MS method was validated by comparison of the results with those of enzymatic assays and an HPLC-UV method; the results were well correlated. Enzymatic methods have the advantage of being very specific, but they require a relatively large sample volume (100–300 μL/assay) and only 5 or 6 metabolic intermediates could be detected. Using the Buchholz LC-MS method, determination of more than 15 intracellular metabolites, including glycolytic intermediates and nucleotides, was possible using minimal sample volumes (10–20 μL) and detection limits between 0.02 and 0.50 mM.

van Dam *et al.*[26] reported the analysis of glycolytic intermediates in the yeast *Saccharomyces cerevisiae* using high-performance anion-exchange chromatography (HPAEC) coupled to negative-ion mode electrospray ionisation tandem MS. A sodium hydroxide gradient was used to separate glycolytic intermediates. Because this is not an MS compatible eluent, a commercially available post-column membrane suppressor (ASRS®, from Dionex) was used to reduce sodium hydroxide by proton exchange and, therefore, avoid ion suppression effects and subsequent loss of sensitivity. Using this method, detection limits as low as 1.5 to 16 nmol g^{-1} dry weight (DW) of *Saccharomyces cerevisiae* cells were achieved for pyruvate and phosphoglycerate (3PG + 2PG), respectively. The intracellular concentrations of glycolytic intermediates such as Fru6P and Glc6P were estimated to be 182 nmol g^{-1} DW and 1282 nmol g^{-1} DW, respectively. The detection limits of this method were at least a factor of 30 lower than those reported by Buchholz *et al.*[25] The van Dam *et al.*[26] LC-MS/MS approach was recently applied by Wittmann *et al.*[27] for the measurement of glycolytic and TCA cycle intermediates during cell-cycle-related oscillation in *Saccharomyces cerevisiae*.

Another interesting example was reported by Wamelink *et al.*[28] who developed an LC-MS/MS method for the analysis of sugar phosphate intermediates of the pentose phosphate pathway (PPP) in bloodspots, fibroblasts and lymphoblasts. LC separation was carried out by means of an ion pair loaded C$_{18}$ column and sugar phosphate detection was performed by negative-ion mode electrospray MS/MS operating in multiple reaction monitoring (MRM) mode. It was shown that all sugar phosphates generate a characteristic fragment ion at m/z 97, and, therefore, the transition of the intact sugar phosphate to m/z 97 was used for MRM analysis. Using this LC method the isomeric pairs ribulose-5-phosphate/xylulose-5-phosphate and Glc6P/Fru6P were not resolved, with each pair eluting as a single unresolved peak. However, the method showed good sensitivity with detection limits ranging from 0.1 to 1.0 μmol L^{-1}, with the exception of erythrose-4-phosphate and 3PG in bloodspots, where the detection limit was found to be 10 μmol L^{-1}. This method was applied to two new inherited defects of metabolism: ribose-5-phosphate isomerase (RPI) and transaldolase (TALDO) deficiency. Cultured cells from patients affected with TALDO showed an increased concentration of sedoheptulose-7-phosphate and cultured cells from patients with RPI deficiency showed a decreased formation of ribose-5-phosphate (Rib5P). In the light of these results, Wamelink and co-workers suggested that the accumulating intracellular sugar phosphate concentrations may play an important role in the pathogenesis of these two newly discovered defects of the PPP.

The LC-MS methods applied to date for the analysis of sugar phosphates and glycolytic intermediates have proven to be efficient for the identification and determination of these metabolic intermediates in complex biological samples. The challenge in LC-MS applications for metabolome analysis still lies in the development of appropriate LC-MS methods based on new column chemistries, using MS compatible mobile phases in order to achieve efficient on-line coupling with MS.

8.2.3.4 CE-MS

Since glycolytic intermediates and sugar phosphates lack a UV-vis chromophore, most work on the analysis of these metabolites by capillary electrophoresis (CE) uses indirect UV detection with various background electrolytes (BGEs). CE was used by Ciringh and Lindsey[29] for the analysis of sugar phosphate standards with indirect UV detection using sorbic acid as the background electrolyte and chromophore. It was shown that, using this CE method, isomers such as Glc1P and Glc6P or ribose-1-phosphate and Rib5P were separated. Separation of a mixture of seven sugar phosphate standards was achieved in 6 minutes and sharp peaks were observed. Another example was reported by Soga and Ross[30] for the simultaneous determination of 82 compounds including pyruvate, glucose, fructose, sucrose and trehalose by CE with indirect UV detection using 2,6-pyridine dicarboxylic acid (PDC) as the BGE. This CE method was further extended by Soga and Imaizumi[31] for the determination of 206 compounds, including sugar phosphates.

However, in order to analyse metabolites from complex mixtures, an increase in sensitivity is required and can be obtained by coupling CE to MS detection. CE provides fast analysis and efficient separations, and MS provides the high sensitivity and selectivity required in metabolome analysis of biological samples.

Recently, CE-MS methods have been developed for the analysis of anionic intermediates extracted from *Bacillus subtilis*. Soga *et al.*[32] developed a simple and reliable CE-ESI-MS method for the comprehensive analysis of intracellular metabolites of the glycolysis and TCA cycle in *Bacillus subtilis* samples. In a first step, a mixture of 32 standards including phosphorylated carboxylic acids, sugar phosphates and glycolytic intermediates were separated by CE and selectively detected by MS using a sheath-flow electrospray ionisation interface. This method was applied to the comprehensive analysis of intracellular metabolites extracted from *Bacillus subtilis*, including glycolytic intermediates and sugar phosphates. Even though migration times of some compounds were very similar (*e.g.*, succinate, malate, 2-oxoglutarate and phosphoenolpyruvate), MS successfully detected and differentiated them. However, isomeric compounds such as Fru6P/Glc6P and 2PG/3PG were not fully resolved by CE, nor identified and quantified by MS. Nevertheless, this CE-ESI-MS method allowed the simultaneous determination of 27 metabolites from *Bacillus subtilis* in less than 25 minutes. The method provides excellent reproducibility and good linearity with relative standard deviations better than 0.4% for migration times and less than 5.4% for peak areas. Detection limits ranged from 9 to 200 fmol at a signal-to-noise ratio of 3.

8.2.4 Data Analysis

In this post-genomic era, bioinformatic tools that allow the integration of qualitative and quantitative data sets have been rapidly developed and applied in metabolomics studies, contributing to a better understanding of the function of unknown genes. The data-processing approaches most commonly used in metabolome analysis include unsupervised methods such as principal component analysis (PCA) and cluster analysis (see Chapter 11).

8.3   Metabolic Disorders of Carbohydrate Metabolism

There are currently just two inborn errors of metabolism (IEMs) routinely screened for in 99% of newborn babies in the UK: phenylketonuria (PKU) and congenital hypothyroidism (CHT). (For more information about the UK newborn screening programme centre, see http://www.ich.ucl.ac.uk/newborn/). IEMs are hereditary, permanent biochemical disorders and although many IEMs are relatively uncommon (in the UK 250 diagnoses of PKU and CHT are annually reported from over 600 000 neonatal screens), it is paramount that these disorders receive an early diagnosis so that treatment can be given as early as possible in order to minimise the devastating and irreversible damage which can be caused.

Genetic abnormalities of carbohydrate metabolism include the following disorders: galactosemia, UDP galactose 4-epimerase deficiency, glycogenoses (glycogen storage diseases), fructose intolerance, fructosuria, fructose 1,6-diphosphatase deficiency, pentosuria, pyruvate dehydrogenase complex deficiency, and pyruvate carboxylase deficiency. Unfortunately, many of these carbohydrate metabolic disorders are often incorrectly diagnosed as diabetes, for example L-xylulose in urine is diagnostic of pentosuria and not diabetes, similarly fructose in the urine is diagnostic of fructosuria. These misdiagnoses can waste valuable time in obtaining the correct diagnosis for the patient and implementing the appropriate treatment.

The treatment and relief of symptoms for many of these carbohydrate metabolic disorders can be achieved through strict diet control. However, in some of these disorders, such as Pompe's disease, there is no current treatment available. Most disorders have been studied with an aim to identify biomarkers in the various body fluids: blood, plasma and urine. However, many early studies of these disorders (for example thin-layer chromatography is used in the identification of lysosomal storage diseases, LSDs) have suffered from a lack of specificity and sensitivity, and found that the complex nature of the biological substrate can cause difficulties in the analysis. In the light of these limitations and the requirement for the rapid and accurate study of biomarkers in these disorders, contemporary mass spectrometry has been found to be an essential tool in newborn screening that can make possible the early diagnosis of various metabolic diseases. However, the expense, effort and effect of implementing mass spectrometry, and in particular, MS/MS into current newborn screening programmes is not a small undertaking. Indeed proposals for pilot studies to screen for rare metabolic diseases using MS/MS for screening programmes are still being carried out in many countries.[33,34]

8.3.1 Sampling and Extraction

The most common biofluids used to determine metabolic disorders of carbo-hydrate metabolism are urine (which is the least invasive sample to obtain) and red/white blood cells or whole blood. However, tissue biopsies and MRI scans are required for the diagnosis of some disorders. In the case of fructose 1, 6-diphosphatase deficiency, diagnosis is not *via* body fluid samples but through the use of a fructose challenge test or a needle biopsy of the liver. Although far less invasive than the liver biopsy, the fructose challenge test also poses a significant risk for the individual under test. During a fasting period the individual is given high fructose-content food and their blood glucose and lactate levels are monitored. In affected individuals, glucose levels are low and lactate high, and are the cause of the hypoglycaemia and lactic acidosis that the individual suffers. The individual must be monitored very carefully during this test as blood glucose and lactate may fall or rise, respectively, to dangerous levels.

Metabolic abnormalities exist prior to the onset of the disease, therefore by detection of abnormal metabolites or abnormally elevated metabolites in blood, urine and other body fluids, diagnosis can be made in pre-symptomatic neonates. The IEMs routinely tested for at present in the UK are diagnosed from blood taken from a prick of the heel of babies approximately 1 week old. The droplets of blood are spotted onto a Guthrie card,[35] and the dried blood droplets on these cards may be stored for many months. In addition to the Guthrie test, it has been reported that babies in critical care units require more specific sampling, as blood transfusions, nil by mouth status and administra-tion of heparanised solutions can affect the newborn screen result.[34]

Inborn errors of metabolism occur when an enzyme does not properly catalyse the transformation of one metabolite to another. The levels of meta-bolites are known to be within a certain range for healthy individuals, so when the enzyme doesn't function, levels of the metabolite and its by-products are distinctly elevated in the urine. Urine can, thus, be an effective biofluid sample for the study and analysis of IEMs. Urine may be obtained as a dried sample on filter paper, which is then extracted into distilled water. Alternatively, urine is frozen on dry ice, stored at $-20\,^{\circ}C$ or lower (often with desiccant) and prepared in aliquots to prevent multiple freeze/thawing of samples before/during ana-lysis. Urine samples that are to be directly infused into mass spectrometers are usually centrifuged thoroughly beforehand.

In addition to blood and urine, biomarkers of LSDs have been identified using amniotic fluid.[36] Distinctive protein, oligosaccharide and glycolipid markers have been found to provide unique signature metabolic profiles in many LSD-affected individuals. In the human fetus, renal glomerular filtration starts 9–12 weeks after conception, and this contributes fetal urine to the amniotic fluid, hence oligosaccharide biomarkers are elevated and were detected from control populations by the second trimester. Little information is given in that paper about how the samples were obtained. However, the post-sampling treatment of the amniotic fluid was by lyophilisation and storage at $-20\,^{\circ}C$. The oligosaccharide metabolite biomarkers were then prepared as 1-phenyl-3-methyl-5-pyrazolone PMP derivatives for analysis (see Section 8.3.2).

8.3.2 Sample Handling and Derivatisation

Guthrie cards are generally stored at –20 °C, and in one instance samples were retrieved for further analysis from a bank of Guthrie cards (stored since they were first used for screening) after 23 years.[37] Retrospective analysis such as this is only possible because the cards are stored at low temperature. However, the cards would not be useable for analysis of protein markers, due to the instability of proteins at this temperature. The dried blood spots are punched from Guthrie cards in, usually, ∼ 3 mm discs. The blood is then extracted into solution from the card into an appropriate solution for the next stage of analysis. In this instance derivatisation of the oligosaccharide component of the blood was carried out with 1-phenyl-3-methyl-5-pyrazolone (Scheme 8.1).

A derivative commonly used for monosaccharides to enable GC-MS analysis is the TMS derivative, which produces trimethylsilyl ethers. In the diagnosis of primary hyperoxaluria type II (PH2), heavy excretion of L-glyceric acid occurs because of D-glycerate dehydrogenase deficiency. However, D-glyceric acid is found in urine from patients with D-glycerate kinase deficiency. The absolute configurations of these isomers are, therefore, important, as different optical isomers can originate from separate metabolic pathways which reflect different enzyme defects. The absolute configuration of glyceric acid cannot be determined using TMS derivatisation, thus O-acetyl-(+)-2-butylation with analysis on a DB-5MS column was carried out in this instance.[38] The method was

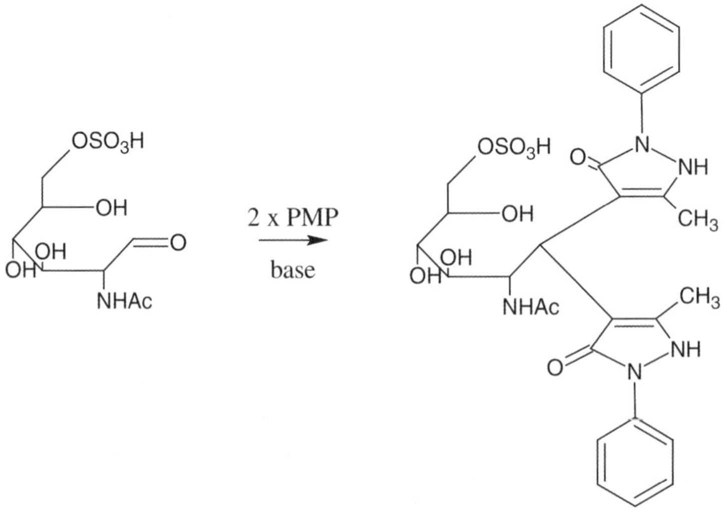

Scheme 8.1   PMP modification of GlcNAc6S; a glycosaminoglycan lysosomal storage biomarker for mucopolysaccharidoses.[39]

shown to have high sensitivity and could separate D- and L-glyceric acid in samples from healthy control subjects (previously the absolute configurations of glyceric acids were only reported from patients with PH2; however, this study identified them in healthy control samples too). In addition, this derivatisation offers greater stability of the derivative group over that of TMS derivatives (samples were re-analysed over a 6-day time course). The method used only 100 μL urine. It is important that when conducting experiments to determine biomarkers for carbohydrate metabolic disorders, all patient samples are analysed concurrently with those of age-matched controls, as much as is reasonably possible.

8.3.3   Analytical Methods

8.3.3.1   Direct Infusion

The direct infusion of a mixture of unknown oligosaccharides is not common practice due to the isomeric nature of the monosaccharides that make up these structures. An element of separation is usually required to separate these isomers and provide a basis for comparison of retention times with those of authentic standards. In addition to this, the electrospray signal suppression is high in biofluid analytes entering the mass spectrometer together, not least because of the many competing analytes present, but also because of salts in the biofluid. There is also the limitation of small sample volumes, particularly in the paediatric setting. A study of LSDs using derivatisation and direct infusion of the oligosaccharide derivatives has been presented.[37] LSDs represent more than 45 genetic diseases (inherited in an autosomal recessive manner) that result in a deficiency of a specific lysosomal protein. This deficiency results in an accumulation of substrates that are normally degraded within lysosomes. The accumulated substrate can be used to group LSDs into categories; these are the mucopolysaccharidoses, lipidoses, glycogenoses and oligosaccharidoses. The clinical similarities between LSDs in these categories are organomegaly, bone abnormalities, coarse hair and facial features (facies). The treatment for such LSDs is bone marrow transplantation and enzyme replacement therapy.

Oligosaccharides resulting from lysosomal storage are very concentrated in the urine compared to plasma, which suggests that the kidneys are very efficient at removing these oligosaccharides from circulation. The derivative used in this direct infusion analysis was 1-phenyl-3-methyl-5-pyrazolone (see Scheme 8.1). The derivatives were directly infused into an ion-spray source on a triple quadrupole mass spectrometer using 50% acetonitrile/0.025% formic acid as solvent at 100 μL min^{-1}. Neutral oligosaccharides were analysed in the positive-ion mode and sulfated oligosaccharides were analysed in the negative-ion mode. The ions generated were then mass analysed using multiple reaction monitoring. Eighty oligosaccharide species were monitored and using MRM of specific oligosaccharide markers, the concentration of these ions could be used to show differences between the LSD disease groups (except for Pompe's disease) and the control groups. Common oligosaccharide markers observed that discriminate diseased patients from control groups are *N*-acetylhexosamine-sulfate (for example in MPS IVA patients) and *N*-acetylhexosamine-hexuronic acid disaccharide (HexNAc-HexA) in Tay-Sachs patients. Storage oligosaccharides in α-mannosidosis are a combination of hexose (Hex) and HexNAc, *e.g.*, $Hex_2HexNAc$, $Hex_3HexNAc$ and $Hex_4HexNAc$, which are all elevated in α-mannosidosis patients. Thus, many LSDs could be identified using this method. However, a number of disadvantages were highlighted and these disadvantages are also pertinent to other similar methods. For large scale screening the derivatisation of oligosaccharides is labour intensive. In addition, due to the low level of oligosaccharide biomarkers in blood, urine is the biofluid of choice. However, using urine as the biofluid requires collection of an additional sample alongside the blood/Guthrie card, which adds expense and further sample handling and storage effort to the screening protocol.

8.3.3.2   GC-MS

GC-MS has been used for the diagnosis of inborn errors of metabolism (IEM) since the 1970s. Quoting from a recent review of GC-MS analyses of IEMs "metabolome analysis by the simplified pre-treatment with urease of urine or the eluates from urine dried on filter paper, stable-isotope dilution, and GC/MS is the most comprehensive approach for the molecular diagnosis of IEMs. A single chromatographic injection by full scan MS and extracted ion chromatograms enables the rapid screening of more than 130 IEMs. The approach is non-invasive, valid, feasible, cost-effective and applicable to patients with IEMs regardless of the presence/absence of clinical symptoms".[40]

Examples of such methods are the use of *O*-acetyl-(+)-2-butylation in the very specific and sensitive analysis of optical isomers of glyceric acid (see Section 8.3.2). In addition to well established GC-MS methods for the analysis of IEMs, authors are still reporting improvements to these methods in the current literature. For example, a stable-isotope dilution method that used ammonia chemical ionisation and negative-selected ion monitoring of 3-hydroxyglutaric acid, a biomarker of glutaric aciduria type 1, is a highly sensitive method. However, workers did not always find ammonia gas easy to handle, especially in routine high-throughput operation. A method of similar sensitivity but with more practical application has since been reported[41] which utilises conventional EI for the measurement of 3HGA as *tert*-butyldimethylsilyl (tBDMS) derivatives. Finally, although not specific to the analysis of IEM biomarkers, common GC-MS derivatives such as TMS have recently been reported using microwave-assisted silylation, which speeds up the derivatisation procedure enormously.[42]

8.3.3.3   LC-MS

The current practice, before enzymatic analysis such as assays for leukocyte or fibroblast enzyme activity, for testing patients suspected of having an LSD, is screening for elevation of storage substrate using thin-layer chromatography. There are limitations to these analyses in that there is difficulty in analysing large numbers of samples and in identification of analytes based just on their retardation factors (R_f). An offline LC-MS analysis for LSDs by analysing amniotic fluid has been presented.[36] Off-line analysis of PMP derivatives of oligosaccharides as biomarkers for LSDs was achieved using C_{18} stationary phase 96-well columns on an automated robotic liquid handling system. The columns were washed with water and dried under vacuum before the addition of $CHCl_3$ to remove excess PMP reagent. The samples were further dried and then derivatised oligosaccharides eluted from the stationary phase using 50% acetonitrile with 0.025% formic acid. Samples were dried under nitrogen and re-aliquoted in a known amount of LC mobile phase before undergoing analysis by ES-MS/MS carried out on a triple quadrupole mass analyser. Neutral oligosaccharides were analysed in positive-ion mode and acidic oligosaccharides in negative-ion mode. Semi-quantitative results were obtained using MRM for 100 ms per MRM pair. The MRM pairs were chosen according to specific fragmentation of PMP derivatives at m/z 175 $(M+H)^+$ and 173 $(M–H)^-$. Semi-quantification of the analytes was achieved using relative peak heights of the internal standards (1 nmol d_3-GlcNAc-6-sulfate for glycosaminoglycans, and 0.5 nmol methyl lactose for all other oligosaccharides).

Pompe's disease is an LSD characterised by lysosomal accumulation of glycogen within cells as a result of the deficiency of the lysosomal enzyme acid α-glucosidase. In infant onset there is massive cardiomegaly (abnormal enlargement of the heart), macroglossia (enlargement of the tongue) and hypotonia (decrease in muscle tone) and death occurs in the first 2 years. Onset later in life is characterised by slower progressive muscle weakness, and death occurs from respiratory failure. Definitive treatment is not available so there are two strategies to treat the disease: enzyme-replacement therapy or gene-transfer of the α-glucosidase gene using viral vectors. Due to early onset and very premature death the requirement for rapid, early diagnosis is evident and is also a requirement for the future monitoring of biomarkers in order to check the efficacy of therapy. Pompe's disease was characterised some years ago by the observation that the concentration of a tetrasaccharide of four glucose residues was elevated in these patients' urine.[43] However, the use of this tetrasaccharide as a biomarker for the disease has not been widely adopted, the likely reasons being that the analytical methods available were complex, and laborious derivatisations for GC- and HPLC-MS analyses and immunoassays. A less laborious method has since been reported which utilises 1-phenyl-3-methyl-5-pyrazolone derivatisation prior to ES-MS analysis of urine,[44] plasma and dried blood spots. The tetrasaccharide concentration was measured against creatinine molarity in urine and was found to increase in both infantile- and adult-onset Pompe individuals, against age-matched controls. Blood spots could not be used to differentiate between controls and affected individuals. The calibration curves generated in relation to the internal standard for quantification were linear up to 8 μmol L^{-1} (LOD 0.32 pmol). This method showed substantial improvement in sensitivity over previous methods that used pulsed amperometric detection (PAD) (LOD 20–50 pmol). In addition, the 96-well solid-phase extraction format allows the analysis of 96 samples in a day compared with, on average, 30 min HPLC run times, making the method applicable for large-scale screening.

Pompe's disease has also been characterised using butyl-*p*-aminobenzoate derivatisation (BAB).[45] BAB labelled oligosaccharides were separated on a C_{18} column and detected with UV absorbance. Fractions were collected offline and analysed using ES-MS/MS. During the analysis of these oligosaccharide BAB derivatives, a co-eluting compound was observed and was shown (using an authentic standard) to be a component of human breast milk. The method was then further modified to separate the Glc_4 BAB derivative (Pompe biomarker) from the breast milk compound. This example underscores one of the most important advantages of using mass spectrometry in clinical HPLC assays over detectors that do not offer the specificity of mass spectrometry.

8.3.3.4   CE-MS

No reports have been found in the literature of the use of CE-MS for the diagnosis or characterisation of inborn errors of metabolism. This probably reflects the generally poor concentration LODs of CE, and the technical difficulties, when interfacing CE with MS, of obtaining reproducible data, especially from samples in complex matrices.

8.3.4   Data Analysis

The common derivatisations of oligosaccharide analytes presented here are not currently pooled into databases built of chromatographic and mass spectrometric data. However, the vast amount of data now present in the literature from metabolomics and metabolite identification has created a need for standardisation in the research community. In addition to standardisation, meaningful comparison of datasets and replication of experiments, user-friendly open access tools for this flood of information are necessary. Although there are numerous databases of information at the genomic, transcriptomic and proteomic level, there is a comparative lack of metabolomic databases, in particular for established GC-MS techniques. Efforts have been made by the plant metabolomics community to agree conventions for their data formats and description of experiments.[46] These issues are only just receiving attention from other workers in metabolome analysis.[47]

8.4   Glucuronidated Metabolites as Biomarkers for Drug and Alcohol Abuse

The detection of a specific drug and its glucuronide metabolite(s) has become of extreme relevance in clinical and forensic toxicology. Being able to estimate metabolite/drug ratios may provide information about route, dose and even time of exposure/uptake. In the majority of cases, a parent drug is biotransformed rapidly and is thus only present in its original form in biological fluids at very low levels, which makes the detection of its metabolite(s) of crucial importance for the identification of the parent drug(s).

Drugs undergo a range of metabolic transformations in the body that act as detoxification mechanisms, and which change the compounds to a less toxic, more water-soluble form that can be rapidly excreted from the body. In general, any drug undergoes a series of enzymatic reactions designated Phase I and Phase II metabolism.[48]

Phase I metabolism is regarded as the enzymatic transformation, such as oxidation, hydroxylation, de-alkylation or reduction of functional groups. Phase II metabolism involves a change in structure of a molecule or its Phase I metabolite *via* conjugation with an endogenous substance (*e.g.*, glucuronide and sulfate). These Phase II reactions often follow Phase I modifications of a drug, but there are examples of xenobiotics that undergo Phase II metabolism without first undergoing Phase I metabolic steps.

The conjugation of a drug with glucuronic acid is probably the most important detoxification reaction and it is found extensively in mammals. The fact that glucuronic acid is readily available from the glucose that is generally stored in the form of glycogen makes this reaction of glucuronidation so common. Additionally, there are several functional groups such as hydroxyl, amino, carboxyl and sulfhydryl to which glucuronic acid can be transferred enzymatically via UDP-glucuronosyltransferase catalysis in the endoplasmic reticulum.[49] The effect of glucuronidation is to produce an acidic compound, more water-soluble than its precursor at physiological pH. Therefore, these compounds are entirely ionised at pH values between 3 and 4. Direct conjugation with glucuronic acid generally removes the pharmacological activity of the drug.

8.4.1 Sampling and Extraction

A range of biological fluids and tissues is used for the determination of drug glucuronide derivatives. These include urine, blood, plasma, serum, cerebrospinal fluid (CSF) and, more recently, even hair.

Analysis of urine has become the most common approach, due to the non-invasive means of sample collection, but also due to the fact that most drug metabolites are found in higher concentrations in urine and in the presence of fewer interfering components, such as proteins and lipids, than other biological samples. Urine can be sampled using several procedures, the most commonly used being the collection of aliquots in sterile containers and subsequent freezing on dry ice and storage at $-20\,°C$, or $-80\,°C$ if available. The sample should normally be stored in several aliquots to prevent loss of the analytes during multiple freeze/thaw cycles. An alternative method that has proven to be quite popular is the use of filter paper onto which the urine sample is collected and dried. Using this method, the sample is easily recovered by extraction with water.

Although reports are to be found in the literature of the analysis of other fluid sample types (blood, plasma, serum and CSF), detail is lacking on the sampling and sample handling approaches used.

Generally, glucuronide drug metabolites are extractable using solid-phase extraction methods with several different chemistries associated with the packed material, the most common being reversed-phase. This extraction step allows reduction of the presence of other metabolites that would mask the presence of the glucuronide metabolites as well as enabling the concentration of these metabolites, which are present in trace amounts in biological fluids and tissues.

8.4.2   Sample Handling and Derivatisation

The determination of glucuronide metabolites has been performed, until recently, by cleavage of the glucuronide moiety with an enzyme (*e.g.*, β-glucuronidase) to yield the parent compound, which may then be detected and quantified. This type of enzymatic approach can suffer from several limitations: (i) incomplete enzymatic hydrolysis due to competitive inhibition of the enzyme used; (ii) difficulties in cleaving some conjugates at specific positions; (iii) possibility of differing levels of hydrolysis of the conjugate with some enzyme preparations. Additionally, there is evidence suggesting that contaminants in the preparation may convert one metabolite into another during the hydrolysis process.[50] Even under very closely controlled conditions, these methods involve extensive sample preparation, which can be time consuming, and the glucuronides of each class of drug require a unique analytical method. Similar arguments can be made for methods that involve derivatisation prior to analysis. All these limitations associated with these procedures reinforce the need for methods that enable the direct detection of these metabolites.

GC-MS analysis of extracts containing these glucuronidated metabolites is generally complicated. Many of the metabolites are not volatile and must be derivatised prior to GC analysis. Trimethylsilylation is widely used as the main derivatisation protocol because it is simple, quick and provides highly volatile derivatives. One drawback of TMS derivatisation is the fact that all samples need to be analysed in a period of 2 hours after the derivative is made, limiting throughput. Several alternative reagents can be found described in the literature, such as the acylating reagents pentafluoropropionic anhydride (PFPA) and heptafluorobutyric anhydride (HFBA), which can provide more stable derivatives but generally fail to provide the necessary volatility for gas chromatography.

8.4.3   Analytical Methods

8.4.3.1   Morphine Glucuronides

Morphine is an extremely potent opioid analgesic used for treatment of moderate to severe pain. Morphine metabolism is achieved mainly via conjugation with glucuronic acid (GlcA) to form morphine-3-glucuronide (M3GlcA) and morphine-6-glucuronide (M6GlcA). M6GlcA contributes to the analgesic potency of morphine. The principal metabolite M3GlcA has no opioid action but seems to contribute to the side effects of morphine. Since morphine is a metabolite of heroin, M3GlcA and M6GlcA are also heroin metabolites.[51] Studies reported in the literature are generally aimed at monitoring therapeutic concentrations in patients, to study opiate abuse, and even to identify causes of intoxication or death in clinical or forensic work.

(a) Direct Infusion. The method of direct infusion into the mass spectrometer has not been found to be used in the analysis of the glucuronide metabolites of morphine although applications of this methodology can be widely found in other metabolite profiling studies. One of the reasons could be because of the highly complex matrix in which the sample is generally obtained, such as urine, blood and serum. These contain high concentrations of other compounds and metabolites, making a chromatographic step necessary to further simplify their analysis.

(b) GC-MS. Only relatively recently has a single method been described for the direct measurement of morphine glucuronides using gas chromatography-mass spectrometry (GC-MS), presumably owing to the thermal lability and involatility of these compounds. In his paper, Leis *et al.*[52] described a procedure for derivatisation and highly sensitive detection and quantitative measurement of morphine glucuronides in human plasma by GC/negative-ion chemical ionisation MS.

Morphine glucuronides, namely M6GlcA and M3GlcA, were derivatised to form their pentafluorobenzyl ester trimethylsilyl ether derivatives. The compounds were then analysed using GC-MS, where a diagnostic fragment ion at m/z 748 was obtained at high relative abundance from both target compounds. Deuterium-labelled morphine glucuronides were used as internal standards. Calibration curves were produced using polynomial fitting for the range 10–1280 and 15–1920 nmol L^{-1} for the M6- and M3-GlcAs, respectively. The method proved to be rapid and robust for batch analysis of morphine glucuronides during pharmacokinetic profiling of the drugs.

(c) LC-MS. LC-MS based methods are widely used in the analysis of morphine glucuronide metabolites and several examples have been reported in the literature. These studies are generally performed using reversed-phase column chemistries, but more recent examples also describe the application of normal-phase columns for LC-MS studies.

Tyrefors *et al.*[53] reported a method using LC–ES-MS in the positive-ion mode for the detection of morphine and its metabolites, M3GlcA and M6GlcA in serum. Separation was achieved on a reversed-phase column using a gradient of 4–70% (v/v) acetonitrile with 0.1% (v/v) formic acid at a flow of 1.0 mL min^{-1}. The compounds were detected in the mass spectrometer by selected-ion monitoring for m/z 286.2 for morphine and m/z 462.2 for both M3GlcA and M6GlcA. A linear range of 5–500 ng mL^{-1} was obtained for M3GlcA and 2–100 ng mL^{-1} for M6GlcA. This method enabled very short analysis times of less than 5 min.

Schanzle *et al.*[54] determined the concentrations of M3GlcA and M6GlcA in serum, urine and cerebrospinal fluid (CSF) of patients or volunteers receiving morphine during a clinical study using LC-ES-MS in the positive ion mode. A reversed-phase separation with a C$_{18}$ column was used with water–acetonitrile–tetrahydrofuran–formic acid (100:1:1:0.1, v/v) as the mobile phase. Retention times were 1.7 and 3.2 min, respectively, for M3GlcA and M3GlcA. The limit of quantitation (LOQ) was reported to be as low as 0.5 ng mL^{-1} for M6GlcA and 2 ng mL^{-1} for M3GlcA when measured in serum or cerebrospinal fluid samples and 25 ng mL^{-1} for M6GlcA and 9 ng mL^{-1} for M3GlcA in urine.

Naidong *et al.*[55] presented an alternative method using normal-phase LC-MS/MS instead of the more generally used reversed-phase column chemistries. In order to obtain a good spray and high sensitivity in MS the authors found that only mobile phases with high organic solvent composition and acidic pH gave the required optimal conditions. Curiously, an extremely fast equilibration time (5–10 min) was achieved with a mobile phase of acetonitrile, water and formic acid, instead of the traditional solvents used in normal-phase liquid chromatography. Retention times for M3GlcA and M6GlcA were 2.4 and 1.9 min, respectively. Linearity for measurements in plasma was achieved over the range 10–1000 ng mL^{-1} for M3GlcA and 1–100 ng mL^{-1} for M6GlcA. The limits of detection (LOD) were reported to be as low as 1 ng mL^{-1}.

Projean *et al.*[56] developed a quick, simple method for the determination of M3GlcA and M6GlcA in rat plasma by LC-ES-MS using selected-ion monitoring (SIM). Chromatographic separation was performed using a phenyl–hexyl column with a step-gradient of acetonitrile and formic acid in water at 1.0 mL min^{-1}. Naloxone was used as the internal standard. The limits of quantification (LOQ) achieved with this method were around 4.88 nM for M3GlcA and M3GlcA.

Whittington and Kharasch developed a method using LC-MS for the detection of morphine and its glucuronides in plasma.[57] The analytes were separated using an isocratic mobile phase consisting of methanol, acetonitrile and formic acid. The LOQ was 0.5 and 5 ng mL^{-1} for M6GlcA and M3GlcA, respectively.

(d) CE-MS. The use of capillary electrophoresis coupled to mass spectrometry (CE-MS) has been recently used for the analysis of morphine glucuronides in human urine.[58] In this study, Wey and Thormann used an aqueous background electrolyte containing 25 mM ammonium acetate and NH$_3$ (pH 9), combined with atmospheric pressure electrospray ionisation mass spectrometry in the positive-ion mode, and showing that injection of untreated or diluted urine would not allow the detection of the morphine glucuronides. Only the incorporation of a solid-phase extraction step using a mixed-mode polymer phase allowed the analysis of the glucuronidated metabolites with limits of detection around 100–200 ng mL^{-1}.

8.4.3.2 Alcohol

Ethyl glucuronide (EtGlcA) (Figure 8.1c) is a non-volatile, water-soluble direct metabolite of ethanol that can potentially serve as a marker of alcohol consumption. EtGlcA was first isolated from rabbit's urine in 1952 by Kamil *et al.*[59] In 1967, Jaakonmaki *et al.*[60] detected the metabolite in human urine for the first time. About 90–95% of blood ethanol is eliminated by hepatic oxidation via alcohol and aldehyde dehydrogenase enzymes. The detoxifying pathway of alcohol elimination via conjugation with glucuronic acid represents about 0.5–1.5% of total ethanol elimination.[61] Despite the fact that it is a minor ethanol metabolite, this compound can be very useful in forensic applications. Previous studies have demonstrated that ethyl glucuronide is highly specific and sensitive as an alcohol marker.

(a) Direct infusion. Again, reports of the use of direct infusion into the mass spectrometer have not been found for the analysis of the metabolite ethyl glucuronide. Again, this is probably due to the highly complex matrix which requires at least a chromatographic step to allow successful analysis.

(b) GC-MS. Janda and Alt[62] have presented a method for the determination of ethyl glucuronide (EtGlcA) in human serum and urine using a combination of solid-phase extraction (SPE) and gas chromatography (GC) with mass spectrometric detection (MS). EtGlcA was isolated from serum and urine using aminopropyl SPE columns after deproteination with perchloric acid and hydrochloric acid, respectively. The chromatographic separation was performed on a DB 1701 fused silica column after conversion of the EtGlcA to its trimethylsilylated derivative. Using GC-MS in selected-ion monitoring mode, LOQs of 173 and 560 ng mL^{-1} and LODs of 37 and 168 ng mL^{-1} were reported for serum and urine, respectively. This clearly shows the sensitivity of the method.

More recently, Jurado *et al.*[63] described a procedure for the detection and quantification of EtGlcA in hair samples. In this method the hair was sequentially washed with water and acetone. The decontaminated sample was then finely cut and a deuterated internal standard (d_5-EtGlcA) and 2 mL of water were added, followed by sonication for 2 hours. The analysis was performed using derivatisation with pentafluoropropionic anhydride (PFPA), following which the acylated derivatives were injected onto the GC-MS system. Linearity over the range of concentrations from 0.050 to 5 ng mg^{-1} was reported by the authors, with LOD and LOQ of 0.025 and 0.050 ng mg^{-1}, respectively.

(c) LC-MS. Wurst *et al.*[64] developed an LC-MS/MS method without the need for derivatisation or sample cleanup prior to analysis. 1 μg of deuterated internal standard was added to 100 μL urine and an aliquot of this mixture was injected onto the LC-MS/MS without further sample preparation. Samples were separated using a reversed-phase C_{18} column with a mobile phase consisting of 1% formic acid in methanol at a flow of 0.18 mL min^{-1}. The detector was a triple–quadruple mass spectrometer operated in multiple-reaction monitoring mode. The detection limit was reported to be 100 ng mL^{-1}.

A simple analytical procedure to routinely screen for urinary EtGlcA was developed by Stephanson *et al.*[65] This method enabled direct injection of urine, diluted and containing a deuterated internal standard, into an electrospray LC-MS. The use of a porous graphitised carbon column (Hypercarb) allowed an isocratic separation with 5–6 min retention times. The mobile phase consisted of 25 mmol L^{-1} formic acid with 5% acetonitrile. The method yielded a linear range of 0.1–1500 µg mL^{-1}. The method was applied to screen for EtGlcA in 252 clinical urine samples.

Weinmann *et al.*[66] recently used LC-MS/MS in negative-ion mode for the detection of EtGlcA in urine that enabled several MS/MS transitions to be monitored (deprotonated molecule $[M–H]^-$/product ions: m/z 75, 85, 113, and optionally also 159). The LOD was found to be 52 ng mL^{-1}.

Politi *et al.*[67] recently presented a negative-ion mode LC-ESI-MS/MS method for the determination of EtGlcA in urine. This assay was applied to several authentic urine samples from social drinkers and to alcoholic beverages. Penta-deuterated EtGlcA was used as internal standard. Two MS/MS reactions were monitored, with the deprotonated molecule as precursor ion: m/z 221 → 75, and m/z 221 → 85. This method was preceded by a simple and rapid sample pre-treatment step (1:50 water dilution and centrifugation of 50 mL of urine). The method was accurate and precise over the linear dynamic range (0.05–10 mg L^{-1}). This study has also shown that EtGlcA is stable in frozen urine for at least 1 month.

(d) CE-MS.   Although there are no examples in the literature of the use of CE-MS for the analysis of EtGlcA, there is a recent method reported by Krivankova *et al.*[68] where the authors used capillary zone electrophoresis (CZE) to analyse EtGlcA in model mixtures and human serum using uncoated and coated fused silica capillaries together with acidic buffers in the pH range between 3.2 and 4.4 and UV detection at 214 nm. In these approaches, separation of EtGlcA from endogenous macro- and micro-components (anionic serum components of high and low concentration, respectively) is based upon transient isotachophoretic stacking, referred to as sample self-stacking. The selection of a favourable buffer co-ion and pH was shown to be crucial for optimised sensitivity. A buffer composed of 10 mM nicotinic acid and epsilon-aminocaproic acid (pH 4.3) was demonstrated to provide a detection limit for EtGlcA in serum of 0.1 mg mL^{-1}.

Further developments will have to be made to allow methods like this to be used in conjunction with mass spectrometry. For a CE-MS method to be successful, it is highly dependent on the use of MS compatible buffers which are frequently not employed in off-line CE separations. In general, CE-MS is still not as robust and is much less user friendly than the routinely used LC-MS methodologies.

8.5 Conclusions

Mass spectrometry, especially hyphenated with GC or LC, is a powerful approach for the characterisation and analysis of glycoconjugate metabolites.

In spite of the theoretical advantages of CE (advantageous flow characteristics, constant solvent composition so that ionisation characteristics do not fluctuate, lack of a requirement for expensive columns, fast separation times and high resolution), this technique is noticeably underused in the analysis of glyco-conjugate metabolites. This probably reflects the technical difficulty of operation in hyphenation with MS, the inherently poor concentration LODs, and the compromises that have to be made as regards separation efficiency when identifying a mass spectrometry compatible background electrolyte. In contrast, GC- and LC-MS(/MS) approaches offer robustness of operation, good resolution, excellent LODs and compatibility of experimental protocols with the complex sample matrices in which these challenging metabolites are generally sampled, and consequently generally represent the approaches of choice for these analyses.

References

1. W. de Koning and K. van Dam, *Anal. Biochem.*, 1992, **204**, 118.
2. G. J. G. Ruijter and J. Visser, *J Microbiol. Methods*, 1996, **25**, 295.
3. H. P. Smits, A. Cohen, T. Buttler, J. Nielsen and L. Olsson, *Anal. Biochem.*, 1998, **261**, 36.
4. U. Shaefer, K. Boos, R. Takors and D. Weuster-Botz, *Anal. Biochem.*, 1999, **270**, 88.
5. N. B. S. Jensen, K. V. Jokumsen and J. Villadsen, *Biotechnol. Bioeng.*, 1999, **63**, 356.
6. B. Moritz, K. Striegel, A. A. de Graaf and H. Sahm, *Eur. J. Biochem*, 2000, **267**, 3442.
7. S. Petersen, C. Mack, A. A. de Graaf, C. Riedel, B. J. Eikmanns and H. Sahm, *Metabol. Eng.*, 2001, **3**, 344.
8. C. van Vaeck, S. Wera, P. van Dijck and J. M. Thevelein, *Biochem. J.*, 2001, **353**, 157.
9. H. Hajjaj, P. J. Blanc, G. Goma and J. François, *FEMS Microbiol Lett*, 1998, **164**, 195.
10. H. Dominguez, C. Rollin, A. Guyonvarch, J. L. Guerquin-Kern, M. Cocaign-Bousquet and N. D. Lindley, *Eur. J. Biochem.*, 1998, **254**, 96.
11. B. Gonzalez, J. François and M. Renaud, *Yeast*, 1997, **13**, 1347.
12. S. R. Hull and R. Montgomery, *Anal. Biochem.*, 1994, **222**, 49.
13. R. R. Swezey, *J. Chromatogr. B*, 1995, **669**, 171.
14. R. P. Maharjan and T. Ferenci, *Anal. Biochem.*, 2003, **313**, 145.
15. S. G. Villas-Bôas, J. Hojer-Pedersen, M. Akesson, J. Smedsgaard and J. Nielsen, *Yeast*, 2005, **22**, 1155.
16. H. C. Lange, M. Eman, G. van Zuijlen, D. Visser, J. C. van Dam, J. Frank, M. J. T. de Mattos and J. J. Heijnen, *Biotechnol. Bioeng.*, 2001, **75**, 406.
17. S. G. Villas-Bôas, S. Mas, M. Akesson, J. Smedsgaard and J. Nielsen, *Mass Specrom. Rev.*, 2005, **24**, 613.
18. C. C. Sweeley, R. Bentley, M. Makita and W. W. Wells, *J. Am. Chem. Soc.*, 1963, **85**, 2497.

19. D. J. Harvey and M. G. Horning, *J. Chromatogr. A*, 1973, **76**, 51.
20. J. M. Halket and V. G. Zaikin, *Eur. J. Mass Spectrom.*, 2003, **9**, 1.
21. J. I. Castrillo, A. Hayes, S. Mohammed, S. J. Gaskell and S. G. Oliver, *Phytochemistry*, 2003, **62**, 929.
22. C. Wittmann and E. Heinzle, *Biotechnol. Bioeng.*, 2001, **72**, 642.
23. S. Strelkov, M. van Elstermann and D. Schomburg, *Biol. Chem.*, 2004, **385**, 853.
24. J. Feurle, H. Jomaa, M. Wilhelm, B. Gutsche and M. Herderich, *J. Chromatogr. A*, 1998, **803**, 111.
25. A. Buchholz, R. Takors and C. Wandrey, *Anal. Biochem.*, 2001, **295**, 129.
26. J. van Dam, M. R. Eman, J. Frank, H. C. Lange, G. W. K. van Dedem and S. J. Heijnen, *Anal. Chim. Acta*, 2002, **460**, 209.
27. C. Wittmann, M. Hans, W. A. van Winden, C. Ras and J. J. Heijnen, *Biotechnol. Bioeng.*, 2005, **89**, 839.
28. M. M. C. Wamelink, E. A. Struys, J. H. J. Huck, B. Roos, M. S. van de Knaap, C. Jakobs and N. M. Verhoeven, *J. Chromatogr. B*, 2005, **823**, 18.
29. Y. Ciringh and J. S. Lindsey, *J. Chromatogr. A*, 1998, **816**, 251.
30. T. Soga and G. A. Ross, *J. Chromatogr. A*, 1999, **837**, 231.
31. T. Soga and M. Imaizumi, *Electrophoresis*, 2001, **22**, 3418.
32. T. Soga, Y. Ueno, H. Naraoka, Y. Ohashi, M. Tomita and T. Nishioka, *Anal. Chem.*, 2002, **74**, 2233.
33. I. Autti-Ramo, M. Makela, H. Sintonen, H. Koskinen, L. Laajalahti, R. Halila, H. Kaariainen, R. Lapatto, K. Nanto-Salonen, K. Pulkki, M. Renlund, M. Salo and T. Tyni, *Acta Paediatrics*, 2005, **94**, 1126.
34. K. G. Balk, *Neonatal Network*, 2005, **24**, 39.
35. R. Guthrie and A. Susi, *Pediatrics*, 1963, **32**, 338.
36. S. L. Ramsay, I. Maire, C. Bindloss, M. Fuller, P. D. Whitfield, M. Piraud, J. J. Hopwood and P. J. Meikle, *Mol. Genet. Metab.*, 2004, **83**, 231.
37. P. J. Meikle, E. Ranieri, H. Simonsen, T. Rozaklis, S. L. Ramsay, P. D. Whitfield, M. Fuller, E. Christensen, F. Skovby and J. J. Hopwood, *Pediatrics*, 2004, **114**, 909.
38. Y. Inoue, T. Shinka, M. Ohse and T. Kuhara, *J. Chromatogr. B*, 2005, **823**, 2.
39. S. L. Ramsay, P. J. Meikle and J. J. Hopwood, *Mol. Genet. Metab.*, 2003, **78**, 193.
40. T. Kuhara, *Mass Spectrom. Rev.*, 2005, **24**, 814.
41. Y. Shigematsu, I. Hata, Y. Tanaka, G. Tajima, N. Sakura, E. Naito and T. Yorifuji, *J. Chromatogr. B*, 2005, **823**, 7.
42. C. Deng, J. Ji, L. Zhang and X. Zhang, *Rapid Commun. Mass Spectrom.*, 2005, **19**, 2974.
43. P. Hallgren, G. Hansson, K. G. Henriksson, A. Hager, A. Lundbald and S. Svensson, *Eur. J. Clin. Invest.*, 1974, **4**, 429.
44. T. Rozaklis, S. L. Ramsay, P. D. Whitfield, E. Ranieri, J. J. Hopwood and P. J. Meikle, *Clin. Chem.*, 2002, **48**, 131.
45. Y. An, S. P. Young, S. L. Hillman, J. L. K. Van Hove, Y. T. Chen and D. S. Millington, *Anal. Biochem.*, 2000, **287**, 136.
46. H. Jenkins, N. Hardy, M. Beckmann, J. Draper, A. R. Smith, J. Taylor, O. Fiehn, R. Goodacre, R. J. Bino, R. Hall, J. Kopka, G. A. Lane,

B. M. Lange, J. R. Liu, P. Mendes, B. J. Nikolau, S. G. Oliver, N. W. Paton, S. Rhee, U. Roessner-Tunali, K. Saito, J. Smedsgaard, L. W. Sumner, T. Wang, S. Walsh, E. S. Wurtele and D. B. Kell, *Nature Biotechnol.*, 2004, **22**, 1601.

47. N. Schauer, D. Steinhauser, S. Strelkov, D. Schomburg, G. Allison, T. Moritz, K. Lundgren, U. Roessner-Tunali, M. G. Forbes, L. Willmitzer, A. R. Fernie and J. Kopka, *FEBS Lett.*, 2005, **579**, 1332.

48. M. Shipkova and E. Wieland, *Clin. Chim. Acta*, 2005, **358**, 2.

49. B. A. Rashid, G. W. Aherne, M. F. Katmeh, P. Kwasowski and D. Stevenson, *J. Chromatogr. A*, 1998, **797**, 245.

50. L. D. Bowers and Sanaullah, *J. Chromatogr. B*, 1996, **687**, 61.

51. A. Dienes-Nagy, L. Rivier, C. Giroud, M. Augburgeer and P. Mangin, *J. Chromatogr. A*, 1999, **854**, 109.

52. H. J. Leis, G. Fauler, G. Raspotnig and W. Windischhofer, *J Mass Spectrom.*, 2002, **37**, 395.

53. N. Tyrefors, B. Hyllbrant, L. Ekman, M. Johansson and B. Langstrom, *J. Chromatogr. A*, 1996, **729**, 279.

54. G. Schanzle, S. Li, G. Mikus and U. Hofmann, *J. Chromatogr. B*, 1999, **721**, 55.

55. W. Naidong, J. W. Lee, X. Jiang, M. Wehling, J. D. Hulse and P. P. Lin, *J. Chromatogr. B*, 1999, **735**, 255.

56. D. Projean, T. M. Tu and J. Ducharme, *J. Chromatogr. B*, 2003, **787**, 243.

57. D. Whittington and E. D. Kharasch, *J. Chromatogr. B*, 2003, **796**, 95.

58. A. B. Wey and W. Thormann, *J Chromatogr A.*, 2001, **916**, 133.

59. I. A. Kamil, N. J. Smith and R. T. Williams, *Biochem. J.*, 1952, **51**, 32.

60. P. I. Jaakonmaki, K. L. Knox, E. C. Horning and M. G. Horning, *Eur. J. Clin. Pharmacol.*, 1967, **1**, 63.

61. P. Droenner, G. Schmitt, R. Aderjan and H. Zimmer, *Forensic Sci. Int.*, 2002, **126**, 24.

62. I. Janda and A. Alt, *J. Chromatogr. B*, 2001, **758**, 229.

63. C. Jurado, T. Soriano, M. P. Gimenez and M. Menendez, *Forensic Sci Int.*, 2004, **145**, 161.

64. F. M. Wurst, C. Kempter, J. Metzger, S. Seidl and A. Alt, *Alcohol*, 2000, **20**, 111.

65. N. Stephanson, H. Dahl, A. Helander and O. Beck, *Ther. Drug Monit.*, 2002, **24**, 645.

66. W. Weinmann, P. Schaefer, A. Thierauf, A. Schreiber and F. M. Wurst, *J. Am. Soc. Mass. Spectrom.*, 2004, **15**, 188.

67. L. Politi, L. Morini, A. Groppi, V. Poloni, F. Pozzi and A. Polettini, *Rapid Commun. Mass Spectrom.*, 2005, **19**, 1321.

68. L. Krivankova, J. Caslavska, H. Malaskova, P. Gebauer and W. Thormann, *J. Chromatogr. A*, 2005, **1081**, 2.

CHAPTER 9

Matrix Assisted Laser Desorption Ionisation Mass Spectrometric Imaging – Principles and Applications

CAROLINE J. EARNSHAW,[1] SALLY J. ATKINSON,[1] MICHAEL BURRELL[2] AND MALCOLM R. CLENCH[1]

[1] Biomedical Research Centre, Sheffield Hallam University, Howard Street, Sheffield S1 1WB, UK
[2] Department of Animal and Plant Sciences, The University of Sheffield, Western Bank, Sheffield S10 2TN, UK

9.1 Introduction

Mass spectrometric imaging using matrix assisted laser desorption ionisation mass spectrometry (MALDI-MSI) is a technique pioneered by the group of Richard Caprioli.[1] In the most common variant of the technique, the sample is imaged by moving it by set increments under a stationary laser. At each position the laser is fired for a pre-selected time or number of shots and a mass spectrum is acquired. Images are obtained by plotting the x and y coordinates *versus* the abundance of a selected ion or ions, which is represented as a grey or colour scale (Figure 9.1a). This is often referred to as the mass microprobe mode of MALDI imaging. Using this approach images are typically acquired with a spatial resolution of 50–200 μm (this being limited by the laser spot size/resolution available for sample target movement).

An alternative to the mass microprobe approach, the mass microscope approach has been pioneered by Ron Heeran and co-workers at FOM in Amsterdam.[2] In this approach, rather than the laser beam being highly focussed a mass spectrometer that accepts a 150–300 μm diameter ion beam is used to map a magnified image of the spatial distribution of a selected m/z

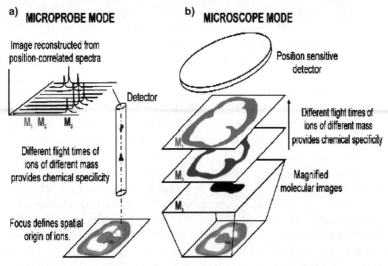

Figure 9.1 Schematic of the two different approaches to macromolecular imaging mass spectrometry. (a) Microprobe imaging collects mass spectra from an array of designated positions to reconstruct a molecular image after completion of the experiment. (b) In microscope imaging magnified images of the ion distributions are directly acquired using a two-dimensional detector. (Reproduced from Luxemborg S. L *et al.*, *Analytical Chemistry* 2004, **76**, 5339–5344 with kind permission of the publishers © 2004 The American Chemical Society.)

value onto a two-dimensional detector. This approach is shown in Figure 9.1b. To image larger areas than the diameter of the laser beam, adjacent images are "stitched" together in software and, by using ion gates, images of differing m/z values can be obtained. Using such instrumentation a spatial resolution of 4 μm has been demonstrated. Whilst such instrumentation holds enormous promise for the future of MALDI-MSI, to date there are currently no commercial systems available and hence in this article we will concentrate on the more widely used microprobe approach.

9.2 Practical Aspects of MALDI Mass Spectrometric Imaging (MALDI-MSI)

9.2.1 Instrumentation for MALDI-MSI

MALDI-MSI in microprobe mode has been performed on a wide variety of instrumentation. Most widespread has been the use of a conventional time of flight mass spectrometer and of note is the availability of software for Perceptive Biosystems/Applied Biosystems MALDI-TOF-MS instrumentation, as free downloads (http://www.maldi-msi.org) which both facilitates the acquisition of MALDI-MSI data and provides tools for its processing. MALDI-TOF-MS instruments with imaging options are, however, available from a range of other commercial vendors.

Whilst conventional MALDI-TOF-MS instrumentation is the most appropriate for the original MALDI-MSI application of protein profiling, there are limitations to its use for small molecule work. Conventional MALDI-TOF-MS instruments are of axial configuration; that is the sample and the mass spectrometer are in direct line of sight. The time of flight of an ion through such an instrument and hence the measured mass can be affected by both the surface topography of the sample and by surface charging if an insulating surface, such as a sample of biological tissue, is analysed. Sputter coating of sample surfaces with gold has been proposed as a method to overcome this effect. For small molecule MALDI-MSI, however, the use of orthogonal MALDI ion sources on hybrid quadrupole time of flight (QqTOF) instruments offers both good mass measurement stability and the potential to use tandem mass spectrometry to increase the specificity of analyses (discussed in Section 9.3.2). Such instruments are now also available commercially from a number of vendors.

In order to simply couple MALDI to mass analysers other than TOF based systems, atmospheric pressure MALDI (AP-MALDI) has been proposed. Here, as the name implies, ions are generated at atmospheric pressure and transferred into the mass spectrometer, either through a capillary inlet forming part of an electrospray or other atmospheric pressure ionisation source or through a nozzle skimmer arrangement. For imaging MALDI-MSI applications, AP-MALDI would clearly have the advantage that the tissue could be kept at standard laboratory conditions rather than under vacuum. Although still in its infancy and reported to suffer from sensitivity limitations, Oktem *et al.*[3] have demonstrated the feasibility of such an approach and produced images of the distribution of phospholipids in rat brain.

9.2.1.1   UV Lasers for MALDI-MSI

The most widely employed lasers for MALDI-MSI are the standard UV lasers, *i.e.* the nitrogen laser ($\lambda = 337$ nm) and a frequency tripled Nd:YAG laser ($\lambda = 355$ nm). Whilst the low cost N_2 laser is the most widely used laser for MALDI applications, solid state Nd:YAG lasers have become popular for MALDI-MSI owing to their high repetition rate (rates up to 1 kHz having been employed compared to 20 Hz on a standard N_2 laser) and extended lifetimes ($\sim 1 \times 10^9$ shots compared to $2 \times 10^7 - 6 \times 10^7$ shots for an N_2 laser). A recent paper, however,[4] has discussed some of the limitations of Nd:YAG lasers for MALDI applications. The narrow beam profile leads to poor sensitivity and a reduction in the number of MALDI matrices that can be employed. The same paper goes on to describe the development of a modified Nd:YAG laser to overcome these limitations.

9.2.1.2 Matrices for UV MALDI-MSI

The commonly used organic acid based UV-MALDI matrices such as sinapinic acid (SA), alpha cyano-4-hydroxycinnamic acid (αCHCA) and 2,5-dihydroxy-benzoic acid (2,5-DHB) have all found application for MALDI-MSI, although the use of 2,5-DHB is somewhat hampered by the large crystals it forms. The requirement for an energy absorbing matrix where a UV laser is employed is a limitation of the technique for metabonomic profiling. This is particularly the case where the common organic acid-type matrices are used. This is due to the high chemical background observed in the low mass (<500 Th) region of the mass spectrum. Ionic liquids, synthesised by reacting a conventional organic acid matrix with a base such as tributylamine or pyridine, have been proposed as an alternative type of organic matrix that yield homogenous samples with a decreased low mass background. Lemarie *et al.*[5] have shown that a solid ionic matrix prepared by the reaction of αCHCA and aniline yielded both higher sensitivity and reduced fragmentation in the MALDI-MSI examination of proteins, peptides and phospholipids in rat brain tissue.

The initial publication on MALDI produced by Tanaka *et al.*[6] did not in fact use an organic acid based matrix at all. In this work, 30 nm particles of cobalt suspended in glycerol were used as the matrix. Such "particle suspension matrices" are particularly suited to low mass work owing to the low background of ions obtained below 500 Th. Schürenberg *et al.*[7] studied a range of particle suspension matrices for the analysis of peptides and proteins and obtained best results using 35 nm TiN particles in glycerol. Crecelius *et al.*[8] investigated the use of such matrices for thin-layer chromatography mass spectrometry (TLC-MS), a technique which when carried out using MALDI instrumentation is closely related to tissue imaging. In this case the best results were obtained using 1–2 μm graphite in ethylene glycol. All reports on particle suspension matrices, however, comment on the transient nature of signals obtained and the poor sensitivity compared to conventional MALDI.

Perhaps one of the most interesting developments in small molecule MALDI is desorption ionisation on porous silica (DIOS).[9] Porous silicon is a UV absorbing semi-conductor with a large surface area and is produced through electrochemical anodisation or chemical etching of crystalline silicon. It is proposed that the structure of porous silica provides a scaffold for retaining solvent and analyte molecules and that its UV absorptivity provides the mechanism for the transfer of laser energy. A comparison of the use of DIOS and 9-aminoacridine for metabolite profiling in *Escherichia coli* was presented recently;[10] this work demonstrated that the sensitivity for such analyses obtainable using DIOS was superior owing to the reduced chemical background.

9.2.1.3 MALDI with Infrared Lasers

An alternative to the use of UV lasers for MALDI-MSI applications is the use of lasers operating in the infrared region. The use of such lasers for conventional

MALDI was first demonstrated by the group of Franz Hillenkamp in 1990.[11] The major attraction of the use of such a system for the study of particularly small molecules in biological tissue is the strong absorption band of water (ice) at $\sim 3\,\mu m$, offering the potential of "matrix free" MALDI from plant and animal tissue sections by using residual water (in the form of ice) as an endogenous matrix. The Hillenkamp group have reported the use of ice as a matrix for IR-MALDI of proteins.[12] Exploitation of the huge promise of this technique has been hampered, however, by the availability of suitable IR lasers. Lasers that appear suited to this application have now started to appear on the market and IR-MALDI-MSI of native tissue sections has been demonstrated by both the Hillenkamp group and that of Vertes at a range of conferences in 2006.

9.2.2   Sample Preparation Procedures

The methodology utilised for the incorporation of the MALDI matrix into the imaging experiment is the key to success. Several approaches have been proposed. Caprioli and co-workers developed the imaging MALDI technique originally to identify and profile peptides and proteins in biological tissue. In their early work they demonstrated that optimum results were achieved by spraying the required laser energy absorbing MALDI matrix onto the samples. Such an approach minimised analyte spreading and allowed the mapping and localisation of specific compounds within tissue sections. This is shown diagrammatically in Figure 9.2. Thin sections of tissue (typically $\sim 15\,\mu m$) are cut using a cryostat. Chaurand et al.[13] have proposed that for protein imaging these sections are "fixed" and "seeded" prior to matrix coating by soaking them in a matrix containing ethanol solution. Sections were then carefully mounted onto the MALDI target plate for matrix coating which was generally achieved by multiple spray passes (for protein imaging) of a solution of 20 mg/ ml sinnapinic acid in 50:50:0.1 acetonitrile:water:trifluoroacetic acid.

The use of spray coating of tissue sections with organic acid based matrices is still the most widely used methodology for tissue preparation; however, recently the use of matrix spotters has been proposed. Here an array of discreet matrix spots is produced on the sample surface. This approach has been demonstrated using ink-jet printer technology and also by the use of a novel matrix spotter based around the use of acoustic droplet ejection.[14] It is proposed by its exponents that such an approach to matrix application minimises the risk of analyte spreading caused by the matrix spraying. Such systems also have potential for automation.

Alternative approaches to direct tissue analysis have been developed, including techniques where the tissue under study is blotted onto a suitable membrane or a hydrophobic surface created by using particles of C_{18} solid phase extraction stationary phase. Such surfaces appeared to aid the blotting/capture of proteins.

Methods of sample preparation for MALDI-MSI have been reviewed by Schwartz et al.[15]

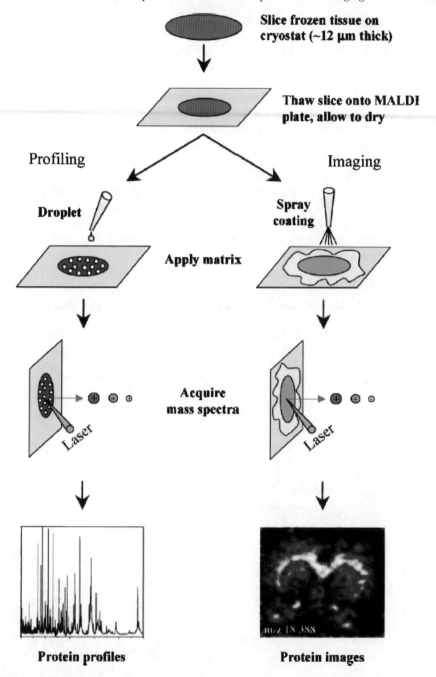

Figure 9.2   Schematic representation of the different steps involved for profiling and imaging mass spectrometry of biological tissue samples. (Reproduced from Chaurand P. *et al.*, *Anal. Chem. A-pages*, 2004, **76**(5), 86A–93A with kind permission of the publishers © 2004 The American Chemical Society.)

9.3   Applications of MALDI-MSI

9.3.1   MALDI-MSI in the Study of Proteins in Diseased and Healthy Tissue

Typical MALDI-MSI data of protein distribution in a 12 μm thick coronal mouse brain section containing a tumour is shown in Figure 9.3 As can be seen, a large number of proteins and polypeptides are observable in these spectra. Multiple molecular images can be produced from such data sets. Figure 9.4 shows MALDI-MSI images generated for 11 different proteins from these data. Histones (H4, H2B1 and H3) were identified in the tumour, consistent with the presence of fast developing tumour cells. Clearly, identification of proteins based on simply determining their relative molecular mass is not reliable. Extraction of the proteins from the tissue followed by identification using standard mass spectrometric based proteomics protocols is required for positive identification.

9.3.2   MALDI-MSI in the Study of Drug/Xenobiotic Distribution

The first demonstration of the use of MALDI to directly study pharmaceutical compounds in animal tissue was published by Troendle *et al.*[16] In this work,

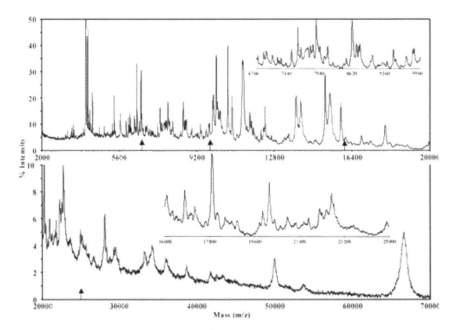

Figure 9.3   MALDI-MS protein profile obtained from a 12 μm thick human grade IV glioma tissue section. (Reproduced from Chaurand P. *et al.*, *Journal of Proteome Research* 2004, **3**, 245–252 with kind permission of the publishers © 2004 The American Chemical Society.)

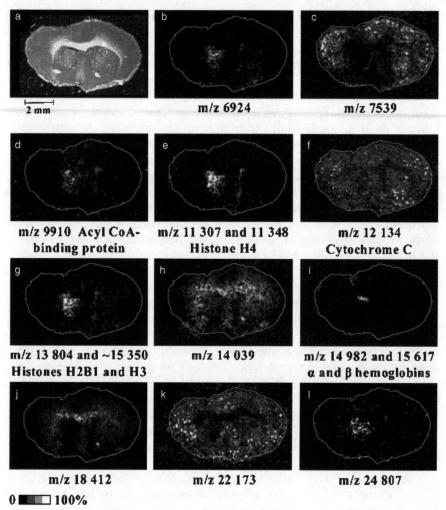

Figure 9.4 MALDI mass spectral image of a 12 µm thick coronal mouse brain section containing a tumour (a) Photomicrograph of the section before matrix application, (b) to (l) ion density maps obtained at different *m/z* values with an imaging resolution of 100 µm. The ion density maps are depicted as pseudo-colour images with white representing the highest protein concentration and black the lowest. (Reproduced from Chaurand P. *et al.*, *Journal of Proteome Research* 2004, **3**, 245–253 with kind permission of the publishers © 2004 The American Chemical Society.)

matrix was applied to the tissue surface by pipetting and electrospraying followed by analysis using MALDI quadrupole ion trap mass spectrometry. The technique was employed to detect the anti-cancer drug paclitaxel in a human ovarian tumour and the anti-psychotic drug siperone in spiked sections of rat liver tissue.

The Caprioli group have themselves begun to study drug distribution using MALDI imaging in collaboration with Schering-Plough. They have reported a study of the distribution of anti-tumour drugs in mouse tumour tissue and rat brains.[17] In these experiments matrix was applied to intact tissue by either spotting small volumes of the matrix in selected areas, or by coating the entire surface by airspraying. MALDI images were created by using the tandem mass spectrometric technique of selected reaction monitoring (SRM) to specifically monitor the drug under study. Such an approach minimises the potential for ions arising from either endogenous compounds or the MALDI matrix to interfere with the analyte signals. Figure 9.5 shows an image of distribution of a candidate anti-cancer drug in mouse tumour tissue created using this methodology. As can be seen, this image indicates that the drug was present over most of the tumour but was concentrated in the outer periphery.

Rohner *et al.*[18] have recently reviewed MALDI imaging and described some of the work being carried out at Novartis. In their article they demonstrate for the first time the use of MALDI-MSI to study drug distribution in a whole body mouse section. An illustration of these data is shown in Figure 9.6, where the optical image of a mouse section is compared with the mass spectral image. As can be seen, the use of the MALDI-MSI technique located the drug to the central nervous system. This pioneering work has been further extended by the Caprioli group[19] who have combined this approach with their own work on protein imaging to produce whole body images showing the location of drug, drug metabolites and endogenous markers for various organs of the body. They then go on to discuss the exciting possibility of imaging markers of efficacy along with the analytes to provide *in vivo* pharmacodynamic data.

Another interesting application of MALDI-MSI in drug and xenobiotic distribution has been the study of trans-dermal absorption.[20] Here it was demonstrated that the absorption of the anti-fungal agent ketoconazole into skin could be examined by MALDI-MSI by the use of an indirect tissue blotting approach. This work has been subsequently extended to show that the absorption of a wide range of xenobiotics into skin can be studied but that the sample preparation technique required, *i.e.* direct or indirect imaging, and the choice of substrate for indirect imaging, is hugely influenced by the analyte.[21] In the same paper this group have also described some preliminary data from a combined solvent assisted transfer/derivatisation approach to sample preparation that may be useful for particularly intractable analytes.

In a paper by Mullen *et al.*, MALDI-MSI was used to study the distribution of agrochemicals in soya plants.[22] Detection and imaging of the herbicide mesotrione (2-(4-mesyl-2-nitrobenzoyl)cyclohexane-1,3-dione) and the fungicide azoxystrobin (methyl (*E*)-2-{2-[6-(2-cyanophenoxy)pyrimidin-4-yloxy]phenyl}-3-methoxyacrylate), on the surface of the soya leaf, and the detection and imaging of azoxystrobin, inside the stem of the soya plant, were achieved using MALDI quadrupole time of flight mass spectrometry. In leaf analysis experiments, the two pesticides were deposited on to the surface of individual soya leaves on growing plants. The soya leaves were removed and prepared for direct and indirect imaging analysis at different periods after initial pesticide

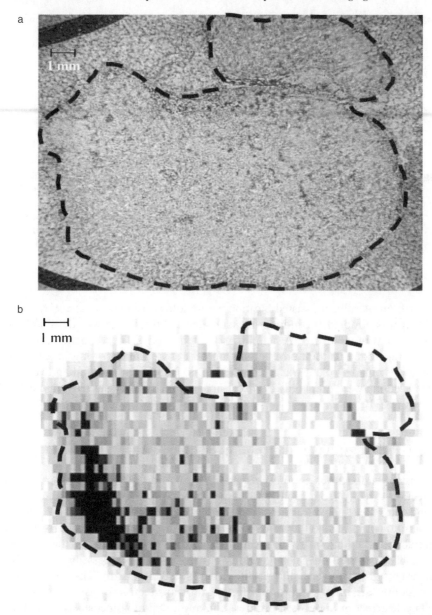

Figure 9.5 Image of the distribution of an anti-cancer drug candidate in mouse tumour tissue. (a) Optical image of the tissue after coating with the MALDI matrix. (b) Mass spectral image showing the distribution of the drug compound in the tumour section. (Reprinted from *Journal of Mass Spectrometry*, M. L. Reyzer *et al.*, "Direct analysis of drug candidates in tissue by matrix assisted laser desorption/ionization mass spectrometry", 2003, **38**, 1081–1092 with permission from John Wiley and Sons Ltd.)

a   b

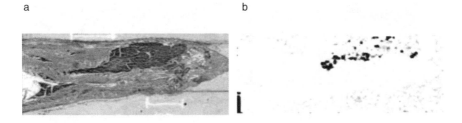

Figure 9.6   Optical image and MALDI mass spectral image of a whole-body mouse
section following dosing with a drug candidate. As can be seen, the
compound under study is located by the mass spectral image into
the central nervous system. (Reproduced from *Mechanism of Ageing
and Development*, Rohner *et al.*, "MALDI mass spectrometric imaging
of biological tissue sections", 2005, **126**, 177–185 reproduced with
permission.)

application. A solvent assisted blotting methodology using acetone as a transfer
solvent was adopted for indirect measurements. Mesotrione was detectable in
negative-ion mode using SA as matrix. In stem analysis experiments, azoxystrobin was added to the nutrient solution of a soya plant growing in a hydro-
ponics system. The plant was left for 48 hours, and then horizontal and vertical
stem sections were prepared for direct imaging analysis by electrospraying the
matrix (αCHCA).

9.3.3   MALDI-MSI and Phospholipids

A class of endogenous compounds that appear to be particularly amenable to
detection directly from biological tissue by MALDI-MSI are phospholipids.
Lipids are key structural/functional components of cells. In their role as
structural components, they provide a platform for membrane protein–protein
interactions (a lipid raft) in addition to providing the physical barrier for
cells. They can also have cellular functions, *e.g.* as second messengers in cellular events such as growth, proliferation and cell death. They and their
metabolites are involved in important signal transduction processes including
cell cycle arrest or apoptosis, proliferation and calcium homeostasis, as well as
cancer development, multidrug resistance, and viral or bacterial infection
processes and their distribution in brain tissue has been studied extensively
by MALDI-MSI.

We have become interested in examining the distribution of phospholipds in
colorectal liver metastasis by MALDI-MSI. Figure 9.7 shows the positive ion
MALDI mass spectra obtained from the three different regions of surface of a
human liver tumour metastasis – namely healthy tissue, tumour and tumour
margin. As can be seen, although the pattern of phospholipids observed is

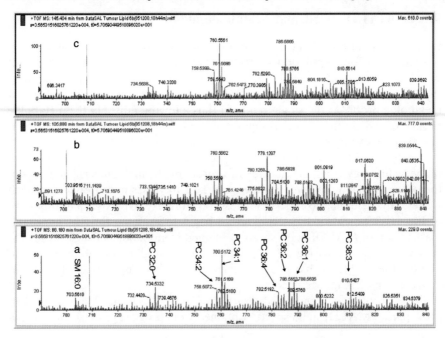

Figure 9.7 Positive-ion MALDI mass spectra obtained from the three different regions of surface of a human liver tumour metastasis namely healthy tissue, tumour and tumour margin.

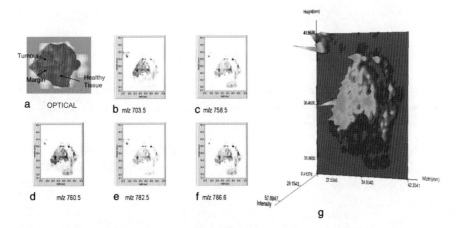

Figure 9.8 A series of MALDI-MSI images from this tissue sample in Figure 9.7 are shown, the agreement with the histology of the sample (a) is remarkable.

similar, there are some differences. One of the most striking differences is the increased expression of sphingomyelin 16:0 in the tumour tissue. A series of MALDI-MSI images from this tissue sample are shown in Figure 9.8; the agreement with the histology of the sample (Figure 9.8a) is remarkable.

9.3.4   MALDI-MSI and Metabonomics

Despite the realisation that spatial resolution of metabolites in complex tissues is required before regulation of metabolism can be understood, this is not possible using existing technologies. MALDI-MSI is an emerging technology that avoids the problems of extraction and has the potential to provide a metabolite profile at the single cell level thus allowing not only the metabolite profile to be determined but also more importantly the known heterogeneity of tissues to be examined.

Sub-cellular MALDI imaging is not possible, however, with the current state of the technology. Animal cells are of the order of 10 μm diameter, plant cells of the order of 50 μm diameter and hence at best the mass microprobe approach would only yield one pixel per cell. However, we have felt it useful at this stage of the development of the technique to begin to examine the low molecular mass region of MALDI mass spectra obtained from biological surfaces for the presence of peaks arising from primary and secondary metabolites and to address the problems that occur in their use for imaging; namely low abundance and matrix interference.[23]

These problems are indicated clearly in Figures 9.9a and 9.9b. These data show partial MALDI mass spectra obtained from a wheat grain section with a

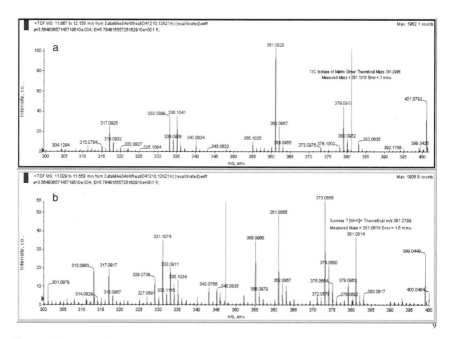

Figure 9.9   Partial MALDI mass spectra from two different regions of a wheat grain acquired during an imaging run. Note the intense αCHCA matrix related ions. Recalibration of data and accurate measurement appears to be essential for the identification of endogenous compounds in such a high matrix background.

typical α-CHCA matrix background spectrum. The spectrum obtained from the wheat grain is dominated by matrix related peaks. Of note are the abundant ions corresponding to alkali metal adducts of the matrix and fragments thereof. We have noted in this work that in fact the $[M + K]^+$ ion for α-CHCA (m/z 228) provides a good marker for the presence of biological material. One obvious and widely used solution to the problem of the low mass background that has been employed in the analysis of xenobiotics in biological tissue is to use an MS/MS approach. In particular the use of selected reaction monitoring has been proposed,[17] where the selected ion of interest in the single stage mass spectrum is subjected to further dissociation by collision with a gas and then the intensity of a characteristic product ion formed by this dissociation is monitored rather than the precursor itself. This approach, however, presupposes that a target compound like approach is being adopted, *i.e.* that the precursor ion m/z value is known or if the instrument automated precursor selection software is used that only a limited number of analytes are being studied.

An approach that potentially allows a wide range of metabolites to be simultaneously monitored without any pre-selection, is the use of accurate mass measurement. In Figure 9.9b a potential $[M + K]^+$ ion for a hex_2 oligosaccharide at m/z 381 is observable. This assignment was confirmed by recalibrating the data using the known accurate masses of matrix peaks and assigning an elemental composition to the 381 peak. The theoretical and actual measured masses were in agreement to 1.5 mmu. Although the assignments made are at this stage tentative, some observations about the data obtained can be made. In wheat grain at 15 days post anthesis, starch is being actively synthesized in the endosperm (Figure 9.10). In the MALDI-MSI data obtained from such a grain, a series of oligosaccharides were apparently detected as their potassium adducts in this region. In contrast, arginine appeared to be more concentrated in the embryo and the seed coat. Similar methodology has been used to detect oligosaccharides in developing wheat stems.[24]

Using αCHCA as matrix, we observed some slightly unusual effects in the analysis of a section of a pea pod (Figure 9.11). Of note from these data are the apparent presence of sucrose as its sodium adduct in the pod (Figure 9.11f) but as its potassium adduct in the pea (Figure 9.11e). This would not appear to make any sense as anything other than a matrix suppression effect.

There are of course alternatives to conventional organic acid matrices such as αCHCA for MALDI. Vermillion, Salsbury and Hercules demonstrated in 2002[25] that 9-aminoacridine could act as a MALDI matrix for the production of negative ions. Edwards and Kennedy have demonstrated recently the detection of 44 metabolites in islets of Langehausen and *Escherichia coli* strain DH5α by using 9-aminoacridine as matrix.[26] We have shown that it is possible to spray coat plant tissue sections with 9-aminoacridine and produce good quality images of the distribution of a range of metabolites.[23] MALDI images of the distribution of hex_1-monophosphate (probably glucose-6-phosphate) and the $[M - H]^-$ for hex_2 sugars are shown in Figure 9.12. The heterogeneity of these images is very striking.

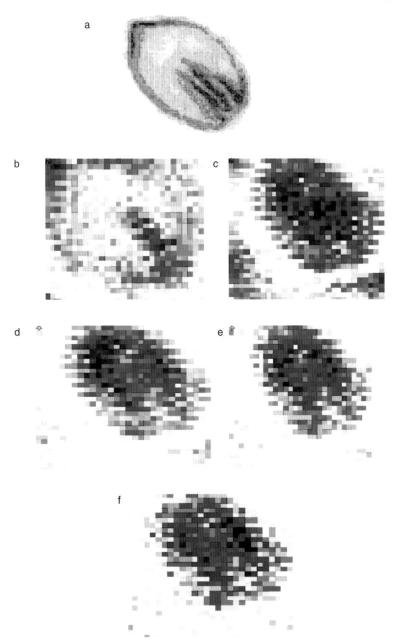

Figure 9.10   MALDI mass spectral images of a developing wheat grain 15 days post anthesis acquired in positive ion mode using αCHCA matrix: (a) optical image; (b) m/z 175 (arginine?); (c) m/z 381 (hex$_2$ oligosaacharide?); (d) m/z 543 (hex$_3$ oligosaacharide?); (e) m/z 705 (hex$_4$ oligosaacharide?); (f) m/z 867 (hex$_5$ oligosaacharide?).

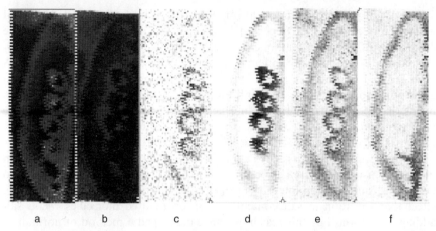

a   b   c   d   e   f

Figure 9.11   Images of (a) m/z 212 (matrix + Na); (b) m/z 228 (matrix + K); (c) m/z 138 (proline?); (d) m/z 175 (arginine?); (e) m/z 381 (hex$_2$ + K); (f) m/z 365 (hex$_2$ + Na), obtained from a section of pea pod.

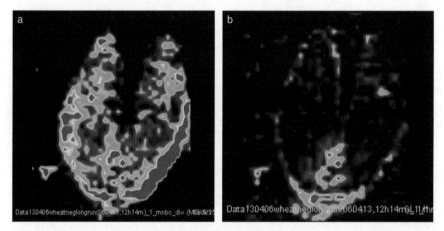

Figure 9.12   MALDI mass spectral images of developing wheat grain 15 days post anthesis acquired in negative ion mode using 9-aminoacridine as matrix. (a) m/z 259 hex-monophosphate (glucose-6-phosphate) $[M-H]^-$; (b) m/z 341 hex$_2$ $[M-H]^-$. (Reproduced from Burrell M *et al.*, "Imaging Matrix Assisted Laser Desorption Ionisation Mass Spectrometry: a technique to map plant metabolites at high spatial resolution", *J. Exp. Bot.* 2007, **58**, 757–763 with kind permission of the publishers © Oxford Univerisity Press 2006.)

9.4   Data Handling and Data Interpretation Issues

The data sets obtained from MALDI-MSI experiments carried out in full scan mode are extremely large files and potentially contain a wealth of information. We have been concerned with how such large data sets are "mined" to find target ions to yield biologically significant images.

   The use of principal components analysis (PCA) in the analysis of MALDI image data sets has been reported by McCombie *et al.*[27] In their work the coefficients of principal components were plotted as images in order to identify regions of tissue that were different. Problems found using raw data in PCA stem from irreproducible ion intensities. This is particularly evident whilst attempting to apply PCA to small molecules where variations in mass and intensity of potential analytes are masked by more significant variations in the type and intensity of matrix peaks. This in turn can be related to homogeneity of matrix coverage and distribution of salts within the tissue.

   We have found that in order to apply PCA to small molecules, normalisation of ion intensities is extremely useful.[21,28] The normalisation must take into account variation in ion intensity in relation to matrix coverage, as well as the effect that variation in salt content has on protonated molecules *versus* salt adduct formation. For this reason we have developed a method of normalisation which functions to normalise each ion species of a compound against the

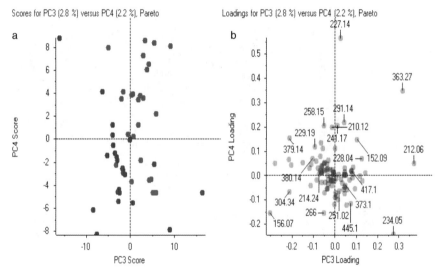

Figure 9.13   Principal components analysis of 25 spectra taken from inside and 25 from outside the drug treated area of porcine skin treated with a commerical 0.1% hydrocortisone cream. The ion intensities in each spectrum were normalised to the corresponding matrix species, *i.e.* peaks presumed to be non-alkali metal containing based on their fractional mass were normalised to the α-CHCA $[M+H]^+$ ion at m/z 190 and others to the $[M+Na]^+$ ion at m/z 212 or the $[M+K]^+$ ion at m/z 228 as appropriate. In the scores plot clear grouping in treated (red spots) and untreated (blue squares) is observed. The loadings plot indicates that m/z 363.28 the $[M+H]^+$ ion for hydrocortisone is a major contributor to the differentiation of the groups. (Reproduced from Prideaux B., *et al.*, "Sample Preparation and Data Interpretation Procedures for the Examination of Xenobiotic Compounds in Skin by Indirect Imaging MALDI-MS", *Int. J. Mass Spectrom.* 2007, **260**, 243–251 with kind permission of the publishers © Elsevier 2006.)

respective ion species of the matrix. PCA carried out on data normalised in this way is also able to give clearer indication of variants than that which is found when using raw data. Figure 9.13a shows the PCA scores plot from analysis of data obtained from an area of drug treated skin. As can be seen, spectra obtained from the treated tissue (red dots) and untreated tissue (blue squares) are clearly differentiated, following the ion species specific normalisation procedure. In the loadings plot (Figure 9.13b), m/z 363.27 is clearly an outlier and hence makes a significant contribution to the difference between the data groups in the scores plot. Therefore it is a clear candidate for imaging despite not being a prominent peak in the original mass spectra.

The use of PCA or a related multi-variant statistical technique to facilitate the detection of hidden variables in image data sets has clear implications in clinical and biomedical research.

9.5 Conclusions

MALDI-MSI is an emerging technology for molecular imaging in biological tissue. Applications of the technology to the study of protein, xenobiotic and phospholipid distribution are rapidly increasing. To date few groups have attempted to apply MALDI-MSI to endogenous metabolites. Its use for this application is hampered by the low mass matrix background commonly encountered in conventional MALDI with organic acid matrices. In this chapter we have described some of the applications of MALDI-MSI that have been carried out to date and attempted to give a brief overview of some of the exciting work that is being carried out to increase its applicability. The first commercial instruments for this technique only became available in 2004 and it is still in a phase of rapid improvement in terms of spatial resolution, sensitivity and applicability. Without question, instrumental and sample preparation innovations over the next 10 years will lead to it becoming a key tool in the increased understanding of cellular processes and their interaction with external stimulants.

Acknowledgements

RSC/ERSRC are gratefully acknowledged for providing Analytical Science PhD studentships for Sally Atkinson and Caroline Earnshaw. The loan of a high repetition rate Nd:YAG laser by Applied Biosystems/MDS Sciex to support our work on MALDI-MSI is also gratefully acknowledged.

References

1. R. M. Caprioli, T. B. Farmer and J. Gile, *Anal. Chem.*, 1997, **69**, 4751–4760.

2. S. L. Luxemborg, T. H. Mize, L. A. McDonnell and M. A. Heeren, *Anal. Chem.*, 2004, **76**, 5339–5344.

3. B. Oktem, V. V. Laiko, V. M. Doroshenko, S. N. Jackson, H-Y. J. Wang and A. S. Woods, Presented at 54th ASMS Conference on Mass Spectrometry and Related Topics Seattle, Washington USA May 28th-June1st 2006 (available from http://www.apmaldi.com/Publications/Posters/2006/AP_MALDI.pdf).

4. A. Holle, A. Haase, M. Kayser and J. Höhndorf, *J. Mass Spectrom.*, 2006, **41**, 705–716.

5. R. Lemaire, J. C. Tabet, P. Ducoroy, J. B. Hendra, M. Salzet and I. Fournier, *Anal. Chem.*, 2006, **78**, 809–819.

6. T. Tanaka, H. Waki, Y. Ido, S. Akita, Y. Yoshida and T. Yoshida, *Rapid Commun. Mass Spectrom.*, 1988, **2**, 151–153.

7. M. Schürenberg, K. Dreisewerd and F. Hillenkamp, *Anal. Chem.*, 1999, **71**, 221–229.

8. A. Crecelius, M. R. Clench, D. S. Richards and V. Parr, *J. Chromatogr. A*, 2002, **958**, 249–260.

9. J. Wei, J. Buriak and G. Siuzdak, *Nature*, 1999, **399**, 243–246.

10. A. Amantionico and R. Zenobi, Presented at 17th International Mass Spectrometry Conference, Prague, Czech Republic, 27th August–1st September 2006.

11. A. Overburg, M. Karas, U. Bahr, R. Kaufmann and F. Hillenkamp, *Rapid Commun. Mass Spectrom.*, 1990, **4**, 293–296.

12. S. Berkenkamp, M. Karas and F. Hillenkamp, *Proc. Natl. Acad. Sci.*, 1996, **93**, 7003–7007.

13. P. Chaurand, S. A. Schwartz and R. M. Caprioli, *J. Proteome Res.*, 2004, **3**, 245–252.

14. A. Aerni, D. S. Cornett and R. M. Caprioli, *Anal. Chem.*, 2006, **76**, 827–834.

15. S. A. Schwartz, M. L. Reyzer and R. M. Caprioli, *J. Mass Spectrom.*, 2003, **38**, 699–708.

16. F. J. Troendle, C. D. Reddick and R. A. Yost, *J. Am. Soc. Mass Spectrom.*, 1999, **10**, 1315–1321.

17. M. L. Reyzner, Y. Hsieh, K. Ng, W. A. Korfmacher and R. M. Caprioli, *J. Mass Spectrom.*, 2003, **38**, 1081–1092.

18. T. C. Rohner, D. Staab and M. Stoeckli, *Ageing and Develop.*, 2005, **126**, 177–185.

19. S. Khatib-Shahidi, M. Andersson, J. L. Herman, T. A. Gillespie and R. M. Caprioli, *Anal. Chem.*, 2006, **78**, 6448–6456.

20. J. Bunch, M. R. Clench and D. S. Richards, *Rapid Commun. Mass Spectrom.*, 2004, **18**, 351–360.

21. B. Prideaux, S. J. Atkinson, V. A. Carolan, J. Morton and M. R. Clench, *Int. J. Mass Spectrom.*, 2007, **260**, 243–251.

22. A. K. Mullen, M. R. Clench, S. Crosland and K. R. Sharples, *Rapid Commun. Mass Spectrom.*, 2005, **19**, 2507–2516.

23. M. Burrell, C. Earnshaw and M. R. Clench, *J. Exp. Bot.*, 2007, **58**, 757–763.

24. S. Robinson, K. Warburton, M. Seymour, M. Clench and J. Thomas-Oates, *New Phytologist.*, 2007, **173**, 438–444.
25. R. Vermillion-Salsbury and D. M. Hercules, *Rapid Commun. Mass Spectrom.*, 2002, **16**, 1575–1581.
26. J. L. Edwards and R. T. Kennedy, *Anal. Chem.*, 2005, **77**, 2201–2209.
27. G. McCombie, D. Staab, M. Stoeckli and R. Knochenmuss, *Anal. Chem.*, 2005, **77**, 6118–6124.
28. S. J. Atkinson, B. Prideaux, J. Bunch, K. E. Warburton and M. R. Clench, *Chim.-Oggi Chem. Today*, 2005, **23**, 5–8.

CHAPTER 10

Plant Metabolomics

THOMAS MORITZ AND ANNIKA I. JOHANSSON

Umeå Plant Science Centre, Dept. Forest Genetics and Plant Physiology, Swedish University of Agricultural Sciences, SE-901 87, Sweden

10.1   Introduction

To date (January 2007) three plant genomes have been fully sequenced: *Arabidopsis thaliana*,[1] *Oryza sativa*[2] (rice) and *Populus trichocarpa*[3] (black cotton wood). The availability of these sequences, and the increasingly abundant information on various other plant genomes, is revolutionising plant biology. The information is an extremely important tool for genetic analysis, and the ability to characterise plants by genome-wide analysis of gene expression (transcriptomics) and alter their phenotypes via specific gene knockouts or over-expression has transformed plant research. However, in order to understand many aspects of plant biology and gene function, broad phenotypic characterisation is also essential. In addition to traditional morphological and anatomical observations, this requires analysis of metabolite profiles, including characterisation of the metabolome, metabolite network analysis and flux determinations.

The aim of this chapter is to provide an overview of metabolomic analysis in plant research, focusing on general methodology and examples of its applications (for reviews see Fiehn[4] and Sumner *et al.*[5] and Table 10.1 for links to various web pages). Ten years before the term metabolomics was coined, Sauter *et al.*[6] used metabolite profiling to unravel the mode of action of herbicides. Now, nearly 20 years later the number of publications related to plant metabolomics and metabolite profiling has increased dramatically. Today, plant metabolomics is used in diverse applications including high-throughput fingerprinting of genetically[7–9] or environmentally modified plants,[10–12] the identification of metabolite quantitative loci,[13] the characterisation of gene functions[14] and the analysis of metabolic differences between genetically modified and parental lines of crop species.[15] Metabolomics is now an integral component of plant functional genomics and its importance will almost

Table 10.1 Links to various plant metabolomics and database sites on the internet.

Max Planck Institute, Golm, Germany	http://csbdb.mpimp-golm.mpg.de/csbdb/gmd/gmd.html
Noble Foundation, USA	http://www.noble.org/PlantBio/MS/metabolomics.html
RIKEN Plant Science, Japan	http://www.psc.riken.go.jp/eng/group/metabolomic/index.html
Rothamsted Research, UK	http://www.metabolomics.bbsrc.ac.uk/
UC Davis, USA	http://fiehnlab.ucdavis.edu/staff/fiehn
Umeå Plant Science Centre, Sweden	http://www.upsc.se/metabolomics.htm
Utrech University, Netherlands	http://www.abc.uu.nl/centersofexcelle/metabolomicscent/29714main.html
TAIR	http://www.arabidopsis.org/index.jsp
KEGG	http://www.genome.ad.jp/kegg/kegg2.html
MetaCyc	http//metacyc.org

certainly increase in coming years with further advances in metabolic profiling techniques. One of the most challenging aspects of plant metabolomics is to characterise the large number of metabolites that occur in plants, which are estimated to produce, across the plant kingdom, up to 200 000 different metabolites (cited in Sumner *et al.*). Although the number of different metabolites in specific species is much lower (a roughly guess is 5000–10 000), there is an urgent need to identify metabolites and create databases with structural and spectral information to fulfil the ultimate aims of metabolomics; to identify and quantify all metabolites in biological samples.[16]

10.2   Analytical Technology

10.2.1   General Considerations

In plant metabolite profiling many different analytical platforms have been used, including GC/MS,[17] LC/MS,[18] CE-MS,[19] HPLC-PDA[20] and NMR.[21] However, since the most widely used approaches are mass spectrometry coupled with GC or LC separation systems, only GC/MS- and LC/MS-based metabolomics will be covered in this chapter. Nevertheless, it should be emphasised that other analytical approaches are also useful for profiling metabolites and they should not be considered to have no value in plant metabolomic analyses. Indeed, appropriate combinations of analytical tools are often required to cover relevant aspects of plant biochemistry and chemistry.

Analysis of metabolites can be divided into several steps:

(i) design of experiment;
(ii) sampling and extraction of metabolites;
(iii) derivatisation of metabolites (especially when GC/MS is used);

    (iv) analysis of metabolites, *e.g.* by GC/MS or LC/MS;
     (v) data processing of MS-files;
    (vi) statistical analysis of data;
   (vii) interpretation of data.

Metabolomic experiments must be carefully designed since the information obtained from them heavily depends on the starting point. Further discussion of the importance of design of experiments (DOE) is provided by various introductions such as the one published by Trygg *et al.*[22] The growth conditions, sampling strategies, numbers of replicates and controls must all be appropriate. In addition, optimised extraction protocols and MS analysis, robust MS-data processing and critical statistical analysis are also crucial for relevant metabolomic analysis.

10.2.2 Sample Preparation

The pre-extraction sampling procedures applied in metabolomic analyses of plant tissues are very important, partly because relevant tissues must be selected for the purposes of the study and (just as importantly) all metabolic processes must be stopped as immediately after sampling as possible. The most frequently used procedure today is to freeze the samples directly in liquid nitrogen. However, it must be emphasised that changes and artefacts in metabolite profiles will inevitably occur during the short period between sampling of a leaf (for instance) and placing it in liquid nitrogen. An alternative is to use freeze-clamping techniques, but they are difficult to apply in large-scale experiments. Therefore, placing sampled tissues immediately in liquid nitrogen is still the best alternative for minimising pre-extraction changes in the metabolome.

   For similar reasons, metabolites should be extracted as rapidly as possible and during the extraction the degradation or modification of metabolites must be avoided (or at least minimised). The metabolites represent many diverse classes of compounds, including amino acids, fatty acids, carbohydrates and organic acids, hence their physico-chemical properties vary widely. This raises problems when choosing extraction protocols, since the optimum extraction conditions differ widely for different types of compounds, and probably for different plant tissues and species. The most common way to extract metabolites is to homogenise the plant tissue (*e.g.* by grinding in a mortar and pestle or using one of a range of mechanical devices), and then to shake the homogenised tissue at high or low temperatures in either a pure organic solvent, or a mixture of solvents. For polar metabolites, methanol, ethanol and water are often used, while for more lipophilic compounds chloroform is the most commonly used solvent. Combinations of chloroform:MeOH:water are also often used, either in a 1:3:1 mixture to avoiding solvent partitioning[23] or in other mixtures that result in the separation of a MeOH/water phase with associated lipophobic metabolites from a chloroform phase and associated lipophilic compounds. Gullberg *et al.*[23] showed, using a design of experiment approach, that a 1:3:1 methanol:water:chloroform mixture is an acceptable compromise for extracting metabolites with diverse chemical properties from *Arabidopsis* plant tissue.

Since only volatile and heat-stable compounds can be analysed by GC/MS, the next step is derivatisation if GC/MS is to be used to analyse metabolites that are insufficiently volatile in their native state. Although there are a number of strategies for derivatising compounds prior to GC/MS analysis, *e.g.* silylation, alkylation, acylation and alkoxyamination, the standard procedure in plant metabolomics is to first derivatise them using methoxyamine (CH_3–O–NH_2) in pyridine to stabilise carbonyl moieties in the metabolites, thereby suppressing keto-enol tautomerism and the formation of multiple acetal or ketal structures. Methoxyamination helps to reduce the numbers of derivatives of reducing sugars, and generates only two forms of the –N=C< derivative (*syn* and *anti* forms). After methoxyamination, functional groups, such as –OH, –COOH, –SH or –NH groups, are converted into TMS-ethers, TMS-esters, TMS-sulfides or TMS-amines, respectively, using a trimethylsilyl (TMS) reagent, usually BSTFA or MSTFA. TMS derivatisation has been thoroughly investigated and shown to be very efficient. However, one should be aware that derivatisation artefacts occur, including multiple derivatives of some compounds, *e.g.* amino acids (see Chapter 1). An alternative to GC/MS is LC/MS analysis, and for such analysis no derivatisation of the metabolites is needed prior to the MS analysis.

10.2.3 GC/MS

In GC/MS-based plant metabolomics and metabolite profiling both electron impact-quadrupole and TOF-instruments have been used (see Table 10.2 for a general overview of mass spectrometry techniques). Due to its fast spectral accumulation (and hence rapid analysis) capacities, GC/TOFMS has become popular within the plant metabolomics community.[9,13,23] A typical GC/TOFMS TIC-chromatogram from an *Arabidopsis thaliana* extract is shown in Figure 10.1. In such an analysis between 300 and 500 peaks can usually be detected. Between 200 and 1000 detected metabolites have been reported from GC/MS-based plant metabolomic studies, but the exact number of verified metabolites (known or unknown) or used in final statistical analysis has not been clearly stated in some of these studies. Nevertheless, in many laboratories 100–200 plant metabolites can now be routinely quantified using GC/MS, including many carbohydrates, fatty acids, amino acids, amines and sterols.

GC/MS profiling of plant samples is now a robust technique from which reproducible data can be obtained using internal standards to track recovery rates, derivatisation efficiency and other potential sources of analytical error.[17,23] The standard deviations of normalised peak areas depend on the chemical nature of the detected compounds, but mean RSD values of 13.8%[23] and 10.8%[24] have been reported. These standard deviations can be reduced by including more isotope-labelled internal standards and matching their peaks

Table 10.2 Common mass spectrometry terms. Note that only the most common ionisation techniques and types of mass analysers are mentioned. For a more thorough description see Chapter 1 of this book and other text books published in this area.

Ionisation techniques	*Processes that produce ions from neutral molecules in the ion source of the mass spectrometer*
Electron impact (EI)	The most commonly used GC/MS ionisation technique. Ions are generated by bombarding gaseous sample molecules with a beam of high energy electrons $(M + e^- \rightarrow M^{+\bullet} + 2e^-)$. Both intact molecular ions and fragments can appear in a mass spectrum.
Chemical ionisation (CI)	Ions are generated by exposing sample molecules to a large excess of ionized reagent gas, *e.g.* CH_4 or NH_3, yielding (inter alia) $[M + H]^+$ ions. Negative ions can also be detected. CI is an alternative to EI ionisation, *e.g.* when molecular weight information is needed.
Electrospray ionisation (ESI)	The most commonly used LC/MS ionisation technique. A liquid containing mobile phase and analyte(s) is pushed through a small charged capillary. An aerosol of charged liquid droplets is then formed. Ionisation takes place at atmospheric pressure. Mainly $[M + H]^+$ or $[M - H]^-$ ions are formed.
Atmospheric pressure chemical ionisation (APCI)	A form of chemical ionisation for LC/MS which takes place at atmospheric pressure. Can be successfully used for many substances that can not be analysed by ESI, *e.g.* carotenoids and some lipids.
Mass analysers	The part of the MS that separates the ions according to their mass-to-charge ratio (m/z). Tandem mass spectrometry (MS/MS) involves multiple steps of mass analysis, usually separated by some form of fragmentation. Example of instruments with MS/MS capabilities are QTOF, triple-quadrupole and ion traps.
Time-of-flight (TOF)	An analyser that measures the flight time of ions with the same kinetic energy over a fixed length.
Quadrupole (Q)	A mass filter consisting of four metal rods that uses oscillating electrical fields to selectively stabilise or destabilise ions passing through a radio frequency quadrupole field. Only ions of a certain m/z will reach the detector for a given ratio of voltages.
Ion-trap	An analyser that produces a three-dimensional symmetrical quadrupole field capable of storing ions with a certain m/z value.
Fourier transform ion cyclotron resonance (FT-ICR)	The m/z is determined by measuring the ions' cyclotron frequency in the presence of a magnetic field. FT-ICR is an ultra high mass resolution technique.

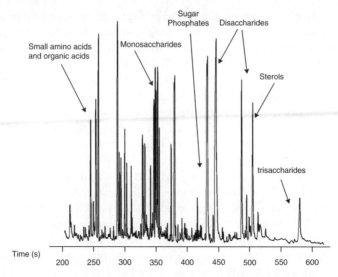

Figure 10.1 Total ion current chromatogram of a methanol:chloroform:water extract of *Arabidopsis thaliana* leaves analysed by GC/TOFMS. The elution order of the main groups of metabolites is indicated in the chromatogram.

against those of corresponding endogenous metabolites.[23] However, a problem with GC/MS analysis that needs to be addressed is the paucity of standard mass spectra libraries, which restricts the scope to identify detected compounds. Large libraries are commercially available, *e.g.* the NIST library, which contains more than 100 000 electron impact mass spectra, but since many of the spectra included are of synthetic compounds, and thus of little general relevance in plant metabolomics, there is a need to create specific libraries for plant metabolomics. The Max Planck Institute in Golm, Germany, has made its mass spectra library, which includes spectra of standard compounds and mass spectra tags from several laboratories, publicly available[25] (http://csbdb.mpimp-golm.mpg.de/csbdb/gmd/gmd.html). This database includes several kinds of information, including mass spectra and corresponding retention indices (RI) that facilitate the identification of metabolites (see Schauer *et al.*[26]). Although this library is a help for many laboratories, the number of identified peaks in an *Arabidopsis* extract analysed by GC/TOFMS is still only 20–30% of all detected peaks (Johansson, Moritz *et al.* unpublished). Further pooling of information from different laboratories, thereby creating common mass spectra databases, would definitely increase the number of identifications.

10.2.4 LC/MS

Since there is no need to derivatise compounds prior to the MS analysis when using LC/MS (as mentioned above), fewer sample preparation steps are required in LC/MS than in GC/MS analyses, and there are generally fewer artefacts. It is also capable of analysing large numbers of compounds that cannot be analysed by GC/MS, including compounds with labile glucosidic bonds such as phenylpropanoid derivatives, carotenoids and many lipids. However, although direct infusion ESI-MS can, in principal, be used for metabolite fingerprinting/footprinting analysis,[27] it cannot be used when metabolites need to be definitively identified, due to problems with suppression effects, its inability to distinguish structural isomers and the need for ultra-high resolution MS detection. Therefore, LC/MS-based metabolomics requires the use of an appropriate separation technique, such as HPLC, prior to the MS analysis. Compared to GC, HPLC has some drawback in the chromatographic performance. However, the recent introduction of ultra-performance liquid chromatography (UPLC) and columns with smaller particle sizes (together with associated improvements in plate number and efficiency) has improved the chromatographic resolution.[28] UPLC also allows higher flow rates than HPLC (since the optimum flow velocity has a broader range) and thus more rapid analysis without loss of resolution. However, despite the improvements provided (*inter alia*) by the introduction of UPLC systems and monolithic columns,[29] one should be aware that the optimal packing materials for separating specific classes of compounds differ widely. For example, highly polar compounds like ATP and GTP are very difficult to retain on a C_{18} column, but for such substances HILIC-type columns can be used instead. Consequently, the LC separation of whole metabolomes is impossible (currently at least) using a single chromatographic system.

Following separation by LC the liquid mobile phase, including compounds of interest, is introduced into the MS and the metabolites are ionised by appropriate techniques, such as electrospray ionisation (ESI) or atmospheric pressure chemical ionisation (APCI) (Table 10.2). To obtain as much information as possible regarding the metabolome both ESI and APCI should be used. For example, carotenoids and many lipids can be satisfactorily detected using APCI-MS, but not ESI-MS systems, while the reverse applies to other compounds. Since the sensitivity of ESI-based systems towards different compounds varies, depending on whether positive or negative ions are detected, each sample should be analysed twice using both positive/negative modes, alternatively using positive/negative switching mode. Since LC/MS analysis usually provides poor structural information ($[M + H]^+$ or $[M - H]^-$ are the main ions produced) high-resolution detection systems are essential, *e.g.* TOF, QTOF, Orbi-trap or FT-MS instruments. LC/MS profiling has been successfully applied to samples from a number of species, *e.g.* tomato[18] and *Arabidopsis;*[30] for an example see the UPLC-ESI/TOF data acquired from an *Arabidopsis* leaf extract shown in Figure 10.2. The number of detectable peaks using the commercial software "MarkerLynx™" (Waters) is estimated to be several thousands, but this includes isotopes, adducts and fragments. The number of true compounds detected is therefore much lower, a fact very seldom mentioned in publications.

Mass spectra libraries for LC/MS analysis are less common than GC/MS libraries, partly because tandem mass spectra vary depending on the instrument

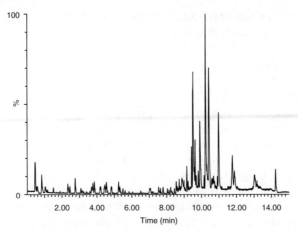

Figure 10.2 Base-peak chromatogram of a methanol:chloroform:water extract of *Arabidopsis thaliana* leaves analysed by UPLC/ESI-TOFMS in the positive-ion detection mode.

and technique used. Therefore, there is a need to compile further metabolite databases (including data on the compounds' mass spectra and structure together with other relevant information, such as LC conditions, retention times, species and sampled tissues) in order to fully exploit the potential of LC/MS in plant metabolomic analysis. One example of such a database that has been made publicly available is the Metabolome Tomato Database[31] (MoTo DB; http://appliedbioinformatics.wurl.nl).

10.3 Applications of Plant Metabolomics

10.3.1 Phenotyping Plants Using Metabolomics

In order to characterise genetically modified plants phenotypically, easily identifiable traits such as morphological features and rates of fertility, lethality and growth are often measured. Alternatively, biochemical approaches, for instance monitoring gene expression or protein profiles or enzyme activities, can often be successfully used to identify mutants and mutated gene functions. However, in many cases phenotypic differences between plants are not solely determined by differences in the expression of a specific gene, instead they are due to the interactive effects of multiple differences in the plants' transcriptomes and metabolomes, especially when the expression levels of genes encoding metabolic enzymes are altered. It is also important to remember that changes in gene expression do not necessarily have linearly proportional effects on protein expression, enzyme activity and/or metabolite levels since complex feed-back and feed-forward loops are involved in the control of plant metabolism.[31]

Metabolite profiling approaches (mainly GC/MS based) have been used to phenotype various plants that have been genetically or environmentally manipulated for more than 5 years. A large number of publications have shown the value of these approaches for phenotyping genetically altered plants, including potato[32] and tomato.[8] More than 100 primary metabolites have been simultaneously identified and quantified, and by comparing the profiles obtained using multivariate statistical techniques like principal component analysis (PCA) and hierarchical cluster analysis (HCA) different genotypes can be clustered according to metabolic variations. Using this approach an overview can be obtained of the relationships between and among the different genotypes or environmentally triggered phenotypes. Interpretation of loading plots from PCA enables metabolites responsible for the differences between clusters (*e.g.* genotypes) to be identified. In all these studies the main focus has been on known metabolites involved in primary metabolism. However, it should be possible to apply a similar approach to all detected metabolites (peaks) using PCA or other multivariate statistical tools. The advantage of this would be that a database could then be created with information on unidentified metabolites (non-annotated mass spectra) and their variations in different genotypes and/or environments, as well as identified metabolites. This would be of great help in the future if and when unknown peaks are identified.

Many organisms that are genetically modified or grown under specific conditions do not show any obvious phenotypic variations, such as differences in growth rate, morphology or related parameters, in comparison to parental lines or control conditions. Such phenotypes, in which there are no apparent phenotypic deviations from controls, are usually called "silent" phenotypes,[33] and can be observed in such diverse types of organisms as yeast and *Arabidopsis*. In some cases even when key genes are mutated the observed plant phenotype does not show any obvious alteration (see Bouche and Bouchez[34]). The generally accepted explanations for the phenomenon are that organisms have a redundancy of gene families, or cope in other ways with the presumed deleterious effect of a mutated key gene. It is also generally believed that metabolomic analysis can help to characterise such silent phenotypes functionally, especially when the primary effect is on enzyme-encoding genes.[31] The underlying rationale for this belief is that fingerprints of the metabolome under a given biological/biochemical situation obtained by metabolomic analyses should provide valuable indications regarding differences in both the levels of specific compounds and the status of associated metabolic networks between the plants compared. Silent phenotypes in yeast have been successfully clustered from WT strains using direct infusion ESI-MS[27] combined with multivariate statistical analysis.

In contrast to the work in yeast, Weckwerth *et al.*[9] presented another strategy for characterising silent phenotypes of potato based on GC/TOFMS profiling (Figure 10.3). The main advantage of this approach, compared to the direct infusion ESI-MS strategy, was that it enabled the metabolites associated with the clustering to be identified. The study objects were potato lines transformed with sucrose isoform II (SS2) in antisense orientation. The role of sucrose

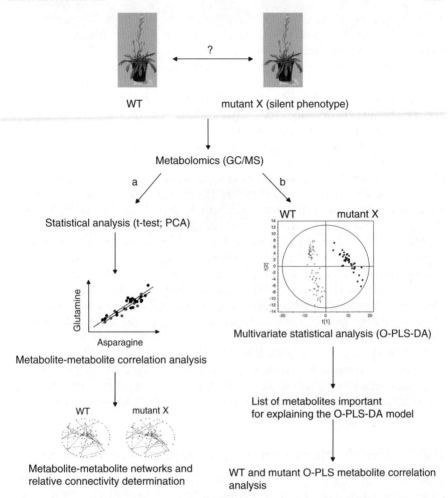

Figure 10.3 (a) Methodology used by Weckwerth *et al.*[9] for identifying metabolic differences between wild type and a mutant with a silent phenotype. (b) Suggestion for an alternative approach for the analysis of silent phenotypes.

synthase is to catalyse, enzymatically and reversibly, the cleavage of sucrose into UDP-glucose and fructose. Sucrose synthases are very important regulators of plant metabolism and have functions in various processes, including phloem unloading and plant development. However, since the expression patterns of different isoforms of sucrose synthase overlap both temporally and spatially, their specific roles are very difficult to determine. Silencing specific sucrose synthases by antisense or RNAi approaches may therefore result in silent phenotypes. In the study published by Weckwerth *et al.*[9] the aims were to investigate the potential of metabolomic analysis to discriminate the antisense line from the parental line and to identify the effect of down-regulation of *SS2* on primary metabolism. The results showed that the antisense SS2 potato line did not have any obvious visible phenotypic deviations from the parental line with regards to leaf and tuber morphology, or selected enzyme activities. Furthermore, t-tests and PCA did not detect any major differences in the GC/MS-acquired metabolite profiles of leaves and tubers between the antisense and parental line. To obtain more detailed knowledge about the biochemical differences between the lines the cited authors then applied metabolic correlation and metabolite network topology analyses. The purpose of these analyses was to investigate the hypothesis that primary carbohydrates like sucrose and glucose are interconnected to multiple pathways, and thus changes in their levels might have an impact on overall networks.[35] One of the problems with such studies is that the large number of pairwise correlations involved complicates attempts to overview the data. The authors therefore chose to use a sub-network of mainly polar metabolites for their connectivity topology ranking. Interestingly, using this strategy they were able to detect subtle metabolic differences between the modified and unmodified lines by determining the number of "metabolite connectivities" in the different genotypes. At any given relevance level the average number of correlations per metabolite was found to be higher in the SS2 tubers and lower in SS2 leaves compared to the parent line. The metabolites that showed differences in connectivity between genotypes were also identified. Although this interesting approach seems to give valid and interpretable results, it would be intriguing to compare it with the use of supervised multivariate projection methods like PLS-DA and O-PLS-DA (Figure 10.3; see Trygg et al.[22] for an introduction to chemometrics). These methods can handle large data sets, give interpretable results and are not dependent on univariate statistics, which causes problems when variables (metabolites) are correlated.

10.3.2   Genetic Metabolomics: Identification of Metabolic QTLs

There is a need to develop improved plant breeding tools that facilitate earlier selection based on desirable traits in (inter alia) commercially important crops, fruit trees and forest trees. Recently, an example of the use of metabolite profiling for identifying important traits, in tomato, was published by Schauer et al.[13] The cited authors phenotyped tomato ingression lines containing chromosome segments of wild species in a background of a cultivated variety. Using this approach they identified more than 800 so-called quantitative metabolic loci (metabolite QTLs) and more than 300 loci associated with yield-associated traits. About 50% of the metabolic loci were associated with quantitative trait loci. The authors conclude that this approach, combining metabolite profiling and detailed morphological analysis, is useful for analysing traits that are potentially important for crop breeding. An interesting study, that is similar in some ways to the tomato study, was recently published in which flavonoid profiles were detected by HPLC-PDA in apical tissues of two full-sib families of Populus deltoids × P. nigra and P. deltoides × P. trichocarpa. Metabolite correlation followed by QTL analyses were used to detect the point of metabolic control of flavonoid biosynthesis.[36]

In forest breeding one of the major problems is the lack of tools for early selection of important traits. Traditionally used traits such as growth rate and biomass, together with wood characteristics, such as lignin content and fibre length, are commercially important and there is a need to accelerate the selection of trees since the demand for forest trees will increase in the future. Although the number of studies in which metabolite profiling has been applied in forest tree related research is limited, GC/MS profiling has been successfully applied to identify differences between wild type *Populus tremula* × *P. alba* and two transgenic lines with modified lignin monomer composition.[37] PCA was able to distinguish the two lines irrespectively of whether the samples were taken from wood-forming tissues or non-lignifying tissues. An alternative (and perhaps more attractive) approach to find metabolic markers or traits related to specific phenotypic characteristics, is to combine metabolomics analysis of a large number of trees in which the phenotypic characteristic of interest has been quantified and then apply multivariate statistical tools like O-PLS[22,38] to correlate and identify metabolites with the phenotypes. An example of this is shown in Figure 10.4. In a long-term project we are performing metabolomics analysis of xylem scrapings from wood-forming tissues of a large number of trees (both wild type and genetically modified) in which the xylem fibre length and width have been determined. By using multivariate projection methods it is then possible to both classify the trees and correlate the length or width of their fibres with their metabolite profiles.

10.3.3 Plant Metabolomics and Gene Identification

In the post-genomics era the main focus is on identifying gene functions. The classical approach is to use gene cloning, and forward or reversed genetic approaches. In large numbers of cases the sequences of the examined genes (including, unfortunately, many plant metabolic genes) suggest that they encode enzymes involved in secondary metabolism or enzymes with unknown biochemical effects. In such cases metabolomics is likely to be a helpful or perhaps essential complementary approach to obtain relevant information to identify the functions of the genes,[17] although few publications have really demonstrated the value of metabolomics for identifying gene functions *per se*. In the legume *Medicago* both isoflavonoid and triterpine pathways have been elucidated using LC/MS-based metabolomic approaches.[14] However, it must be emphasised that in such cases the use of transcriptomics has also been helpful. Figure 10.5 describes a general approach for using metabolomics to identify a bioactive molecule involved in branching in *Arabidopsis* (see McSteen and Leyser[39]). In this case a number of genes involved in the biosynthesis of the branching regulator have been identified and gene homology data suggest that carotenoids are precursors. By analysing root tissue from the branching mutants (max-mutants) the mass spectra that differ between the mutants and wild type can be identified. Although the phenotypic

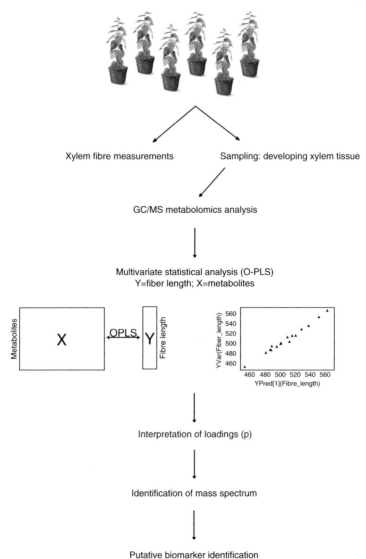

Figure 10.4   Illustration of how metabolomics and multivariate projection methods can be used for identifying metabolic markers related to phenotypic characteristics. A large number of trees are selected, and their fibre lengths and metabolite profiles are determined. Multivariate analysis can then be used to correlate their fibre lengths and metabolite profiles, and thus metabolite markers can be identified.

differences in root tissues between wild type and mutants are small, the metabolic differences are large, therefore the strategy has been to use metabolic profiling (GC/MS) to identify mass spectra that might correspond to a carotenoid cleavage product. Luckily, one of the peaks that appeared to differ

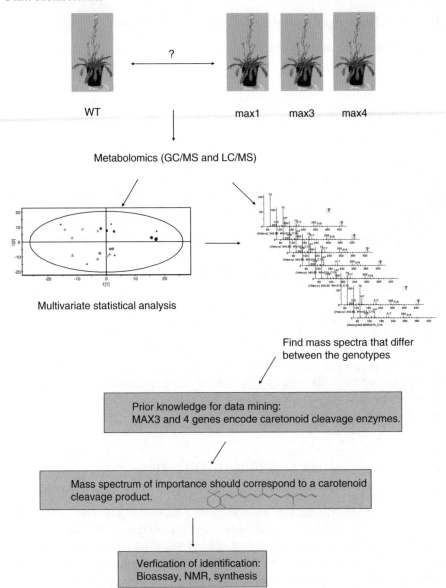

Figure 10.5 Schematic presentation of how a metabolomics approach can be used for identifying gene function, exemplified by the identification of an unknown branching regulator in *Arabidopsis* (see McSteen and Leyser[39]). The key-point is the use of multivariate statistical tools combined with prior knowledge about gene homologies.

between spectra obtained from the wild type and the mutants has a mass spectrum that can be interpreted as a carotenoid cleavage product. The next step is to purify and identify the compound thoroughly by NMR and chemical synthesis.

10.3.4 Other Applications of Plant Metabolomics

In plant biology a common objective is to follow biochemical events associated with environmental changes. Inevitably, metabolomics has gained interest in such studies since changes in metabolite levels can help to elucidate the biological processes underlying phenotypic changes induced by alterations in the environment, *e.g.* diurnal rhythms,[10] sink-source transitions of leaves[40] and acclimatory responses to cold stress.[12]

Metabolomics has also been applied (and will be applied further) to identify unintended effects of genetically modified crops (GM plants; see *e.g.* Rischer and Oksman-Caldentey[41]), based on the hypothesis that potentially unintended effects of genetic modifications on human health or the environment will be connected to changes in metabolite levels in the plants. However, there have been very few studies on this issue, although the general assumption should be that changes in a gene will inevitably cause some metabolic changes too (see the section on silent phenotypes above). Catchpole *et al.*[15] used different metabolite profiling and clustering approaches to determine differences between field-grown GM potato tubers and conventional potato tubers. Although the cited study's general objective was to compare genotypes, the data acquired show that levels of a number of metabolites related to the genetic modification differed between the cultivars, but the general metabolic profiles of the two cultivars were very similar. Although metabolomics approaches, especially when a large proportion of the metabolome is analysed, might be useful for evaluating how GM crops differ from the respective wild-type crops, a problem that still needs to be addressed is to define an "unhealthy" metabolite profile.

10.4 Conclusions and Future Directions

Metabolomics or metabolite profiling is now widely used in plant research. Methods for routinely detecting hundreds of metabolites have been developed and databases containing information from experiments have started to be compiled. To further exploit the potential of plant metabolomics, there is a need to identify more metabolites. Since only a small proportion of the metabolome in plants has been identified, the focus must be on the unknown metabolites. With further advances in the analytical equipment used, such as mass spectrometers, the prospects for determining metabolite profiles in specific cells (metabolomics at the cellular or organellar level) will hopefully be enhanced in the future. Sample fractionation using solid-phase extraction or related techniques, and vapour-extraction methodology[42] may, in the future, also be useful for monitoring metabolite profiles. Furthermore, to obtain further insights into biological and biochemical processes, fluxes of metabolites must also be measured. Since plants use CO_2 in photosynthesis, $^{13}CO_2$ has a potential to be of great help for determine fluxes, as after a certain period all molecules will be labelled with ^{13}C.[43] This can also be a help for identification of unknowns (determination of the number of carbons) and also for creating a true set of internal standards for all metabolites.

References

1. Arabidopsis Genome Initiative, Analysis of the genome sequence of the flowering plant Arabidopsis thaliana, *Nature*, 2000, **408**, 796–815.
2. T. Matsumoto, J. Z. Wu, H. Kanamori, Y. Katayose and M. Fujisawa *et al.*, The map-based sequence of the rice genome, *Nature*, 2005, **436**, 793–800.
3. G. A. Tuscan, S. DiFazio, S. Jansson, J. Bohlmann and I. Grigoriev *et al.*, The genome of black cottonwood, *Populus trichocarpa* (Torr. & Gray), *Science*, 2006, **313**, 1596–1604.
4. O. Fiehn, Metabolomics – the link between genotypes and phenotypes, *Plant Mol Biol.*, 2002, **48**(1–2), 155–171.
5. L. W. Sumner, P. Mendes and R. A. Dixon, Plant metabolomics: large-scale phytochemistry in the functional genomics era, *Phytochemistry*, 2003, **62**, 817–836.
6. H. Sauter, M. Lauer, H. Fritsch, Metabolite profiling of plants – a new diagnostic technique, *Abstr. Pap. Am. Chem. Soc.*, 1988.
7. U. Roessner, A. Luedemann, D. Brust, O. Fiehn, T. Linke, L. Willmitzer and A. R. Fernie, Metabolic profiling allows comprehensive phenotyping of genetically or environmentally modified plant systems, *Plant Cell*, 2001, **13**, 11–29.
8. U. Roessner-Tunali, B. Hegemann, A. Lytovchenko, F. Carrari, C. Bruedigam, D. Granot and A. R. Fernie, Metabolic profiling of transgenic tomato plants overexpressing hexokinase reveals that the influence of hexose phosphorylation diminishes during fruit development, *Plant Physiol.*, 2003, **133**, 84–99.
9. W. Weckwerth, M. E. Loureiro, K. Wenzel and O. Fiehn, Differential metabolic networks unravel the effects of silent plant phenotypes, *Proc. Natl. Acad. Sciences USA*, 2004, **101**, 7809–7814.
10. E. Urbanczyk-Wochniak, C. Baxter, A. Kolbe, J. Kopka, L. J. Sweetlove and A. R. Fernie, Profiling of diurnal patterns of metabolite and transcript abundance in potato (*Solanum tuberosum*) leaves, *Planta*, 2005, **221**, 891–903.
11. F. Kaplan, J. Kopka, D. W. Haskell, W. Zhao, K. C. Schiller, N. Gatzke, D. Y. Sung and C. L. Guy, Exploring the temperature-stress metabolome of Arabidopsis, *Plant Physiology*, 2004, **136**, 4159–4168.
12. M. A. Hannah, D. Wiese, S. Freund, O. Fiehn, A. G. Heyer and D. K. Hincha, Natural genetic variation of freezing tolerance in arabidopsis, *Plant Physiol.*, 2006, **142**, 98–112.
13. N. Schauer, Y. Semel, U. Roessner, A. Gur, I. Balbo, F. Carrari, T. Pleban, A. Perez-Melis, C. Bruedigam, J. Kopka, L. Willmitzer, D. Zamir and A. R. Fernie, Comprehensive metabolic profiling and phenotyping of

interspecific introgression lines for tomato improvement, *Nat. Biotech.y*, 2006, **24**, 447–454.

14. L. Achnine, D. V. Huhman, M. A. Farag, L. W. Sumner, J. W. Blount and R. A. Dixon, Genomics-based selection and functional characterization of triterpene glycosyltransferases from the model legume *Medicago truncatula*, *Plant J.*, 2005, **41**, 875–887.

15. G. S. Catchpole, M. Beckmann, D. P. Enot, M. Mondhe, B. Zywicki, J. Taylor, N. Hardy, A. Smith, R. D. King, D. B. Kell, O. Fiehn and J. Draper, Hierarchical metabolomics demonstrates substantial compositional similarity between genetically modified and conventional potato crops, *Proc. Natl. Acad. Sciences USA*, 2005, **102**, 14458–14462.

16. R. Goodacre, S. Vaidyanathan, W. B. Dunn, G. G. Harrigan and D. B. Kell, Metabolomics by numbers: acquiring and understanding global metabolite data, *Trends Biotech.*, 2004, **22**, 245–252.

17. O. Fiehn, J. Kopka, P. Dormann, T. Altmann, R. N. Trethewey and L. Willmitzer, Metabolite profiling for plant functional genomics, *Nat. Biotech.*, 2000, **18**, 1157–1161.

18. R. J. Bino, C. H. R. de Vos, M. Lieberman, R. D. Hall, A. Bovy, H. H. Jonker, Y. Tikunov, A. Lommen, S. Moco and I. Levin, The light-hyper-responsive high pigment-2(dg) mutation of tomato: alterations in the fruit metabolome, *New Phyt.*, 2005, **166**, 427–438.

19. S. Sato, T. Soga, T. Nishioka and M. Tomita, Simultaneous determination of the main metabolites in rice leaves using capillary electrophoresis mass spectrometry and capillary electrophoresis diode array detection, *Plant J.*, 2004, **40**, 151–163.

20. P. D. Fraser, M. E. S. Pinto, D. E. Holloway and P. M. Bramley, Application of high-performance liquid chromatography with photodiode array detection to the metabolic profiling of plant isoprenoids, *Plant J.*, 2000, **24**, 551–558.

21. J. M. Baker, N. D. Hawkins, J. L. Ward, A. Lovegrove, J. A. Napier, P. R. Shewry and M. H. Beale, A metabolomic study of substantial equivalence of field-grown genetically modified wheat, *Plant Biotech. J.*, 2006, **4**, 381–392.

22. J. Trygg, J. Gullberg, A. I. Johansson, P. Jonsson and T. Moritz, Chemometrics in metabolomics – an introduction In *Plant metabolomics*, K. S. K, R. A. Dixon, L. Willmitzer, Eds., Springer-Verlag, Berlin Heidelberg, 2006, Vol. 57.

23. J. Gullberg, P. Jonsson, A. Nordstrom, M. Sjostrom and T. Moritz, Design of experiments: an efficient strategy to identify factors influencing extraction and derivatization of *Arabidopsis thaliana* samples in metabolomic studies with gas chromatography/mass spectrometry, *Anal. Biochem.*, 2004, **331**, 283–295.

24. W. Weckwerth, K. Wenzel and O. Fiehn, Process for the integrated extraction identification, and quantification of metabolites, proteins and RNA to reveal their co-regulation in biochemical networks, *Proteomics*, 2004, **4**, 78–83.

25. J. Kopka, N. Schauer, S. Krueger, C. Birkemeyer, B. Usadel, E. Bergmuller, P. Dormann, W. Weckwerth, Y. Gibon, M. Stitt, L. Willmitzer, A. R. Fernie and D. Steinhauser, GMD@CSB.DB: the Golm Metabolome Database, *Bioinformatics*, 2005, **21**, 1635–1638.

26. N. Schauer, D. Steinhauser, S. Strelkov, D. Schomburg, G. Allison, T. Moritz, K. Lundgren, U. Roessner-Tunali, M. G. Forbes, L. Willmitzer, A. R. Fernie and J. Kopka, GC-MS libraries for the rapid identification of metabolites in complex biological samples, *FEBS Lett.*, 2005, **579**, 1332–1337.

27. J. Allen, H. M. Davey, D. Broadhurst, J. K. Heald, J. J. Rowland, S. G. Oliver and D. B. Kell, High-throughput classification of yeast mutants for functional genomics using metabolic footprinting, *Nat. Biotech.*, 2003, **21**, 692–696.

28. M. E. Swartz, UPLC (TM): An introduction and review, *J. Liq. Chromatogr. Related Technol.*, 2005, **28**, 1253–1263.

29. V. V. Tolstikov, A. Lommen, K. Nakanishi, N. Tanaka and O. Fiehn, Monolithic silica-based capillary reversed-phase liquid chromatography/electrospray mass spectrometry for plant metabolomics, *Anal. Chem.*, 2003, **75**, 6737–6740.

30. M. Y. Hirai, M. Yano, D. B. Goodenowe, S. Kanaya, T. Kimura, M. Awazuhara, M. Arita, T. Fujiwara and K. Saito, Integration of transcriptomics and metabolomics for understanding of global responses to nutritional stresses in *Arabidopsis thaliana*, *Proc. Natl. Acad. Sciences USA*, 2004, **101**, 10205–10210.

31. W. Weckwerth, Metabolomics in systems biology, *Ann. Rev. Plant Biol.*, 2003, **54**, 669–689.

32. U. Roessner, L. Willmitzer and A. R. Fernie, High-resolution metabolic phenotyping of genetically and environmentally diverse potato tuber systems. Identification of phenocopies, *Plant Physiol.y*, 2001, **127**, 749–764.

33. D. Thorneycroft, S. M. Sherson and S. M. Smith, Using gene knockouts to investigate plant metabolism, *J.f Exp. Bot.*, 2001, **52**, 1593–1601.

34. N. Bouche and D. Bouchez, Arabidopsis gene knockout: phenotypes wanted, *Curr. Opp. Plant Biol.*, 2001, **4**, 111–117.

35. K. Morgenthal, W. Weckwerth and R. Steuer, Metabolomic networks in plants: Transitions from pattern recognition to biological interpretation, *Biosystems*, 2006, **83**, 108–117.

36. K. Morreel, G. Goeminne, V. Storme, L. Sterck, J. Ralph, W. Coppieters, P. Breyne, M. Steenackers, M. Georges, E. Messens and W. Boerjan, Genetical metabolomics of flavonoid biosynthesis in Populus: a case study, *Plant J.*, 2006, **47**(2), 224–237.

37. A. R. Robinson, R. Gheneim, R. A. Kozak, D. D. Ellis and S. D. Mansfield, The potential of metabolite profiling as a selection tool for genotype discrimination in Populus, *J. Exp. Bot.*, 2005, **56**, 2807–2819.

38. J. Trygg and S. Wold, Orthogonal projections to latent structures (O-PLS), *J. Chemometrics*, 2002, **16**(3), 119–128.

39. P. McSteen and O. Leyser, Shoot branching, *Ann. Rev. Plant Biol.*, 2005, **56**, 353–374.

40. M. L. Jeong, H. Y. Jiang, H. S. Chen, C. J. Tsai and S. A. Harding, Metabolic profiling of the sink-to-source transition in developing leaves of quaking aspen, *Plant Physiol.*, 2004, **136**(2), 3364–3375.

41. H. Rischer and K. M. Oksman-Caldentey, Unintended effects in genetically modified crops: revealed by metabolomics? *Trends Biotech.*, 2006, **24**, 102–104.

42. E. A. Schmelz, J. Engelberth, J. H. Tumlinson, A. Block and H. T. Alborn, The use of vapor phase extraction in metabolic profiling of phytohormones and other metabolites, *Plant J.*, 2004, **39**, 790–808.

43. C. Birkemeyer, A. Luedemann, C. Wagner, A. Erban and J. Kopka, Metabolome analysis: the potential of in vivo labeling with stable isotopes for metabolite profiling, *Trends Biotech.*, 2005, **23**, 28–33.

CHAPTER 11

Data Mining for Metabolomics

ANDERS NORDSTRÖM

The Scripps Research Institute, Scripps Center For Mass Spectrometry, BCC007, 10550 North Torrey Pines Road, La Jolla CA 92122, USA

11.1 Introduction

"Data mining" is a broad term and has been described as "the science of extracting useful information from large data sets or databases".[1] The terms metabolomics, metabonomics, and metabolic profiling are also broad and to some extent used differently throughout the literature.[2] They share a common trait of a more or less global profiling approach with an ambition to perform untargeted measurements of metabolites in cells, tissues, fluids, or entire organisms. It is difficult to generalize the concept of data mining in conjunction with all "-omics" approaches and, hence, the data must be mined in a fashion that best answers the particular experimental question. In Figure 11.1 some possible paths of data mining are outlined, covering cases from finding metabolic markers for a certain disease/phenotype or markers for toxicology to more holistic scenarios where gene function and systems biology are addressed. The first three boxes can be considered universal regardless of further downstream data mining. The process starts with a question and an experimental design of some kind. Then follows a data acquisition step which most often in global profiling approaches involves mass spectrometry (MS)[3–7] and/or nuclear magnetic resonance (NMR).[3,4,7–9] It is typically necessary to generate data in a format suitable for further processing – data extraction will be slightly different between MS and NMR due to the different data structures. The actual data mining can take several different routes depending on the question asked. Mining data for potential biomarkers typically will involve statistical analysis, unsupervised such as t-tests, hierarchical cluster analysis, or principal component analysis (PCA) or supervised classification through partial least squares regression (PLS), analysis of variance (ANOVA), or discriminant analysis. Data mining in a more exploratory fashion can involve different types of

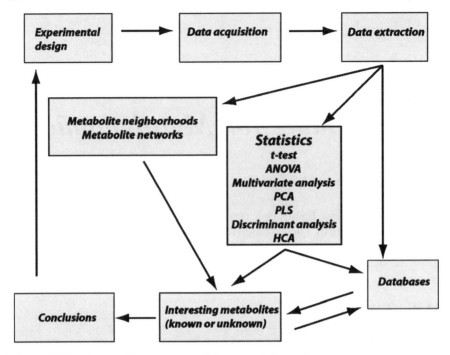

Figure 11.1   Scheme illustrating possible ways of data mining.

correlation studies or direct database queries or browsing. Output of a meta-bolomics data mining is a set of interesting metabolites that can be used to draw direct conclusions and potentially design new experiments. In this chapter we will focus on data mining of mass spectrometry (MS) generated data, but the approaches covered could generally be applied to analysis of data generated by other means (*i.e.* NMR) as long as the data is in a suitable format. Section 11.1 covers primary data extraction and pre-processing strategies and Section 11.2 covers different approaches to perform data mining.

11.1.1   Data Extraction – Generating Variables

The aim of data extraction is to create variables which subsequently can be used for data mining. The challenge in this process is to generate variables which can be compared to each other between samples, that is that variable X represents variable X throughout all observations (samples). For metabolomics data acquired by means of mass spectrometry coupled to some type of chromatography, a generalized data structure is illustrated in Figure 11.2. The data have three dimensions: intensity, m/z, and time, with different observations (replicates and samples) constituting a fourth dimension. To generate comparable variables from this type of data we must first consider how different types of "drift" in the first three dimensions affect our data output when we want to

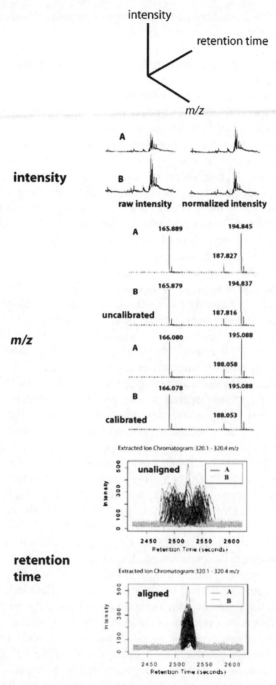

Figure 11.2   General structure for chromatography hyphenated mass spectrometry data. Three aspects of the data have to be aligned (between observations) prior to data mining: intensity, *m/z*, and retention time.

compile data in the fourth dimension. Here are some general considerations for the individual variables:

- m/z Spectral drift, m/z shifts will vary with type of mass spectrometer and resolution. A time of flight (TOF) mass spectrometer will typically have a larger m/z drift than quadrupole due to the flight tube's sensitivity to small temperature changes and, hence, changes in flight length for the ions. A simple calibration of the m/z scale compensates for drift in the m/z dimension. However, quite often peak shape causes small variations in the interpretation of the m/z data when it is converted to discrete tabular values. It can therefore sometimes be beneficial to perform "binning" of spectral data prior to data analysis to compensate for these small variations in the spectral data. Reduction of spectrum resolution to, for example unit mass resolution, will ensure that a specific m/z peak remains in its bin regardless of small spectra shifts.
- Intensity   The intensity dimension is subject to variation from differences in ionization efficiency. Factors such as ionization suppression, ambient temperature, and ion source condition will affect intensity recorded for each ion. All these factors can drift over time and therefore especially in experiments running over longer periods of time, it is important to normalize the intensity response. Considering that ultimately this is the dimension that we will perform data mining on, it is worth monitoring and evaluating how the intensity is normalized. The most robust way to normalize data is to use stable isotope labeled versions of the metabolites of interest. In an untargeted global profiling approach such as metabolomics, this is, however, somewhat difficult since the experimenter basically doesn't have specific target compounds to measure. Using a set of selected stable isotope labeled compounds, representing a verity of potential metabolites and metabolite classes, have proven to be a good substitute strategy.[10,11] Other mathematical approaches for normalization have also been proposed for use in a metabolomics context.[12,13]
- Retention time   Drift in the time domain is arguably the most difficult to normalize or compensate for. The reason for drift in the time domain can be: temperature changes, gradual clogging/contamination of columns, pump/gas pressure fluctuations, *etc.* One aspect of drift in the time domain is that it is typically not a linear phenomenon, which makes it more difficult to compensate for. Calculating retention time index (*i.e.* kovats indices) for the different components is one possible approach, but this strategy involves introduction of standards (typically alkane series for GC) and is potentially problematic to automate. In our lab we have developed XCMS, which is a freely available open-source software.[14] XCMS generates ion chromatograms that are individually processed. It performs filtering, peak detection, matching, and alignment. The non-linear retention time correction is based upon the identification of groups of ions with a conserved retention time from which a non-linear retention time correction filter can be calculated and applied to all the other ions. Other

Table 11.1 Platform independent data extraction tools.

Name	Reference	URL	Comment
XCMS	14	http://metlin.scripps.edu/	Open source, runs in software "R", reads CDF files, peak detection and non-linear retention time correction
MZ-MINE	71	http://mzmine.sourceforge.net/index.shtml	Peak detection, alignment and normalization. GUI for import and processing
MET-IDEA	72	http://www.noble.org/PlantBio/MS/	Input is a list of ion/retention time pairs typically generated manually in software or through AMDIS
H-MCA (Hierarchical multivariate curve resolution)	21		Generates a table with metabolites (and their relative concentration) that differs between samples
LCMSWARP	73		Aligns both mass and time aspect of both LC-MS and LC-MS/MS data
MS-resolver		http://www.prs.no/	Commercial, platform-independent
metAlign		http://www.pri.wur.nl/UK/	Commercial, platform-independent

strategies have been developed, both commercial and free, that address data extraction and time alignment and they are summarized in Table 11.1.

The data extraction strategy will depend on the type of data acquisition. Some parameters to consider before choice of data extraction protocol are the following.

(1) What is the super structure of the data – direct infusion or chromatographic step prior to mass spectrometry analysis.
(2) Chromatographic resolution.
(3) Spectral information, fragmentation (electron ionization, EI, or electrospray ionization, ESI).

For direct infusion type of metabolomics where the data consist of an intensity and m/z domain, the mass spectra can be directly exported in a table format containing raw spectral data[15] or as discrete m/z values.[16] These tables are subsequently subject to further evaluation. Binning of data (as discussed in the section above) can facilitate data mining of exported spectral tables.

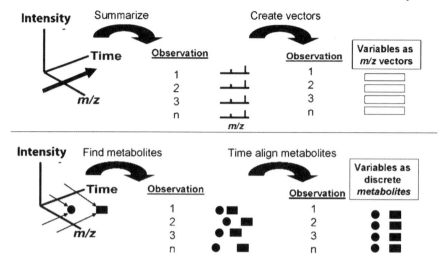

Figure 11.3   The two principal strategies for generating variables for further data mining.

It is more common to use a chromatographic step prior to the mass spectrometer. Chromatography can facilitate the detection of isomers and reduce ion suppression during the ionization, which enhances detection limits. By using chromatography a time domain is introduced. The two principal options for data extraction are illustrated in Figure 11.3. In the first case data dimensionality is reduced to m/z and intensity by summation over the time domain. This process creates variables in the form of m/z vectors and the output is similar to that of direct infusion mass spectrometry.[11,17] The other option is to find the components (metabolites) through deconvolution[18–21] or curve resolution of the total ion chromatogram. The variables then become discrete metabolites (known or unknown). After metabolites have been deconvoluted, they must be aligned creating comparable variables. How well the deconvolution algorithm can perform largely depends on (1) amount of spectral information and (2) chromatographic resolution. Electron ionization (EI) typically causes more fragmentation of the molecules in the ionization process than electrospray ionization (ESI). Furthermore, gas chromatography (GC) is achieving higher resolution (narrower chromatographic peak profiles) than liquid chromatography (LC). Combined, these two facts make GC data easier to deconvolute. Recent developments in LC (*e.g.* ultrahigh performance liquid chromatography, UPLC) paired with new extraction software (see Table 11.1) will facilitate deconvolution strategies for LC generated metabolomics data.

11.1.2   Data Pre-treatment

Before subjecting extracted data to further analysis, it is worth considering some inherent properties of the data variables. Many statistical operations such as a *t*-test are valid on the assumption that the data is normally distributed. It might therefore be worthwhile to check some variables as histogram plots to make sure that data appear normally distributed. With mass spectrometry (depending on data extraction approach) the number of variables can range from a few hundred to several thousand, but browsing through some 20 to 50 variables will give a general impression of how data is distributed. One strategy that can be used to address skewed data (data that is not normally distributed) is to perform a transformation of the entire data set. The transformation would typically be a power transformation (for example the square root) or a logarithmic transformation, with logarithmic transformations being the most applied. Another aspect of data structure to consider is that many unsupervised (see section below) classification tools such as principal component analysis (PCA) will project data according to their maximum variation. This means that variables that vary with smaller amplitude in absolute terms between observations (or potential classes) will be less important to the model than variables with a higher variation (in absolute terms). Most often it is the proportional or relative variation which is more interesting, therefore we would like to compensate data through "scaling" or "weighing" of some kind to give variables equal weight in the data analysis. The most common technique to standardize the variance is to perform "unit variance scaling" in which the inversed standard deviation of a certain variable is multiplied by that particular variable. Another standard procedure is to eliminate the offset of a given variable through centering the data around zero instead of the mean. This is performed by calculating the average for a certain variable, and subsequently subtracting that average from each observation of that variable. Several different approaches for scaling and transformation of metabolomics data has been evaluated by van den Berg and colleagues.[22]

11.2 Data Mining

The output of a data mining routine can be a list of metabolites ranked in a certain order, for example concentration, fold change, *p* value (from a *t*-test), score value (from a multivariate evaluation), or loading value (from a multivariate evaluation). The purpose of the investigation can also be more explorative in its nature, in these cases the output might be best mined visually as a relational "neighbor" networks or a hierarchical cluster (heat map). The choice of strategy for data mining will depend on the question that we want to answer. With data extraction and data pre-treatment, we have now produced a data set containing variables in a format suitable for further data analysis. In the paragraphs below are described a few scenarios of data mining, including an example from our laboratory. It is important to have a well-defined question; for example, do we want to be able to classify a certain sample as "healthy" or "diseased", or we want to find/understand the underlying metabolite difference (and biochemistry) between "healthy" and "diseased". In the first case it might be enough to establish a data mining scheme that can recognize the metabolite pattern of a certain condition and classify samples accordingly. It could also involve finding one or a set of metabolites that have a significantly different concentration level associated with a certain condition such as a disease. In more exploratory types of data mining such as elucidation of gene function where biochemical redundancy has to be accounted for, it might not be sufficient only to rank interesting metabolites according to their concentration differences. Here, ranking of interesting metabolites could potentially be more relevant in terms of altered concentration ratios towards other metabolites. Initial data mining can frequently result in "unknown" compounds among the interesting metabolites. Verification of metabolite significance is important before proceeding to identification since this can be a time consuming effort especially if the list of interesting metabolites is long. This can involve reverting to raw data for verification of data extraction and/or re-analyzing samples.

The origin of what we refer to as metabolomics/metabonomics can be found in metabolite profiling. Profiling of blood, for example, has resulted in the discovery of several biomarkers that are now routinely used in screening of newborn babies for inborn metabolite disorders.[23] Finding biomarkers for disease diagnosis is a very important application of metabolomics and putative markers for conditions like myocardial ischemia[24] (loss of blood flow to the heart muscle) and influenza-associated encephalopathy[25] have recently been identified. For an experimental design where samples can be classified either as "A" or "B", assuming that there is no metabolic difference between them (null hypothesis), it might be sufficient to compare and rank the list according to p values calculated by performing a t-test to determine if the means (of A and B) are equal or distinct. This "simple" type of data mining is aimed at discovering concentration changes between metabolites of the two classes "A" and "B". The next step would typically be to identify, and subsequently perform, targeted analysis on a few top metabolite candidates from the produced "rank list" and use these as biomarkers for a specified condition or state.

11.2.1   Multivariate Data Mining

Often an experimental design might be more complex than of the type "A" and "B" samples. The design can for example involve a series of pathological progression states[26] or samples taken in a time series[27] which generates data sets containing several variables in several observations placed in several sub-classes such as time points or progression states. Statistical tools such as t-tests might not be sufficient to rationally find underlying variable–observation correlations. Multivariate statistics is a set of different procedures that deal with the observation and analysis of more than one variable at the time. This section will cover some of these procedures that have found application within metabolomics. Multivariate analysis can be divided into non-supervised classification and supervised classification methods. Non-supervised classification methods analyze the data set as it is by methods such as principal component analysis (PCA)[28–31] and hierarchical cluster analysis (HCA).[32–34] In supervised classification, additional external information relating to the data set, such as class or certain quantitative or qualitative responses, is used for making predictions through multivariate calibration or by performing discriminant analysis. Supervised classification methods include partial least square analysis (PLS),[28,35] discriminant analysis,[33] analysis of variance (ANOVA),[33,36] and projections to latent structures by means of partial least squares discriminant analysis (PLS-DA).[37]

HCA has the advantage of being easy to visualize in tree diagrams (dendrograms). The similarity or difference between different observations is calculated by, for example, mean Euclidean distances. With a distance/similarity scale in place, observations can step by step form clusters of similarity until all observations are gathered in one cluster (dendrogram). In metabolomics/metabonomics investigations HCA has been used for data mining of endogenous plant compounds[38] and to classify toxicological effects.[39] Projection models such as PCA are widely used for data mining and evaluation in metabolomics experiments.[10,22,30] When performing a principal component analysis, the information in a data matrix is compressed to a few principal components (latent variables) by means of matrix decomposition (single value decomposition, SVD, or non-linear iterative partial least squares, NIPALS). Each principal component consists of one score vector (t) and one loading vector (p). The scores are linear combinations (eigenvectors) of the original data and the loading is the link between the original data and the score. The compression should explain as much variation as possible in the original data set. By reducing dimensionality of the data it becomes easier to get an overview over trends, groups, and outliers. The result can be viewed graphically showing how observations relate to one and other (score plots) and also perhaps more importantly what variables are contributing to these relations (loading plots). In PLS-DA the latent variables (score and loading pairs) are calculated to maximize the covariance between the data and the class assignment. This is a powerful technique that can be used for predictions[26] and for finding variables that explain class belonging.[40]

In our laboratory we wished to validate our metabolomics protocol (data acquisition and data extraction) for its capacity to find minor differences and small concentration changes between different samples. We set up an experimental design involving spiking of known quantities of small organic molecules into human serum (Figure 11.4), which were subsequently analyzed using LC-MS. Data extraction was performed using XCMS[14] software. The resulting exported data tables from XCMS were further analyzed using the SIMCA software (UMETRICS, Umeå, Sweden). The "A" serum was not spiked, "B" serum was spiked with 19 compounds of various concentrations, and the "C" serum was spiked with a slightly modified version of the "B" spiking mixture where two compounds were completely removed, two compounds were increased 25% in concentration, two compounds were decreased 25%, two compounds were increased 50%, and two were decreased 50% in concentration (Figure 11.4). Each spiking mixture was injected in 5 replicates, generating 15 observations in total. The XCMS data extraction and alignment resulted in

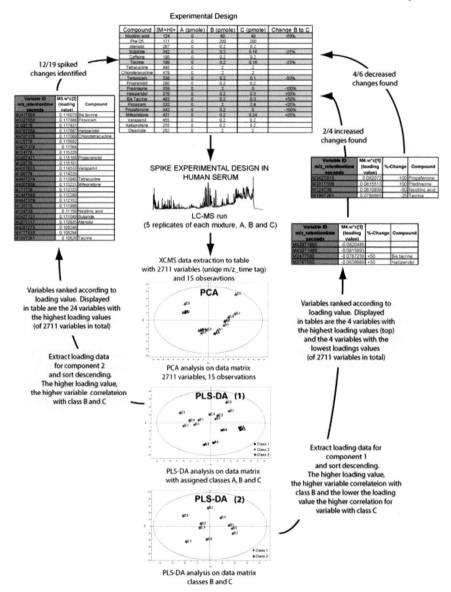

Figure 11.4 Evaluation of data extraction performance and subsequent data mining.

2711 unique ion features (unique *m/z* at a unique retention time). The data matrix was initially mined with PCA (Figure 11.4), which did not reveal any groups or clusters. In fact, the most dominant feature in the resulting PCA score plot turned out to be the sample run order. The next step was to perform supervised data mining using PLS-DA. Assignment of class belonging A, B and C resulted in two distinguishable clusters, one consisting of observations "A" and the other of a mixture of "B" and "C" observations – PLS-DA score plot

(1) in Figure 11.4. To evaluate which variables contributed to the group formation, the loading values for component 2 were tabulated and sorted in descending order. Inspection of the variables revealed that 12 out of 19 spiked compounds in "B" could be identified amongst the 24 variables with the highest loading values (Figure 11.4). The "A" observations were removed and classes "B" and "C" were modeled. The same procedure was repeated with the loading values for component 1 – PLS-DA plot (2) in Figure 11.4. Variables associated with a higher loading value are more correlated with class "B" and vice versa for class "C". The resulting loading value tables are shown in Figure 11.4 and it was evident that 4 out of 6 of the compounds that were decreased between "B" and C" could be found amongst the 4 variables with the highest loading values, and 2 out of 4 of the compounds that were increased in "C" could be identified in the same way. The conclusion of the validation was that XCMS provided a reliable way for extracting and aligning data and that the subsequent multi-variate data mining approach could detect differences close to the analytical variance ($\sim 10\text{–}20\%$) (Figure 11.4).

11.2.2 Exploratory Data Mining

Metabolomics studies which are more exploratory in nature might need several different type of statistical and visualization tools. Experiments intended to explore how different metabolites relate to each other, or how protein or gene expression under certain experimental conditions effects the metabolome, calls for a more iterative data mining strategy where output from one analysis triggers another mining strategy or a new experiment (Figure 11.1). The more complex the experimental design is (different perturbations or different pheno-types) and the more levels of data we wish to compare (metabolome or proteome or genome), the more complex the outcome of the data analysis will be. Visualization of data in score plots (PCA) or clusters (HCA) are strategies to facilitate intuitive understanding of how different data sets relate to each other and what variables contribute to these particular relations. When com-paring several layers of data (genes or proteins or metabolites), the outcome of a metabolomics experiment can be placed in the context of metabolic pathways such as the ones displayed in the KEGG database.[41] Visualization software that links metabolite concentration data to protein and gene expression data is available which can provide a basis for better understanding of changes in the metabolome.[42,43] There is, however, a certain debate over what a metabolic pathway actually is and metabolites (intermediates) might be "exchanged" between different "pathways". This fact has promoted the concept to visualize metabolite neighborhoods[42,44] or metabolic networks[45,46] as a tool to under-stand which metabolites are correlated and thus better understand metabolites in context. This approach has, for example, been used to understand metabolic changes in silent plant phenotypes,[47] phenotypes that due to pleiotropic effects and gene redundancy don't display any morphological or biochemical altera-tions to wild type. Other software that can be used for visualization and mining of large networks include PAJEK,[48,49] MetNet3D,[50] and paVESy[51] which allows the researcher to explore gene expression and metabolic pathway data simultaneously. A simple organism, *Saccharomyces cerevisiae*, has had the entire metabolic network reconstructed *in silico*.[52] This is possible with the vast array of genomic, biochemical, and physiological information available for this well-studied species. Furthermore, this type of modeling made it possible to correctly predict, for 70–80% of the considered conditions, the subsequent phenotypic observations.[53] A data mining approach modeling observed metabolomics data against *in silico* predictions of phenotypes is obviously only possible for less complex and well-characterized organisms like yeast. However, it opens up a very interesting scenario for fundamental studies integrating gene, protein, and metabolite data.

11.2.3 Databases

Databases can be a powerful tool at several different levels of the data mining. Mendes has classified databases in a metabolomics context into five different categories.[54]

(1) Databases storing detailed metabolite profiles, including raw data and detailed metadata (information about the data).
(2) Databases storing metabolite profiles obtained for a single biological species.
(3) Databases collecting diverse metabolite profile data from many biological species and at many different physiological states.
(4) Databases listing all known metabolites (metabolome) for each biological species.
(5) Databases representing established facts.

Most available databases today represent category 5 and, to some extent, category 4 (see Table 11.2). The initial use of a database can be to identify metabolites that after initial data mining remain "unknown". Several of the databases listed in Table 11.2 can for example be used to search for compounds by mass (molecular weight or exact mass) if data is acquired using a mass spectrometer (*e.g.* METLIN,[55] KEGG,[41] ChemDB,[56] ZINC,[57] PubChem,[58] ChemBank[59] and KNApSAcK[60]). This can result in a list of potential compound candidates which subsequently can be purchased or synthesized for evaluation. If metabolite identity is already established, the database can be used to place metabolite concentration changes in a biological or biochemistry context by overlaying gene expression and protein data with databases such as KEGG,[41] BioCyc,[61] or BRENDA[62] (see Table 11.2). Modeling results in this fashion can result in hypothesis generation and design of new experiments (see Figure 11.1). Using the highest levels of databases (categories 1, 2, and 3), extracted data can be directly mined against already performed experiments. This requires certain standards for reporting data so that results may be interchangeable between different researchers much like the MIAME[63,64]

Table 11.2 Some databases useful for metabolomics data mining.

Name	Reference	URL	Number of small molecules	Organism (if applicable)	Comment
KEGG Lignad	41	http://www.genome.ad.jp/kegg/	Compounds 14 203 Drugs 4050 Glycans 10 951	Several	Pathway maps, genes, enzymes
BioCyc open chemical database	61	http://biocyc.org/	3558	Several	Portal for several interconnected databases. Pathways, metabolite, protein and genes. Compounds search by name
chEBI	74	http://www.ebi.ac.uk/	8705		Compounds search by name, links to other databases
UM-BBD	75	http://umbbd.msi.umn.edu/	964	microorganisms	Microbial biocatalytic reactions and biodegradation pathways
BRENDA	62	http://www.brenda.uni-koeln.de/	47 630 Substrate/products	several	Extensive enzyme database with many search possibilities
PubChem	58	http://pubchem.ncbi.nlm.nih.gov/	~4000 000		Molecular properties, toxicology/bioactivity listed for several entities
ChemBank	59	http://chembank.broad.harvard.edu	~330 000		Bioactivity reported where known
ZINC	57	http://blaster.docking.org/zinc/	4600 000		Created by compilation of vendor catalogs, possible to purchase compounds directly. Contains 3D structures that can be used for docking experiments

Table 11.2 (*Continued*)

Name	Reference	URL	Number of small molecules	Organism (if applicable)	Comment
ChemDB	56	http://cdb.ics.uci.edu/CHEM/Web/	~4000 000		Various molecular descriptors are provided
METLIN	55	http://metlin.scripps.edu/	~4000		Also contains LC-MS, MS/MS and FT-MS spectra. Many drug metabolites are listed
AraCyc	76	http://www.arabidopsis.org/biocyc/index.jsp	1085	Arabidopsis	Links to enzymes and genes if known
Lipid Maps	77	http://www.lipidmaps.org/	8900	mammalian	Extensive database on lipids
Glycan Database		http://www.functionalglycomics.org/static/consortium/main.shtml	7500	Several	Linked to references, also contains a database for glycan binding proteins
Human Metabolome Database		http://www.hmdb.ca/	2100	Human	Contains links to 5500 protein and DNA sequences linked to the metabolite entities. Possible to search MS/MS and MS data. Related physiological data
MoTo DB	78	http://appliedbioinformatics.wur.nl/	~110	Tomato	Contains *de novo* identifications of some compounds. Also contain MS/MS and DAD data as well as literature references
The Golm Metabolome Database	79	http://csbdb.mpimp-golm.mpg.de/csbdb/gmd/gmd.html			GC-MS data stored with metadata. Possible to search via spectra or compound name

Name	URL				Description
SDBS	http://www.aist.go.jp/RIODB/SDBS/cgi-bin/cre_index.cgi		32 438		Possible to search MS (EI) spectra and NMR data.
NIST Chemistry webbook	http://webbook.nist.gov/chemistry/		Mass spectra 15 000 Thermo chemical data for 7000 Ion energetics data for 16 000		Physical data and structures. Also IR, UV/VIS and GC data for many compounds
Tumor Metabolome	http://www.metabolic-database.com/				Not really a compound searchable database. Source for tumor related biochemistry
K KNApSAcK	http://kanaya.naist.jp/KNApSAcK/	60	11 059	Several	Contains bioactivity if known and references. Possible to search mass spectra with different adduct options
Dictionary of Natural Products	http://www.chemnetbase.com/		170 000		(Commercial) Source data, physical properties, possible to search via drawn structure
NIST Standard Reference Database 1	http://www.nist.gov/srd/nist1.htm		~160 000 compounds with EI spectra ~5000 MS/MS spectra ~25 000 retention indices		(Commercial) Well evaluated database
Wiley Registry of Mass Spectral Data	http://www.wileyregistry.com/		~310 000		(Commercial) Largest mass spectra database
MS/MS Spectral Libraries	http://www.highchem.com/		~1000	Human	(Commercial) ESI MS/MS spectral database

standards for micro array data which makes it possible to query or browse databases such as ARRAY EXPRESS[65] and GEO[58] for gene indexed expression profiles. Standards like MIAME are underway in metabolomics as well[66,67] which will facilitate the creation of similar high level databases for use in a metabolomics context.

11.3 Conclusions

The scope and procedures of a metabolomics/metabonomics data mining scheme is determined by the questions that we want to answer. Any data mining strategy, and hence the conclusions that are drawn, will be based on the metabolites actually measured. It is important to stress (see Figure 11.1) that the choice of data acquisition system is critical. A factor to consider is if there is any discrimination towards certain metabolites (polar or hydrophobic), size or a certain functional groups. For example, GC-MS typically discriminates against thermally labile and non-volatile compounds and LC-ESI-MS can discriminate against only weakly acidic or basic compounds. The data extraction step creates variables of some type that must be comparable between observations. For some experimental designs a *t*-test might be sufficient for mining differences, while other more complex designs are better mined using both non-supervised and supervised multivariate statistics or more exploratory strategies such as metabolite networks. Typically the data mining effort results in a list of metabolites that are ranked according to a certain property for example concentration difference.

More holistic concepts in biology such as "systems biology",[68–70] which aims to understand phenotypic variation and build comprehensive models of cellular/organ/organism organization and function, must use data mining tools on different levels including multivariate statistics, correlation studies, and databases, and has already started to produce some very encouraging system functional models.[69] The establishment of reporting standards for metabolomics is creating a foundation for high level databases to which data can be directly uploaded and mined against. This will allow comparisons of metabolomics data between different experiments, developmental stages, and organisms much like how gene expression data has been mined.

Acknowledgement

The author wish to acknowledge the Swedish Research Council (VR) for financial support through a postdoctoral fellowship.

References

1. D. J. Hand, H. Mannila and P. Smyth, *Principles of Data Mining*, MIT Press, Cambridge, MA, 2001.

2. R. Goodacre, S. Vaidyanathan, W. B. Dunn, G. G. Harrigan and D. B. Kell, Metabolomics by numbers: acquiring and understanding global metabolite data, *Trends Biotechnol.*, 2004, **22**, 245–252.

3. J. van der Greef, P. Stroobant and R. van der Heijden, The role of analytical sciences in medical systems biology, *Curr. Opin. Chem. Biol.*, 2004, **8**, 559–565.

4. W. B. Dunn, N. J. Bailey and H. E. Johnson, Measuring the metabolome: current analytical technologies, *Analyst*, 2005, **130**, 606–625.

5. S. G. Villas-Boas, S. Mas, M. Akesson, J. Smedsgaard and J. Nielsen, Mass spectrometry in metabolome analysis, *Mass Spectrom. Rev.*, 2005, **24**, 613–646.

6. E. J. Want, B. F. Cravatt and G. Siuzdak, The expanding role of mass spectrometry in metabolite profiling and characterization, *Chembiochem.*, 2005, **6**, 1941–1951.

7. I. Nobeli and J. M. Thornton, A bioinformatician's view of the metabolome, *Bioessays*, 2006, **28**, 534–545.

8. M. E. Bollard, E. G. Stanley, J. C. Lindon, J. K. Nicholson and E. Holmes, NMR-based metabonomic approaches for evaluating physiological influences on biofluid composition, *NMR Biomed.*, 2005, **18**, 143–162.

9. I. Pelczer, High-resolution NMR for metabomics, *Curr. Opin. Drug Discov. Devel.*, 2005, **8**, 127–133.

10. O. Fiehn, J. Kopka, P. Dormann, T. Altmann, R. N. Trethewey and L. Willmitzer, Metabolite profiling for plant functional genomics, *Nat. Biotechnol.*, 2000, **18**, 1157–1161.

11. P. Jonsson, J. Gullberg, A. Nordstrom, M. Kusano, M. Kowalczyk, M. Sjostrom and T. Moritz, A strategy for identifying differences in large series of metabolomic samples analyzed by GC/MS, *Anal. Chem.*, 2004, **76**, 1738–1745.

12. W. Wang, H. Zhou, H. Lin, S. Roy, T. A. Shaler, L. R. Hill, S. Norton, P. Kumar, M. Anderle and C. H. Becker, Quantification of proteins and metabolites by mass spectrometry without isotopic labeling or spiked standards, *Anal. Chem.*, 2003, **75**, 4818–4826.

13. Y. I. Shurubor, U. Paolucci, B. F. Krasnikov, W. R. Matson and B. S. Kristal, Analytical precision, biological variation, and mathematical normalization in high data density metabolomics, *Metabolomics*, 2005, **1**, 75–85.

14. C. A. Smith, E. J. Want, G. O'Maille, R. Abagyan and G. Siuzdak, XCMS: processing mass spectrometry data for metabolite profiling using nonlinear peak alignment, matching, and identification, *Anal. Chem.*, 2006, **78**, 779–787.

15. K. O. Boernsen, S. Gatzek and G. Imbert, Controlled protein precipitation in combination with chip-based nanospray infusion mass spectrometry. An approach for metabolomics profiling of plasma, *Anal. Chem.*, 2005, **77**, 7255–7264.

16. A. Aharoni, C. H. Ric de Vos, H. A. Verhoeven, C. A. Maliepaard, G. Kruppa, R. Bino and D. B. Goodenowe, Nontargeted metabolome analysis by use of Fourier Transform Ion Cyclotron Mass Spectrometry, *Omics*, 2002, **6**, 217–234.

17. R. S. Plumb, C. L. Stumpf, M. V. Gorenstein, J. M. Castro-Perez, G. J. Dear, M. Anthony, B. C. Sweatman, S. C. Connor and J. N. Haselden, Metabonomics: the use of electrospray mass spectrometry coupled to reversed-phase liquid chromatography shows potential for the screening of rat urine in drug development, *Rapid Commun. Mass Spectrom.*, 2002, **16**, 1991–1996.

18. S. E. Stein, An integrated method for spectrum extraction and compound identification from gas chromatography/mass spectrometry data, *J. Am. Soc. Mass Spectr.*, 1999, **10**, 770–781.

19. W. Windig, J. M. Phalp and A. W. Payne, A noise and background reduction method for component detection in liquid chromatography mass spectrometry, *Anal. Chem.*, 1996, **68**, 3602–3606.

20. H. Idborg-Bjorkman, P. O. Edlund, O. M. Kvalheim, I. Schuppe-Koistinen and S. P. Jacobsson, Screening of biomarkers in rat urine using LC/electrospray ionization-MS and two-way data analysis, *Anal. Chem.*, 2003, **75**, 4784–4792.

21. P. Jonsson, A. I. Johansson, J. Gullberg, J. Trygg, A, J., B. Grung, S. Marklund, M. Sjostrom, H. Antti and T. Moritz, High-throughput data analysis for detecting and identifying differences between samples in GC/MS-based metabolomic analyses, *Anal. Chem.*, 2005, **77**, 5635–5642.

22. R. A. van den Berg, H. C. Hoefsloot, J. A. Westerhuis, A. K. Smilde and M. J. van der Werf, Centering, scaling, and transformations: improving the biological information content of metabolomics data, *BMC Genomics*, 2006, **7**, 142.

23. D. H. Chace, Mass spectrometry in the clinical laboratory, *Chem. Rev.*, 2001, **101**, 445–477.

24. M. S. Sabatine, E. Liu, D. A. Morrow, E. Heller, R. McCarroll, R. Wiegand, G. F. Berriz, F. P. Roth and R. E. Gerszten, Metabolomic identification of novel biomarkers of myocardial ischemia, *Circulation*, 2005, **112**, 3868–3875.

25. H. Kawashima, M. Oguchi, H. Ioi, M. Amaha, G. Yamanaka, Y. Kashiwagi, K. Takekuma, Y. Yamazaki, A. Hoshika and Y. Watanabe, Primary biomarkers in cerebral spinal fluid obtained from patients with influenza-associated encephalopathy analyzed by metabolomics, *Int. J. Neurosci.*, 2006, **116**, 927–936.

26. J. T. Brindle, H. Antti, E. Holmes, G. Tranter, J. K. Nicholson, H. W. Bethell, S. Clarke, P. M. Schofield, E. McKilligin, D. E. Mosedale and D. J. Grainger, Rapid and noninvasive diagnosis of the presence and severity of coronary heart disease using ^{1}H-NMR-based metabonomics, *Nat. Med.*, 2002, **8**, 1439–1444.

27. O. Teahan, S. Gamble, E. Holmes, J. Waxman, J. K. Nicholson, C. Bevan and H. C. Keun, Impact of analytical bias in metabonomic studies of human blood serum and plasma, *Anal. Chem.*, 2006, **78**, 4307–4318.

28. L. Eriksson, H. Antti, J. Gottfries, E. Holmes, E. Johansson, F. Lindgren, I. Long, T. Lundstedt, J. Trygg and S. Wold, Using chemometrics for

navigating in the large data sets of genomics, proteomics, and meta-bonomics (gpm), *Anal. Bioanal. Chem.*, 2004, **380**, 419–429.

29. J. E. Jackson, *A User's Guide to Principal Components*. Wiley-Interscience, Hoboken, NJ, 2003.

30. J. J. Jansen, H. C. Hoefsloot, H. F. Boelens, J. van der Greef and A. K. Smilde, Analysis of longitudinal metabolomics data, *Bioinformatics*, 2004, **20**, 2438–2446.

31. J. van der Greef and A. K. Smilde, Symbiosis of chemometrics and metabolomics: past, present, and future, *J. Chemometr.*, 2005, **19**, 376–386.

32. M. B. Eisen, P. T. Spellman, P. O. Brown and D. Botstein, Cluster analysis and display of genome-wide expression patterns, *Proc. Natl. Acad. Sci. USA*, 1998, **95**, 14863–14868.

33. S. K. Kachigan, *Multivariate Statistical Analysis: A Conceptual Introduction*, Radius Press, New York, 1991.

34. J. Quackenbush, Computational analysis of microarray data, *Nat. Rev. Genet.*, 2001, **2**, 418–427.

35. S. Wold, M. Sjostrom and L. Eriksson, PLS-regression: a basic tool of chemometrics, *Chemometr. Intell. Lab.*, 2001, **58**, 109–130.

36. A. K. Smilde, J. J. Jansen, H. C. Hoefsloot, R. J. Lamers, J. van der Greef and M. E. Timmerman, ANOVA-simultaneous component analysis (ASCA): a new tool for analyzing designed metabolomics data, *Bioinformatics*, 2005, **21**, 3043–3048.

37. M. Sjostrom, S. Wold, A. Wieslander and L. Rilfors, Signal peptide amino-acid-sequences in *Escherichia coli* contain information related to final protein localization – a multivariate data analysis, *EMBO J.*, 1987, **6**, 823–831.

38. Y. Tikunov, A. Lommen, C. H. de Vos, H. A. Verhoeven, R. J. Bino, R. D. Hall and A. G. Bovy, A novel approach for nontargeted data analysis for metabolomics. Large-scale profiling of tomato fruit volatiles, *Plant Physiol.*, 2005, **139**, 1125–1137.

39. O. Beckonert, M. E. Bollard, T. M. D. Ebbels, H. C. Keun, H. Antti, E. Holmes, J. C. Lindon and J. K. Nicholson, NMR-based metabonomic toxicity classification: hierarchical cluster analysis and k-nearest-neighbour approaches, *Anal. Chim. Acta*, 2003, **490**, 3–15.

40. S. Bijlsma, I. Bobeldijk, E. R. Verheij, R. Ramaker, S. Kochhar, I. A. Macdonald, B. van Ommen and A. K. Smilde, Large-scale human metabolomics studies: a strategy for data (pre-) processing and validation, *Anal. Chem.*, 2006, **78**, 567–574.

41. M. Kanehisa, S. Goto, S. Kawashima and A. Nakaya, The KEGG databases at GenomeNet, *Nucleic Acids Res.*, 2002, **30**, 42–46.

42. X. J. Li, O. Brazhnik, A. Kamal, D. Guo, C. Lee, S. Hoops and P. Mendes, Databases and visualization for metabolomics. *Metabolic Profiling its role in biomarker discovery and gene function analysis*, G. G. Harrigan and R. Goodacre, Norwell, MA, Kluwer Academic Publishers Groupm 2003, pp. 293–309.

43. S. M. Paley and P. D. Karp, The Pathway Tools cellular overview diagram and Omics Viewer, *Nucleic Acids Res.*, 2006, **34**, 3771–3778.

44. A. de la Fuente, N. Bing, I. Hoeschele and P. Mendes, Discovery of meaningful associations in genomic data using partial correlation coefficients, *Bioinformatics*, 2004, **20**, 3565–3574.

45. R. Steuer, J. Kurths, O. Fiehn and W. Weckwerth, Observing and interpreting correlations in metabolomic networks, *Bioinformatics*, 2003, **19**, 1019–1026.

46. R. Steuer, Review: on the analysis and interpretation of correlations in metabolomic data, *Brief Bioinform.*, 2006, **7**, 151–158.

47. W. Weckwerth, M. E. Loureiro, K. Wenzel and O. Fiehn, Differential metabolic networks unravel the effects of silent plant phenotypes, *Proc. Natl. Acad. Sci. USA*, 2004, **101**, 7809–7814.

48. http://vlado.fmf.uni-lj.si/pub/networks/pajek/.

49. V. Batagelj and A. Mrvar, Pajek - Analysis and visualization of large networks, *Lect. Notes Comput. Sci.*, 2002, **2265**, 477–478.

50. Y. Yang, L. Engin, E. S. Wurtele, C. Cruz-Neira and J. A. Dickerson, Integration of metabolic networks and gene expression in virtual reality, *Bioinformatics*, 2005, **21**, 3645–3650.

51. A. Ludemann, D. Weicht, J. Selbig and J. Kopka, PaVESy: Pathway Visualization and Editing System, *Bioinformatics*, 2004, **20**, 2841–2844.

52. J. Forster, I. Famili, P. Fu, B. O. Palsson and J. Nielsen, Genome-scale reconstruction of the *Saccharomyces cerevisiae* metabolic network, *Genome Res.*, 2003, **13**, 244–253.

53. I. Famili, J. Forster, J. Nielson and B. O. Palsson, *Saccharomyces cerevisiae* phenotypes can be predicted by using constraint-based analysis of a genome-scale reconstructed metabolic network, *Proc. Natl. Acad. Sci. USA*, 2003, **100**, 13134–13139.

54. P. Mendes, Emerging bioinformatics for the metabolome, *Brief Bioinform.*, 2002, **3**, 134–145.

55. C. A. Smith, G. O'Maille, E. J. Want, C. Qin, S. A. Trauger, T. R. Brandon, D. E. Custodio, R. Abagyan and G. Siuzdak, METLIN: a metabolite mass spectral database, *Ther. Drug Monit.*, 2005, **27**, 747–751.

56. J. Chen, S. J. Swamidass, Y. Dou, J. Bruand and P. Baldi, ChemDB: a public database of small molecules and related chemoinformatics resources, *Bioinformatics*, 2005, **21**, 4133–4139.

57. J. J. Irwin and B. K. Shoichet, ZINC – a free database of commercially available compounds for virtual screening, *J. Chem. Inf. Model.*, 2005, **45**, 177–182.

58. D. L. Wheeler, T. Barrett, D. A. Benson, S. H. Bryant, K. Canese, V. Chetvernin, D. M. Church, M. DiCuccio, R. Edgar, S. Federhen, L. Y. Geer, W. Helmberg, Y. Kapustin, D. L. Kenton, O. Khovayko, D. J. Lipman, T. L. Madden, D. R. Maglott, J. Ostell, K. D. Pruitt, G. D. Schuler, L. M. Schriml, E. Sequeira, S. T. Sherry, K. Sirotkin, A. Souvorov, G. Starchenko, T. O. Suzek, R. Tatusov, T. A. Tatusova, L. Wagner and E. Yaschenko, Database resources of the National Center for Biotechnology Information, *Nucleic Acids Res.*, 2006, **34**, D173–180.

59. R. L. Strausberg and S. L. Schreiber, From knowing to controlling: a path from genomics to drugs using small molecule probes, *Science*, 2003, **300**, 294–295.

60. Y. Shinbo, Y. Nakamura, M. Altaf-Ul-Amin, H. Asahi, K. Kurokawa, M. Arita, K. Saito, D. Ohta, D. Shibata and S. Kanaya, KNApSAcK: A Comprehensive Speciies-Metabolite Relationship Database, *Biotechnology in Agriculture and Forestry*, eds. K. Saito, R. A. Dixon and L. Willmitzer, Springer, 2006, **57**, pp. 165–181.

61. P. D. Karp, C. A. Ouzounis, C. Moore-Kochlacs, L. Goldovsky, P. Kaipa, D. Ahren, S. Tsoka, N. Darzentas, V. Kunin and N. Lopez-Bigas, Expansion of the BioCyc collection of pathway/genome databases to 160 genomes, *Nucleic Acids Res.*, 2005, **33**, 6083–6089.

62. I. Schomburg, A. Chang and D. Schomburg, BRENDA, enzyme data and metabolic information, *Nucleic Acids Res.*, 2002, **30**, 47–49.

63. A. Brazma, P. Hingamp, J. Quackenbush, G. Sherlock, P. Spellman, C. Stoeckert, J. Aach, W. Ansorge, C. A. Ball, H. C. Causton, T. Gaasterland, P. Glenisson, F. C. Holstege, I. F. Kim, V. Markowitz, J. C. Matese, H. Parkinson, A. Robinson, U. Sarkans, S. Schulze-Kremer, J. Stewart, R. Taylor, J. Vilo and M. Vingron, Minimum information about a microarray experiment (MIAME)-toward standards for microarray data, *Nat. Genet.*, 2001, **29**, 365–371.

64. http://www.mged.org/Workgroups/MIAME/miame.html.

65. A. Brazma, H. Parkinson, U. Sarkans, M. Shojatalab, J. Vilo, N. Abeygunawardena, E. Holloway, M. Kapushesky, P. Kemmeren, G. G. Lara, A. Oezcimen, P. Rocca-Serra and S. A. Sansone, ArrayExpress – a public repository for microarray gene expression data at the EBI, *Nucleic Acids Res.*, 2003, **31**, 68–71.

66. H. Jenkins, N. Hardy, M. Beckmann, J. Draper, A. R. Smith, J. Taylor, O. Fiehn, R. Goodacre, R. J. Bino, R. Hall, J. Kopka, G. A. Lane, B. M. Lange, J. R. Liu, P. Mendes, B. J. Nikolau, S. G. Oliver, N. W. Paton, S. Rhee, U. Roessner-Tunali, K. Saito, J. Smedsgaard, L. W. Sumner, T. Wang, S. Walsh, E. S. Wurtele and D. B. Kell, A proposed framework for the description of plant metabolomics experiments and their results, *Nat. Biotechnol.*, 2004, **22**, 1601–1606.

67. J. C. Lindon, J. K. Nicholson, E. Holmes, H. C. Keun, A. Craig, J. T. Pearce, S. J. Bruce, N. Hardy, S. A. Sansone, H. Antti, P. Jonsson, C. Daykin, M. Navarange, R. D. Beger, E. R. Verheij, A. Amberg, D. Baunsgaard, G. H. Cantor, L. Lehman-McKeeman, M. Earll, S. Wold, E. Johansson, J. N. Haselden, K. Kramer, C. Thomas, J. Lindberg, I. Schuppe-Koistinen, I. D. Wilson, M. D. Reily, D. G. Robertson, H. Senn, A. Krotzky, S. Kochhar, J. Powell, F. van der Ouderaa, R. Plumb, H. Schaefer and M. Spraul, Summary recommendations for standardization and reporting of metabolic analyses, *Nat. Biotechnol.*, 2005, **23**, 833–838.

68. J. van der Greef, E. Davidov, E. Verheij, J. Vogels, R. van der Heijden, A. S. Adourian, M. Oresic, E. W. Marple and S. Naylor, The Role of Metabolomics in Sytems Biology. *Metabolic Profiling its role in biomarker*

discovery and gene function analysis, G. G. Harrigan and R. Goodacre. Norwell, MA, USA, Kluwer Academic Publishers, 2003, pp. 171–198.

69. C. B. Clish, E. Davidov, M. Oresic, T. N. Plasterer, G. Lavine, T. Londo, M. Meys, P. Snell, W. Stochaj, A. Adourian, X. Zhang, N. Morel, E. Neumann, E. Verheij, J. T. Vogels, L. M. Havekes, N. Afeyan, F. Regnier, J. van der Greef and S. Naylor, Integrative biological analysis of the APOE*3-leiden transgenic mouse, *Omics*, 2004, **8**, 3–13.

70. A. R. Fernie, R. N. Trethewey, A. J. Krotzky and L. Willmitzer, Metabolite profiling: from diagnostics to systems biology, *Nat. Rev. Mol. Cell Biol.*, 2004, **5**, 763–769.

71. M. Katajamaa, J. Miettinen and M. Oresic, MZmine: toolbox for processing and visualization of mass spectrometry based molecular profile data, *Bioinformatics*, 2006, **22**, 634–636.

72. C. D. Broeckling, I. R. Reddy, A. L. Duran, X. Zhao and L. W. Sumner, MET-IDEA: data extraction tool for mass spectrometry-based metabolomics, *Anal. Chem.*, 2006, **78**, 4334–4341.

73. N. Jaitly, M. E. Monroe, V. A. Petyuk, T. R. W. Clauss, J. N. Adkins and R. D. Smith, Robust Algorithm for Alignment of Liquid Chromatography-Mass Spectrometry Analyses in an Accurate Mass and Time Tag Data Analysis Pipeline, *Anal. Chem.*, 2006, **78**(21), 7397–7409.

74. C. Brooksbank, G. Cameron and J. Thornton, The European Bioinformatics Institute's data resources: towards systems biology, *Nucleic Acids Res.*, 2005, **33**, D46–53.

75. L. B. Ellis, C. D. Hershberger, E. M. Bryan and L. P. Wackett, The University of Minnesota Biocatalysis/Biodegradation Database: emphasizing enzymes, *Nucleic Acids Res.*, 2001, **29**, 340–343.

76. L. A. Mueller, P. Zhang and S. Y. Rhee, AraCyc: a biochemical pathway database for *Arabidopsis*, *Plant Physiol.*, 2003, **132**, 453–460.

77. D. Cotter, A. Maer, C. Guda, B. Saunders and S. Subramaniam, LMPD: LIPID MAPS proteome database, *Nucleic Acids Res.*, 2006, **34**, D507–510.

78. S. Moco, R. J. Bino, O. Vorst, H. A. Verhoeven, J. de Groot, T. A. van Beek, J. Vervoort and C. H. de Vos, A liquid chromatography-mass spectrometry-based metabolome database for tomato, *Plant Physiol.*, 2006, **141**, 1205–1218.

79. J. Kopka, N. Schauer, S. Krueger, C. Birkemeyer, B. Usadel, E. Bergmuller, P. Dormann, W. Weckwerth, Y. Gibon, M. Stitt, L. Willmitzer, A. R. Fernie and D. Steinhauser, GMD@CSB.DB: the Golm Metabolome Database, *Bioinformatics*, 2005, **21**, 1635–1638.

CHAPTER 12

Metabonomics and Global Systems Biology

IAN D WILSON[1] AND JEREMY K. NICHOLSON[2]

[1] Department of Drug Metabolism and Pharmacokinetics, AstraZeneca, Mereside, Alderley Park, Macclesfield, Cheshire SK10 4TG, UK
[2] Department of Biomolecular Medicine, Faculty of Medicine, Imperial College London, South Kensington, London SW7 2AZ, UK

12.1   Introduction

"Systems biology" is a term that has a relatively recent origin and currently means many different things to different investigators. The ideas encompassing the term systems biology have arisen as a result of the development of the "omics" technologies such as genomics, proteomics or metabonomics/metabolomics. In these fields of study large amounts of quantitative (or semiquantitative) data are being derived, at a variety of levels of biomolecular organisation, from genes through proteins down to metabolites. One of the expectations of systems biologists is that, in some way, such data can be integrated to give a holistic picture of the state of the "system" that provides insights that are not available by other, more directed, methods, ultimately enabling a more fundamental understanding of biology to be obtained via networks of interactions at the molecular level.

This may, or may not, be a realistic ambition but, successful or not, such work may greatly aid in efforts to deliver the "Personalised Healthcare Solutions" so desired by the practitioners of 21st century medicine. Such therapeutic approaches, tailored to the exact biology (or biological state) of an individual, clearly require methods of patient evaluation that enable the clinician to select the most appropriate combinations of drugs, dosages and treatment regimens before commencing therapy. In an ideal world this process would maximise therapeutic benefit and minimise adverse drug events. Attempts at this type of sub-classification of individuals (patient stratification) are beginning to be performed and are currently most often attempted using some particular genetic feature.

Moving away from disease, such concepts could easily be extended to more general lifestyle paradigms aimed at minimising the propensity of an individual, found to have gene-level risk factors, to acquire a disease later in life by optimising lifestyle (nutrition and exercise, *etc.*). Given the current cost of providing such detailed information on an individual it is difficult to believe that personalised medicine will be delivered via a systems biology approach in the near future (at least to large populations). However, this does not mean that systems approaches may not be valuable in identifying better diagnostics and, paradoxically, many of the insights that will illuminate "personalised medicine" may well come from omics-based epidemiological studies of populations.

If it is taken as a given that the state of any biological system, be it cell, organ or whole organism, is a function of a combination of factors such as genotype, physiological state (*e.g.* age), disease state, nutritional state, environment (both current and historical), *etc.*, the complexity faced by such investigations is clearly enormous (an attempt to illustrate this is given in Figure 12.1). It is arguable that metabonomics,[1,2] because it measures the outputs of the system rather than potential outcomes, offers the most practical approach to measuring global system activity via accessing the metabolic profiles that are determined by these combinations of genetic and environmental factors.[3,4] This set of assumptions provides the basis for the following discussion of the use of global metabolic profiling in systems approaches.

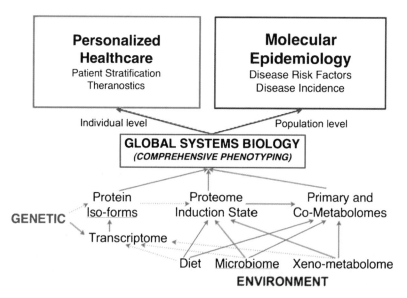

Figure 12.1 Relationships between systems biology, personalised healthcare and molecular epidemiology. Dotted lines indicate indirect connections or influences. Modified from ref. 45.

12.2 Biological Fluids as a Window on the Global "System"

In attempting to perform studies at a systems level on a complex, multi-cellular, multi-organ, organism such as an experimental animal or human there are obvious problems associated with sampling (what, when and how much). From a practical standpoint biological fluids, such as urine and plasma, provide both a convenient sample for analysis and a window on the metabolic state of the subject, be it rat or man, and can be obtained using non-, or minimally, invasive procedures (unlike organ biopsies). Other, currently underexploited, sources of non-invasive metabolic monitoring include breath and possibly saliva, though clearly there are more problems in obtaining such samples from animals compared to humans.

Urine is particularly useful because it can be collected in a completely non-invasive manner from both experimental animals and human subjects and such collections can be repeated over periods of hours, days, weeks (or even years) on the same test subjects to observe the effects of treatment and determine the time course of changes in the "system". Indeed, the utility of urine as a suitable sample type for performing metabonomic studies has been amply demonstrated in innumerable studies covering basic physiology, toxicology and clinical investigations (reviewed in refs 5–10, for example). Whilst somewhat more invasive, blood samples can also be obtained from both experimental animals and humans and, as with urine, there are examples of the successful application of metabonomics using this approach. Blood and urine are, however, not equivalent and provide the investigator with different levels of information on the metabolic status of the whole organism. Thus, the excretory function of urine makes it particularly appropriate for monitoring the state of the global system as it provides a "sum of histories" of what has taken place in the whole animal since the last sample was voided. Such information can be immensely subtle, providing genetic, environmental, nutritional and physiological inform-ation, not only on the host but also on the commensal/symbiotic bacteria and other microorganisms in the gut that form the microbiome. Also, given that one of the functions of the kidney is to maintain homeostasis, it is arguable that early markers of disease, or toxicity, are likely to be more easily detected in the urine rather than plasma, where they would become readily detectable only when the ability of the excretory mechanisms to maintain homeostasis became overwhelmed and unable to adequately compensate for the disease-driven dysbiosis, or themselves became damaged and defective. The urinary metabolic profile therefore provides a metabolic "phenotype" for the animal or human being studied and we have termed this the "metabotype".[11]

Plasma, in contrast to urine, provides a system-level read out of the physio-logical state of the organism under study at the time the sample is taken. Obviously integrating data from both plasma and urine samples taken at the same times greatly enhances the investigators ability to extract meaningful information from the study.

12.3   Examples of the Use of Metabonomics in Normal and Disease Models

As discussed above, the metabolic profile detected in biological fluids or tissues/ extracts represents the outcome of the various potentials inherent in the expressed genome and proteome. Examination of these metabolic profiles can illuminate many biological process such as, for example, normal aging and development. A recent example of this is provided by studies on the urinary metabolome of normal male Wistar-derived rats performed using both ^{1}H NMR spectroscopy and HPLC-MS to obtain profiles on samples collected from these animals every 2 weeks (post weaning at 4 weeks up to 20 weeks of age).[12] The advantage of using several analytical techniques is that a much wider coverage of the metabolome is obtained than could be provided by either alone. (It is a continuing problem for those trying to obtain global metabolite profiles is that there is no one analytical method that will provide comprehensive determination of all the components of the sample.) Multivariate statistical analysis of the resulting analytical data allowed clusters to be visualised within the data set that were dependent on the age of the animals. This analysis of the data found that urine collected at 4 and 6 weeks of age, shortly after weaning, showed the greatest differences in composition compared to later samples. Some of the age-related markers, which increased as the animals aged (detected using ^{1}H NMR spectroscopy), included creatinine, taurine, hippurate and resonances associated with amino acids/fatty acids. Others that decreased in amount with age included citrate and glucose/myoinositol. Analysis using HPLC-MS, using both positive and negative electrospray ionisation, in contrast detected a different range of compounds, some of which were only present in urine samples at 4 weeks of age. Based on these data, models were constructed that enabled the urinary metabolite profiles to be used to predict the age of the animals. These predictions were made by PLS-regression modelling and clearly demonstrated an age-related trend from between 4 and 12 weeks for HPLC-MS, and 4 to16 weeks for NMR spectroscopy, after which the trend levelled out indicating a degree of stability (or at least a slower rate of change) in urine composition with maturity. Examples of these plots are shown in Figure 12.2 for NMR and Figure 12.3 for HPLC-MS.

Such results reveal a number of possibilities, and raise a number of questions. Firstly, there is clearly the potential to use of these techniques to study the aging process and determine the age-related changes in biofluid composition in both the rat and other species, including man. The second is the basic biological question of why some of these changes occur at all. What, for example, is the biology that underlies the steady increase in taurine concentrations in the urine of these rats noted by ourselves and others?[12,13] Another obvious result of these studies is that, if profiles change with age, then it is essential to have suitable control animals in pharmacological and toxicological studies, as an animals predose urine cannot be used as the control profile. Indeed, the sensitivity of these analytical techniques combined with multivariate statistical analysis is

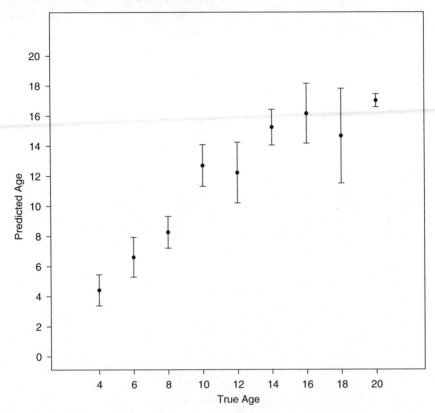

Figure 12.2 ^{1}H NMR spectroscopy and PLS regression. Cross-validated age predictions for male rats for the period from 4 to 20 weeks of age made using a
PLS model with three components. Data expressed as mean ± standard
deviation. (From ref. 12.)

such that we have seen measurable changes in metabolic profiles in periods of a
few days that are probably related to aging and development in young rats.[14] If
not recognised, looked for, monitored and controlled, such effects could easily
make nonsense of poorly designed studies (and such comments can probably be
made with equal force for the other omic techniques as well).

   We have also extended this approach to a metabonomics examination of the
development of the male Zucker (fa/fa) obese rat, an important experimental
model of Type II diabetes, comparing the global urinary metabolite profiles
with those of the Wistar-derived animals.[15] As seen in the previous study, both
^{1}H NMR spectroscopy and HPLC-MS revealed changes in the composition of
the samples with age. Strain-related differences were also observed and even at
4 weeks of age the Zucker and Wistar-derived animals could readily be
differentiated. These differences became more pronounced with time and by 8
to 10 weeks a clear divergence in metabolic trajectories was evident (Figure
12.4), continuing up to the end of the study at 20 weeks. The "markers"
identified in the urine of the Zucker (fa/fa) obese rats via ^{1}H NMR

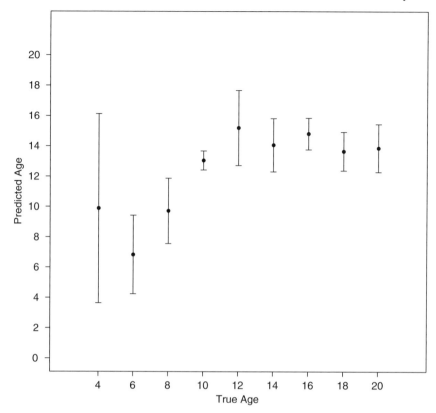

Figure 12.3   HPLC-MS, negative ion mode and PLS regression. Cross-validated age
predictions for male rats for the period from 4 to 20 weeks of age made
using a PLS model with three components. Data expressed as
mean ± standard deviation. (From ref. 12.)

spectroscopy by the 20 week time point included increased protein and glucose,
changes in Krebs cycle intermediates and reduced concentrations of taurine
compared to the Wistar-derived strain. Markers were also seen using HPLC-
MS and a number of ions were observed that increased by 20 weeks of age in
the Zucker (fa/fa) obese animals compared to the Wistar-derived rats. These
included those at m/z 71.0204, 111.0054, 115.0019, 133.0167 and 149.0454
(negative-ion ESI) and m/z 97.0764 and 162.1147 (positive-ion ESI). Ions
observed to decrease in intensity relative to the normal strain included those
at m/z 101.026 and 173.085 (negative-ion ESI) and m/z 187.144 and 215.103
(positive-ion ESI). Possible identities of some of these ions included fumarate,
maleate, furoic acid, ribose, suberic acid, carnitine and pyrimidine nucleoside.
    As noted above, in the case of the normal Wistar-derived strain the urinary
metabolite profile eventually became relatively stable and from 16 to 20 weeks
showed relatively little change. For the Zucker (fa/fa) obese rats, in contrast,
the urine composition continued to change with age and the metabolite profiles

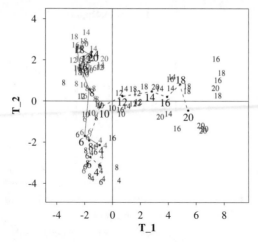

Figure 12.4 Age-related PCA trajectories, derived from ^{1}H NMR spectroscopic data, showing time-related changes for two groups of male rats: Wistar-derived (red) and Zucker (fa/fa) obese (blue) animals. The data point for each individual animal is displayed as a number representing the time point of sample collection (week) whilst the mean data point for each strain at each time point is displayed as a dot, labelled with the week number. (From ref. 15.)

did not stabilise, no doubt reflecting the progress of the disease and an increasingly abnormal state of health of the Zucker animals.

Again, the results of these global metabolite profiling efforts show both the utility of the techniques employed to study the onset and progression of a pathophysiological state in a non-invasive manner in an animal model and also highlight a number of questions. Thus, whilst the appearance of glucose in the urine of some of the Zucker animals could have been anticipated, the reasons for the observed fall in taurine concentrations are less obvious. This is clearly not a function of the normal aging process as, as described above, in the Wistar-derived rats the concentrations of taurine increased from 4 weeks to 20 weeks, with the resonances for this metabolite eventually becoming one of the dominant features of the urinary spectra of these animals. Interestingly, at early time points, the concentration of taurine in the urine of the Zucker rats was much higher than that for the Wistar-derived strain, as illustrated in Figure 12.5. The relationship, if indeed there is one, between the development of insulin resistance and the decline in taurine concentrations is not clear. Taurine, an abundant and essential non-protein amino acid, appears to have many functions, including *e.g.* conjugation of bile acids, cytoprotection against myocardial and hepatic necrosis and pulmonary fibrosis. Insofar as the compound has effects in diabetes, taurine also appears to affect glucose and insulin concentrations in the blood[16] and a role in the function and integrity of pancreatic β cells has also been suggested. In addition, taurine has been found to decrease body weight in hyperglycaemic obese mice and suppress the

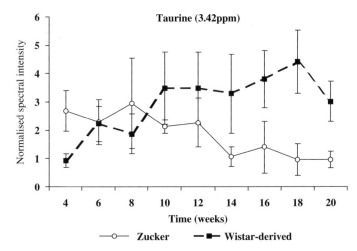

Figure 12.5 Normalised spectral intensity of the spectral region arising from taurine (3.40–3.44 ppm) obtained from ¹H NMR spectra of urine samples collected from male Zucker and Wistar-derived rats between 4 and 20 weeks old. Data expressed as mean ± standard deviation. (From ref. 15.)

development of atherosclerotic lesions in mice.[17,18] When the plasma of the 20-week-old animals examined in the metabonomic study described above was examined, the concentrations of taurocholate detected were elevated in the Zucker (fa/fa) obese animals compared with Wistar-derived rats.[19] Given the possibility therefore that taurine may have some role the development of the condition it is possible that the decline in urinary taurine concentrations that precedes an elevation in urinary glucose reflects some fundamental process in the development of diabetes. Equally, there may only be some casual relationship. However, causal or casual, the metabonomic study performed here does allow a hypothesis to be generated that taurine is in some way important, and allows studies to be designed to investigate this. It is also noteworthy that not all of the animals progressed towards disease at the same rate (glucose was only elevated in the urinary profile of 4 out of 6 of the Zucker (fa/fa) obese rats), reflecting normal biological variation in the animals. This type of disease progression, or response to a toxic insult *etc.*, is readily monitored using non-invasive metabonomics and allows such variability to be taken into account.

As well as glucose and taurine, other changes, such as increased concentrations of acetate (also seen for patients with type 2 diabetes by NMR,[20] with the authors suggesting that this was the result of impaired renal function rather than metabolic alteration) were noted, together with alterations in the ratios of citric acid cycle intermediates. These metabolites have been suggested to be acting as ligands for orphan G-protein-coupled receptors (GPCR) (specifically GPR91 which acts as a receptor for succinate and GPR99 for α-keto-glutarate) implying a link between energy homeostasis/metabolic status and haemodynamic regulation.[21] Perturbed concentrations of these citric acid cycle intermediates have been associated with the molecular pathology of diseases such as hypertension, atherosclerosis and diabetes,[21] and thus changes in the ratio of these metabolites may reflect a signalling response occurring in these animals as the disease state progresses.

However, another valuable feature of metabonomics is that it allows access to effects on the metabolic profile that result from interactions of the host organism and its symbiotic consortium of gut microbiota. This is discussed in more detail in the next section, but it is noteworthy, in the context of disease models and insulin resistance, that in recent studies on the differential induction of insulin resistance (IR) and non-alcoholic fatty liver disease in two strains of mice fed high-fat diets it was shown that the disease progression was linked to critical gut microfloral metabolic events leading to choline conversion to trimethylamines in IR susceptible animals.[22]

12.4 The Microbiome – an Integral Part of the Global System

The influence of the gut microflora on the metabolic profiles represents another neglected area of research in systems biology;[4] however, as alluded to above, the gut microflora make a significant contribution to the metabolic profiles determined for urine (*e.g.* see Figure 12.6).[23–26] The recent concentration of

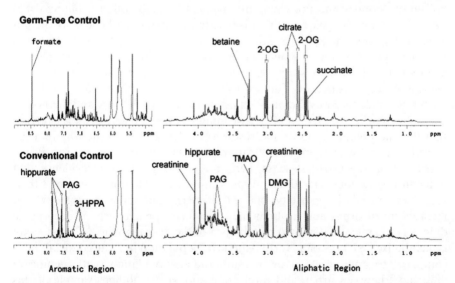

Figure 12.6 ^{1}H NMR spectra of urine samples obtained from conventional (lower) and germ-free (upper) rats showing the differences in metabolic profile resulting from the presence or absence of gut microbiota. Key: 2-OG, 2-oxoglutarate, PAG, phenylacetylglycine, 3-HPA, 3-hydroxyphenyl propionic acid, TMAO, trimethylamine *N*-oxide, DMG, dimethylglycine.

attention on obtaining genome sequences, by focussing solely on the host, has further tended to obscure the fact that most organisms have a symbiotic relationship with the microorganisms inhabiting their guts that is fundamental to their health.[27,28] The microbiome, which in effect represents an external "organ" is, self-evidently, not represented in the gene sequence of rats or humans, for example, and this is to the detriment of simplistic genome-based attempts to understand the biology of the host. This very close symbiotic relationship between the gut flora and the host is such that the two can perhaps be best envisaged as a "superorganism".[29] To put this into context, there are approximately 10^{12} parenchymal cells in an average human (excluding blood cells and neurons) compared to the microbiome which comprises some 10^{14} microbes (weighing > 1 kg). In humans the majority of the intestinal bacteria are obligate anaerobes and include species of the genera *Bacteriodes*, *Clostridium*, *Lactobaccillus*, *Eschericia* and *Bifidobacteria*, amongst others. In addition, there are a variety of yeasts and other microorganisms and all of these co-exist in a dynamic ecological equilibrium. Current estimates are that there are more than 1000 species of gut microorganisms, but many more may yet be discovered. As indicated above, the interaction between the host and the "forgotten organ" represented by the gut microflora provides an interesting challenge for global systems biology as the there are a huge range of interactions between host and gut flora involving metabolic exchange and co-metabolism of many substrates. Thus these microbial symbionts stimulate the development of intestinal microvilli, provide the host with significant amounts of energy (via the fermentation of non-host digestible dietary fibre and the anaerobic metabolism of peptides and proteins), are involved in xenobiotic (including drug) metabolism, and are part of the host's defence against pathogens at the gut level. In addition, the gut microflora are involved in the development the host's immune system. There is increasing evidence that these "metabolome–metabolome" interactions may have significant implications in the development a variety of diseases. The acquisition of the gut microbiome begins at birth but is influenced by factors such as the method of delivery and method of feeding in newborns, diet, *etc*. Once weaning has taken place, there is evidence that the microbiome is stable and resistant to change. Given the way that the microbiome is, at least in part, "transmitted" via the mother at birth it will clearly be, in the broadest sense, heritable, and there is good evidence that different populations have different microfloral compositions. Because of their very large metabolic capacity, the gut bacteria can influence toxicity and metabolism of drugs and, in humans, there have been reports of geographic differences in microbial-related metabolism of the drug digoxin.[30] Thus digoxin is metabolised by 36% of a North American population to reduced metabolites compared to only *ca.* 14% by a South Indian population (with significant differences between urban and rural populations).[30] Another example of this type of effect is provided by studies of the metabolic fate of isoflavones and lignans in germ-free rats and in those inoculated with human gut microbiota.[31] The germ-free animals were found to excrete substantial amounts of isoflavones in urine following feeding with a soy-isoflavone-containing diet but the isoflavone metabolites equol, *O*-desmethylangolensin and the lignan entero-lactone were not detected. However, following colonisation by with human-derived microbiotia these pharmacologically active compounds, which will almost certainly have modulating effects on host metabolism, and therefore metabolic profiles, were detectable in urine. Of particular interest, however, was that the ability to produce these metabolites depended on the human donor, as equol excretion by the animals was limited to those inoculated with bacteria from subjects who themselves produced equol. Thus it seems evident that some of the variability in drug response, or toxicity, between individuals may result from differences in the microbiome, rather than host genetic factors. Thus genome-based studies for effects such as these are likely to prove poorly predictable, whilst metabonomic investigations, which do take into account both host genetic and environmental factors such as the status of the microbiome, will capture such effects.

In our own metabonomic studies in the rat, we have shown large micro-biome-derived variations in the urinary metabolic profile for a number of common aromatic species derived from the metabolism of dietary components such as chlorogenic acid.[23] In the presence of a full complement of gut flora this compound can be metabolised in a variety of ways, in a complex interaction between the microbiome and the host, depending upon the composition of the gut microbiotia. One route leads to the excretion of 3-(hydroxyphenyl)propionic acid (and related compounds) whilst another route results in the formation of benzoic acid (which is converted by the host to hippuric acid). Interventions, such as dietary changes, can result in the microbiota switching from one route to another,[23] with profound changes to the aromatic metabolite profiles, but such changes can also occur when the diet is kept constant. As an example, in a recent study the initial urinary profiles of control animals were those of a 3-(hydroxyphenyl)propionic acid metabotype.[14] However, as the study progressed the profile changed to that of a hippurate excretor. It is not unknown, in our experience, for some of these animals to then revert to the 3-(hydroxyphenyl)propionic acid metabotype as the study progresses (unpublished observations). As the diet supplied to the animals and other environmental conditions were constant the observed changes must presumably have reflected a shift in the gut microenvironment resulting from other factors such as, *e.g.* stress, *etc.* Clearly it is essential to monitor and understand such background variation as it might affect the interpretation of the study and these gut microbial effects on metabolite profile. As a simple example, we have noted effects on the aromatic profiles of rats administered antibiotics as part of a series of studies on nephrotoxicity.[32] These observations were, however, more likely to have been the result of "pharmacological" effects of the drug on the gut microbiota rather than any effect on the kidney. The need to consider effects such as these certainly increases the complexity of biomarker interpretation, as the effects being observed in a study could be the result of changes in either the mammalian metabolic process or the microbiome or, indeed, both. In the extreme case, where the gut bacteria have been effectively eliminated by rearing the animals in germ-free colonies, the effect on the urinary metabolic profile can be large. This is illustrated in Figure 12.6, where typical ^{1}H NMR profiles of normal and axenic rats are shown, revealing, amongst other things, differences in the profiles of the aromatic metabolites present in the samples. Once exposed to the external environment, such germ-free animals rapidly begin to acquire gut microbiota, but the process is not instantaneous and indeed takes several weeks before a "normal" metabotype is observed.[25]

We have termed the products of microbial and host–micrbobial metabolism sym-xenobiotics and would contend that the metabolic signature provided by these compounds in biofluids such as urine offers a unique, and non-invasive, insight into the often dynamic interaction between the gut microbiome and the host organism.[3] The global system in this instance, therefore, is a metabolic continuum including all of the mammal–microbiome interactions formed from a combination of the hosts "primary" metabolome (effectively endogenous metabolites and metabolic processes under host genome control) and the microbial co-metabolome. The co-metabolome can result in a supply of metabolites to the host, both useful (such as energy-rich metabolites, vitamins *etc.*) and microbial waste products for processing and disposal by the host. This "combinatorial metabolism" greatly increases the number of metabolites that will be encountered in the analysis of biofluid samples. Thus the microbiome interacts with the host via absorption of metabolites from the gut whilst the various host cell types are also able to "communicate" with the microflora by a metabolic axis via the enterohepatic circulation (bile plus gut secretions into the enteron). In this way microbial-derived compounds can flow from the gut into the circulation (with or without pre-processing by the gut wall or liver) and compounds can also flow from the gut to the liver (and other tissues) and be recycled to the gut for excretion or further microbial processing and structural modification. The latter type of metabolic co-processing is well characterised, bile acids for example, and occurs on a large mass scale.

12.5   Pharmaco-metabonomics

"Pharmaco-metabonomics" is an approach that aims to enable the prediction of the effects of drugs (efficacy, toxicity and drug metabolism) based on mathematical models constructed from pre-dose metabolic profiles. This has been demonstrated in a "proof-of-concept" study in the rat that used the model hepatotoxins galactosamine, allyl alcohol and acetaminophen (paracetamol), which act via different mechanisms.[33] These investigations showed that the pre-dose urinary profiles encoded information that predicted the degree of toxicity that resulted from the administration of the toxins (or indeed to predict the metabolism of the drug itself). Thus, in the case of galactosamine, the responder/non-responder pattern of liver damage at 24 hours post-dosing was reflected in the pre-dose metabolic profile of the urine analysed using a simple PCA model. For acetaminophen a more complex study was performed with animals administered a "threshold" toxic dose of the drug that produced a wide range of liver toxicity between individuals. The resulting data were analysed using a supervised approach, projection-to-latent structure (PLS) but, once again, there was a significant association between pre-dose metabolic profile and post-dose outcome with respect to both degree of toxicity and metabolic fate with the acetaminophen to acetaminophen glucuronide excretion ratio strongly correlated to the pre-dose urinary metabolite profiles. The potential to apply such metabotyping to screening patients for drug trials is obvious, but clearly requires significant efforts to extend our knowledge on the relationships between endogenous metabolic status and drug metabolism outcomes.

12.6   Combining Omics

Whilst there is no doubt that metabonomics on its own can provide many new and valuable insights into biological processes there are a variety of reasons why performing simultaneous gene and protein level analysis of the system under study are intellectually attractive. Not the least of these is the assumption that having information from more than one (preferably all three) of the major levels of biomolecular organisation on the organism under study should enable more complete understanding and allow more robust models to be derived. Now this view is somewhat simplistic as, as we have discussed elsewhere,[34] responses to stimuli may evolve at different rates depending upon the level of biomolecular organisation, with effects dispersed over time (*e.g.* gene transcription fast, protein synthesis and post transcriptional modification slower, metabolite synthesis last, by which time gene activity may be past), and have different half lives. Nonetheless, true systems biology approaches need to be able, wherever possible, to combine the data obtained by different "omic" technologies to obtain a single, global, view.

There are two basic approaches that can be taken to combining such multi-omic data. In the first, the data at each level is statistically analysed, independently of the other levels, and the major changes correlating with the observed effects of treatment on the organism are discovered. Following this analysis, the investigators attempt to combine the results from the genomic, proteomic and metabonomics levels in such a way as to link them into some structure that makes narrative sense. Hypothesises are then constructed and tested. There are a number of examples of this type of "combinatorial omics" of the first kind, from various groups,[35-42] including our own. These include investigations on acetaminophen (paracetamol)[36] and, most recently, the drug methapyrilene (a histamine antagonist).[35] In the rat, administration of high doses of methapyrilene causes periportal liver necrosis. The mechanism of toxicity is ill-defined, probably involving reactive metabolites, and in this study we attempted to perform an integrated systems approach to understanding the toxic mechanisms by combining gene expression profiling, proteomics and metabonomics profiling. Male rats were dosed with methapyrilene for 3 days at either 150 mg/kg/day to induce liver necrosis, 50 mg/kg/day as a "therapeutic", subtoxic, dose, or as a control, undosed, group. Urine, blood plasma and liver were obtained from the rats and analysed by the various gene transcript, proteome and metabonome analytical platforms as appropriate. The resulting data showed changes occurring in signal transduction and metabolic pathways as a result of methapyrilene hepatotoxicity, and revealed changes in the expression levels of genes and proteins associated with oxidative stress and a change in energy usage that was reflected in both gene/protein expression patterns and metabolites. The interrogation of genes, proteins and metabolites thus provided, to some extent, an integrated picture of the response of the liver to methapyrilene-induced hepatotoxicity with mutually supporting and mutually validating evidence arising from each biomolecular level that correlated with conventional histopathology and clinical chemistry. Not unexpectedly there were a number of examples of genes and proteins, either encoded by the same gene or by other genes within the same pathway, that were co-regulated in response to the drug-induced toxicity, and sometimes this was in concert with an associated metabolic product. This example demonstrates the synergy that might be expected from combining these parallel omic approaches. However, as well as providing a different insight into the hepatotoxic effects of this particular drug, the study highlighted a number of areas that require addressing in approaches to systems biology employing a multi-omic approach. For example, alterations in expression of genes or enzyme levels, or the modification of proteins, while suggestive do not either prove a potential target of toxic effects or show that function or activity must be altered. This means that, from gene and proteomic profiling alone, the distinction between causative and casual effects cannot be made. On the other hand, an alteration to a metabolic profile does indeed reflect a functional change and thus can aid in interpretation of the changes at other levels of biomolecular organisation. So, whilst the metabolic profile provides a window on function, the changes seen in gene and protein expression may be interpreted as the homeostatic response to the current metabolic "challenge". There is of course the issue of experimental design in such multi-omic studies. In the case of methapyrilene, it may well have been the case that sampling at time points where the toxicity was already well established is not helpful in obtaining a clear understanding of the temporal dynamics of the mechanism for the reasons described above concerning the time scales of changes at the gene, protein and metabolite level. This could lead to highly non-linear relationships between the concentrations of various species at the different levels of biomolecular organisation and experiments that allowed the time course of the changes to be determined would clearly be of benefit. In addition, using the approach described here, the analysis of such complex data sets is both challenging and, to an extent, involves a degree of subjectivity. It may thus be very easy to miss more subtle relationships present in these data linking the genes, proteins and metabolites that might nevertheless be highly significant.

Another complexity is due to the way in which the analytical data are generated in multi-omic studies. So, whilst the DNA microarray methods measure the expression of individual genes, and a single expression numerical value can be derived for each, the same is not true for the proteomic and metabonomic data. The latter methodologies instead provide a measure of the overall change in the profile, where the variables are expressed in terms of differential responses or relative changes. This disparity represents a significant modelling challenge and is one reason why latent variable methods dominate the metabonomics literature. In the case of methapyrilene, hundreds of gene expression changes were detected with a much smaller number seen for both proteomic and metabonomic profiling and it is quite conceivable that insufficient detail was obtained at each biomolecular level to elaborate fully on the mechanism of methapyrilene toxicity. However, at each of three levels examined (gene expression, protein expression and global metabolic profiles) a range of dose-dependent changes were seen relating to broad areas such as stress responses and to changes in metabolic pathways involving lipid, glucose and choline metabolism and the urea cycle. With respect to mechanism of toxicity, previous studies have suggested covalent binding of a reactive metabolite of methapyrilene may be responsible for the observed toxicity.[43,44] This binding has been associated with mitochondrial proteins and the mechanism is therefore likely to be related to mitochondrial disfunction. In this study we observed that two mitochondrial proteins, pyruvate carboxylase and carbamoyl phosphate synthase, underwent modification to negatively charged species, possibly (though this was not conclusively demonstrated in this work) as a result covalent adduct formation. Whilst the exact mechanism of toxicity was not unambiguously defined, what was not in doubt, however, was the fact that many of the metabolic processes that were apparently affected by methapyrilene toxicity (such as gluconeogenesis, fatty acid β-oxidation, urea cycle, choline metabolism), in the main, reside within the mitochondria. Mitochondrial dysfunction causes oxidative stress, the hallmarks of which were observed in the gene and protein stress response, and eventual cell death either through apoptosis or necrosis. This study also provides an example of the information that can be obtained on intact tissue using the technique of high resolution magic angle spinning (HR-MAS) NMR spectroscopy, which is ideally suited to solid and semi-solid samples. Figure 12.7 shows a series of ^{1}H MAS NMR spectra of liver samples obtained from each treatment group, clearly illustrating the dose-related changes in fatty acid composition, elevations in trimethylamine *N*-oxide concentrations and the decrease in liver glucose and glycogen with increasing dose. Interestingly, this pattern of changes was also observed in studies on the mouse following administration of sub-toxic and hepatotoxic doses of acetaminophen, where major effects on energy production were also indicated, suggestive of mitochondrial toxicity.[36]

An alternative to this approach of measure, evaluate and "manually" integrate the different levels of biomolecular organisation, is to use statistical methods that exploit the variability inherent in any biological experiment to discover connections between metabolites and proteins, for example. An example of this is the investigation of cancer biomarkers in the plasmas of nude mice implanted with a human-derived prostate PC3 tumour line.[38] In this work, parallel 2D-DIGE proteomic and ^{1}H NMR metabolic profile data were collected on blood plasma from mice with the xenograft and from matched control animals. At the individual omic level the metabonomics-detected changes in the plasma metabolite profile were made up of decreased amounts

Group Mean Normalised Standard Spectra for Each Treatment

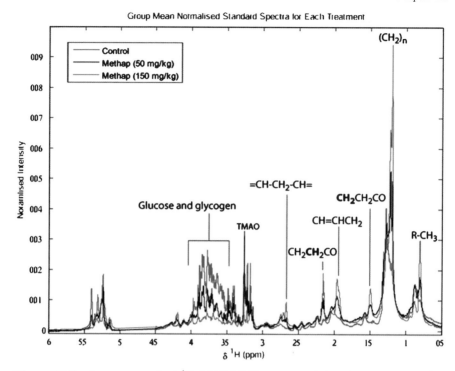

Figure 12.7   Average standard ^{1}H MAS NMR spectra of liver from methapyrilene treated animals at 0, 50 and 150 mg/kg day for 3 days showing dose-related elevations and composition changes in fatty acid species as indicated by the change in shape of the CH_2 and CH_3 peaks in the regions of 1.3 and 0.9 ppm, respectively. TMAO levels are also elevated in high-dose animals. There is also a clear decrease in liver glucose and glycogen levels. (From ref. 35.)

of amino acids such as valine, isoleucine, glutamine, leucine, lysine, tyrosine and phenylalanine combined with increased quantities of glucose, 3-D-hydroxybutyrate and acetate. Similarly, clear differences were also detected in the plasma protein profile between PC3 and control mice, including increased amounts of gelsolin precursor, serotransferrin precursor, α-enolase/β-2-glycoprotein 1 precursor, plasminogen precursor/fibrinogen gamma polypeptide, and complement C4 precursor. These increased concentrations of proteins were also accompanied by decreased amounts for major urinary protein 1 precursor and complement factor H precursor in PC3 tumour-bearing animals. As discussed in more detail elsewhere, these changes in metabolites and proteins can be related back to various aspects of tumour biology.[38]

However, as well as this attempt to interpret the data using a "one omic at a time" approach, we also used a statistically integrated proteometabonomics method to interrogate the xenograft-induced differences in plasma profiles. This was performed via the use of multivariate statistical algorithms, including orthogonal projection to latent structure (O-PLS), which were applied to generate models characterising the disease profile. Two approaches to integrating metabonomic data matrices, based on O-PLS algorithms, were used to provide a framework for generating models relating to the specific and common sources of variation in the metabolite and protein data matrices that can be directly related to the disease model. Patterns of correlation between ^{1}H NMR data and 2D-DIGE data were initially explored by visualisation of the data as a correlation map, to provide an overview of similarities between variables in the two data sets (Figure 12.8) which can be used to aid the identification of proteins associated (correlated) to specific metabolite signals and vice versa. Then, in order to further investigate and confirm the relationships observed between NMR regions and protein spots, O-PLS was used. The O-PLS models were constructed using the 2D-DIGE data spots and individual ^{1}H NMR data peaks, giving the highest discriminatory power between the control and tumour-bearing animals. This approach provides a means of adding further confidence in correlations between NMR and DIGE variables since we are able to apply cross-validation for the O-PLS models. The separate O-PLS models were built by *e.g.* regressing all NMR variables against a single DIGE variable or the reverse, an NMR-detected metabolite against all of the DIGE spots. For example, the metabolite 3-D-hydroxybutyrate was found to be present in significantly higher concentrations in the plasma of mice carrying the PC3 xenograft. Interpretation of the regression coefficients for a model where the DIGE data were regressed against the 3-D-hydroxybutyrate NMR signal identified a number of protein spots with both high positive and negative regression coefficients against this metabolite. In the same way an O-PLS model generated for tyrosine, which was down-regulated in the tumour-bearing animals, also showed associations with a number of proteins.[38]

The correlation maps generated linking the NMR and DIGE data sets show associations between many proteins and metabolites (Figure 12.9) and provide leads for further analysis and modelling. These correlations may also be used to generate hypotheses on biological relationships or pathway activity that can be further tested experimentally *in vivo* or *in vitro*. However, because of the current state of our knowledge of the proteome, many of the proteins that were detected as significant in this study remain, as yet, unidentified. This clearly limits and complicates any attempt to provide a robust biological interpretation of correlations between metabolites and proteins.

This approach differs significantly from that employed in the methapyrilene study described above in that, as indicated above, the subjectivity of selecting for significant metabolites and proteins, and then trying to find connections, was eliminated. What was particularly interesting in this study, however, were the multiple correlations seen between metabolites and proteins that were not obviously connected by the same pathways. These included statistical connections between the variations in the protein serotransferrin precursor and both tyrosine and 3-D-hydroxybutyrate. Additionally, a correlation between decreased excretion of tyrosine and increased presence of gelsolin was also observed for PC3 tumour-bearing mice compared to the controls.

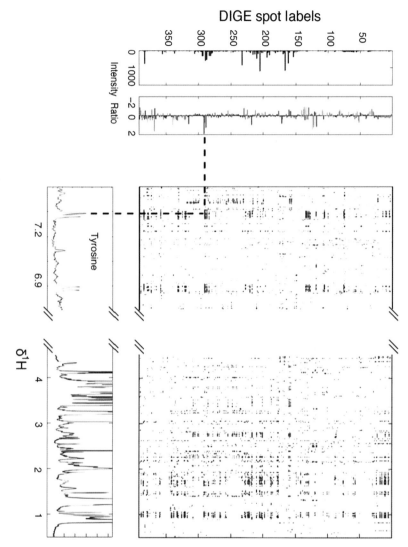

Figure 12.8 Visualisation of correlations between NMR variables (*x*-axis) and DIGE variables (*y*-axis) in the form of a correlation map (correlations > ± 0.77 are shown in the figure). Red coloured areas indicate positive correlations and blue coloured area indicates negative correlations between NMR and DIGE variables. The region corresponding to the tyrosine resonance in the ¹H NMR spectrum is expanded in the left most part along the *x*-axis. The dashed lines shows and example of how the correlation map may be used to identify proteins associated with a specific metabolite and vice versa. In this case the marked DIGE spot, which is negatively correlated with the ∼7 ppm tyrosine NMR signal, is identified as serotransferrin precursor/fibrinogen A alpha polypeptide. (From ref. 38.)

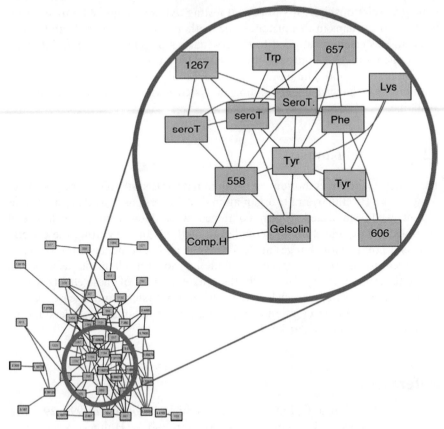

Figure 12.9 Visualisation of correlations between selected DIGE spots and NMR data points with a correlation to class of > 0.77. Edges in the network are present between nodes (NMR/DIGE variables) where the correlation is > 0.85. Key: Blue node = NMR variable, red node = DIGE spot, red edge = positive correlation > 0.85, blue edge = negative correlation < –0.85). (From ref. 38.)

Clearly, where biological sense can be made out of co-varying proteins and metabolites, the results are more robust since changes in protein expression, to some extent, validate the changes seen in metabolites and *vice versa*, providing the opportunity to obtain increased information from the study and improve interpretation. Self evidently not all changes in the amounts or activities of proteins will result in a change in metabolite concentrations and not all changes in the metabolome will connect exactly with obvious changes in the protein expression profile but, where they do, the observation may turn out to be of great value. This approach for examining, interrogating and integrating complex data sets, derived from different analytical platforms, does provide one means of deriving statistical relationships between metabolites and proteins. These relationships can then be used to construct hypotheses regarding biological relationships for subsequent testing. Although applied only to meta-bolites and proteins in this instance, the method is of general applicability and could just as easily be applied to the co-analysis of gene expression data with proteomic data. This approach thus seems able to provide enhanced recovery of combination candidate biomarkers across multi-omic platforms, potentially enabling an enhanced understanding of systems where multiple omic data is available.

12.7 Conclusions

The potential for metabonomics to illuminate the system-level responses of complex multi-cellular, multi-organ animals as diverse as rodents and humans is now well established. When combined with data from other levels of biomolecular organisation and coupled to methods of integrating these meta-bolic end points with changes at the proteome and genome level, for example, there seem to be good prospects for obtaining novel mechanistic insights into the systems under study. Hopefully the outcome of all of these efforts will be better biomarkers in areas such as the detection of toxicity and disease, patient stratification and efficacy monitoring and a more fundamental understanding of biology at the molecular level.

References

1. J. K. Nicholson, J. C. Lindon and E. Holmes, *Xenobiotica*, 1999, **29**, 1181.
2. J. K. Nicholson, J. Connelly, J. C. Lindon and E. Holmes, *Nature Rev. Drug Discov.*, 2002, **1**, 253.
3. J. K. Nicholson and I. D. Wilson, *Nature Drug Discov.*, 2003, **2**, 668–676.
4. J. K. Nicholson, E. Holmes and I. D. Wilson, *Nature Rev. Microbiol.*, 2005, **3**, 2.
5. J. K. Nicholson and I. D. Wilson, *Prog. NMR Spectrosc.*, 1989, **21**, 449.
6. J. C. Lindon, E. Holmes and J. K. Nicholson, *Prog. NMR Spectrosc.*, 2004, **45**, 109.
7. J. C. Lindon, J. K. Nicholson and E. Holmes, in *Metabonomics in Toxicity Assessment*, Eds. D. G. Robertson, J. C. Lindon, J. K. Nicholson, E. Holmes, CRC Press, Boca Raton, 2005, pp. 105.
8. M. E. Bollard, E. G. Stanley, J. C. Lindon, J. K. Nicholson and E. Holmes, *NMR Biomed.*, 2004, **9**, 1.
9. D. G. Robertson, M. D. Reily, J. C. Lindon, E. Holmes, J. K. Nicholson, in *Comprehensive Toxicology*, Eds. J.P. Vanden Heuvel, G. J. Perdew, G. J. Mattes and P. Greenlee, Elsevier Science BV, 2002, **14**, 583–610.
10. J. C. Lindon, E. Holmes and J. K. Nicholson, *Prog. Nuclear Magn. Res. Spectrosc.*, 2004, **45**, 109.
11. C. L. Gavaghan, E. Holmes, E. Lenz, I. D. Wilson and J. K. Nicholson, *FEBS Lett.*, 2000, **484**, 169–174.

12. R. E. Williams, E. M. Lenz, J. Lowden, M. Rantalainen and I. D. Wilson, *Mol. Biosystems*, 2005, **1**, 166–180.
13. J. D. Bell, P. J. Sadler, V. C. Morris and O. A. Levander, *Mag. Res. Med.*, 1991, **17**, 414.
14. E. M. Lenz, J. Bright, R. Knight, I. D. Wilson and H. Major, *Analyst*, 2004, **129**, 535.
15. R. E. Williams, E. M. Lenz, M. Rantalainen and I. D. Wilson, *Mol. BioSyst.*, 2006, **2**, 193.
16. E. C. Kulakowski and J. Matura, *Biochem. Pharmacol.*, 1984, **33**, 2835.
17. M. Zhang, L. F. Bi, J. H. Fang, X. L. Su, G. L. Da, T. Kuwamori and S. Kagamimori, *Amino Acids*, 2004, **26**, 267.
18. C. Li, L. Cao, Q. Zeng, X. Liu, Y. Zhang, T. Dai, D. Hu, K. Huang, Y. Wang, X. Wang, D. Li, Z. Chen, J. Zhang, Y. Li and R. Sharma, *Cardiovasc. Drugs Ther.*, 2005, **19**, 105.
19. R. Williams, E. M. Lenz, A. J. Wilson, J. Granger, I. D. Wilson, H. Major, C. Stumpf and R. Plumb, *Mol. Biosyst.*, 2006, **2**, 174.
20. I. Messana, F. Forni, F. Ferrari, C. Rossi, B. Giardina and C. Zuppi, *Clin. Chem.*, 1998, **44**, 1529.
21. W. He, F. J. Miao, D. C. Lin, R. T. Schwandner, Z. Wang, J. Gao, J. L. Chen, H. Tian and L. Ling, *Nature*, 2004, **429**, 188.
22. M. E. Dumas, R. H. Barton, A. Toye, O. Cloarec, C. Blancher, A. Rothwell, J. R. Fearnside, Tatoud, V. Blanc, J. C. Lindon, S. Mitchell, E. Holmes, M. I. McCarthy, J. Scott, D. Gauguier and J. K. Nicholson, *Proc. Natl. Acad. Sci., USA*, 2006, **103**, 12516.
23. A. N. Phipps, J. Stewart, B. Wright and I. D. Wilson, *Xenobiotica*, 1998, **28**, 527–537.
24. R. E. Williams, H. W. Eyton-Jones, M. J. Farnworth, R. Gallagher and W. M. Provan, *Xenobiotica*, 2002, **32**, 783–794.
25. A. W. Nicholls, R. J. Mortishire-Smith and J. K. Nicholson, *Chem. Res. Toxicol.*, 2003, **16**, 1395.
26. L. C. Robosky, D. F. Wells, L. A. Egnash, M. L. Manning, M. D. Reily and D. G. Robertson, *Toxicol. Sci.*, 2005, **87**, 277.
27. M. J. G. Farthing, *Best Practice & Res. Clin. Gastroent.*, 2004, **18**, 233.
28. L. V. Hooper and J. L. Gordon, *Science*, 2003, **299**, 1999.
29. J. Lederberg, *Science*, 2000, **288**, 287.
30. V. I. Mathan, J. Wiederman, J. F. Dobkin and J. Lindenbaum, *J. Gut*, 1989, **30**, 971.
31. E. Bowey, H. Adlercreutz and I. Rowland, *Food Chem. Toxicol.*, 2003, **41**, 631.
32. E. M. Lenz, J. Bright, R. Knight, F. R. Westwood, D. Davies, H. Major and I. D. Wilson, *Biomarkers*, 2005, **10**, 173.
33. T. A. Clayton, J. C. Lindon, H. Antti, C. Charuel, G. Hanton, J. P. Provost, J. L. Le Net, D. Baker, R. J. Walley, J. E. Everett and J. K. Nicholson, *Nature*, 2006, **440**, 1073.
34. J. K. Nicholson, E. Holmes, J. C. Lindon and I. D. Wilson, *Nature Biotech.*, 2004, **22**, 1268.

35. A. Craig, J. Sidaway, E. Holmes, T. Orton, D. Jackson, R. Rowlinson, J. Nickson, R. Tonge, I. Wilson and J. K. Nicholson, *Proteome Res.*, 2006, **5**, 1586.
36. M. Coen, S. U. Ruepp, J. C. Lindon, J. K. Nicholson, F. Pognan, E. M. Lenz and I. D. Wilson, *J. Pharm. Biomed. Anal.*, 2004, **35**, 93.
37. T. G. Kleno, B. Kiehr, D. Baunsgaard and U. G. Sidelmann, *Biomarkers*, 2004, **9**, 116.
38. M. Rantalainen, O. Cloarec, O. Beckonert, I. D. Wilson, D. Jackson, R. Tonge, R. Rowlinson, S. Rayner, J. Nickson, R. W. Wilkinson, J. D. Mills, J. Trygg, J. K. Nicholson and E. Holmes, *J. Proteome Res.*, 2006, **5**, 2642.
39. M. Y. Hirai, M. Yano, D. B. Goodenowe, S. Kanaya, T. Kimura, M. Awazuhara, M. Arita, T. Fujiwara and K. Saito, *Proc. Natl. Acad. Sci. USA,*, 2004, **101**, 10205.
40. J. L. Griffin, S. A. Bonney, C. Mann, A. M. Hebbachi, G. F. Gibbons, J. K. Nicholson, C. C. Shoulders and J. Scott, *J. Physiol. Genomics*, 2004, **17**, 140.
41. W. H. Heijne, R. J. Lamers, P. J. van Bladeren, J. P. Groten, J. H. J. van Nesselrooij and B. van Ommen, *Toxicol. Pathol.*, 2005, **33**, 425–433.
42. C. E. Thomas and G. Ganji, *Curr. Opinion Drug Discov. Dev.*, 2006, **9**, 92.
43. G. S. Ratra, S. Cotterell and S. Powell, *Toxicol. Sci.*, 1998, **46**, 185.
44. G. S. Ratra, W. A. Morgan, J. Mullervy, C. J. and M. C. Wright, *Toxicology*, 1998, **130**, 79.
45. J. K. Nicholson, *Molecular Systems Biology*, 2006, **2**, 52.

Subject Index